Periods of Quaternionic Shimura Varieties. I.

CONTEMPORARY MATHEMATICS

762

Periods of Quaternionic Shimura Varieties. I.

Atsushi Ichino
Kartik Prasanna

AMERICAN MATHEMATICAL SOCIETY

2020 *Mathematics Subject Classification.* Primary 11F41, 11F27, 11G18.

Library of Congress Cataloging-in-Publication Data

Names: Ichino, Atsushi, 1976– author. | Prasanna, Kartik A., 1976– author.
Title: Periods of quaternionic Shimura varieties. I. / Atsushi Ichino, Kartik Prasanna.
Description: Providence, Rhode Island : American Mathematical Society, [2021] | Series: Contemporary mathematics, 0271-4132 ; volume 762 | Includes bibliographical references.
Identifiers: LCCN 2020044411 | ISBN 9781470448943 (paperback) | ISBN 9781470464189 (ebook)
Subjects: LCSH: Shimura varieties. | Quaternions. | AMS: Number theory – Discontinuous groups and automorphic forms – Automorphic forms on | Number theory – Discontinuous groups and automorphic forms – Theta series; Weil representation; theta correspondences | Number theory – Arithmetic algebraic geometry (Diophantine geometry) – Arithmetic aspects of modular and Shimura varieties
Classification: LCC QA242.5 .I24 2021 | DDC 516.3/5–dc23
LC record available at https://lccn.loc.gov/2020044411
Contemporary Mathematics ISSN: 0271-4132 (print); ISSN: 1098-3627 (online)
DOI: https://doi.org/10.1090/conm/762

∞ The paper used in this book is acid-free and falls within the guidelines
established to ensure permanence and durability.
Visit the AMS home page at `https://www.ams.org/`

10 9 8 7 6 5 4 3 2 1 26 25 24 23 22 21

Contents

Introduction

In this paper and its sequels [**31**], [**32**], we study periods of automorphic forms on quaternionic Shimura varieties. Specifically, the periods that we focus on are the Petersson inner products of Hilbert modular forms and of their Jacquet–Langlands lifts to quaternionic Shimura varieties. This subject was pioneered by Shimura, who proved many results on algebraicity of ratios of Petersson inner products and made a precise general conjecture ([**70**] Conjecture 5.10) that predicts a large number of algebraic relations in the $\overline{\mathbb{Q}}$-algebra generated by such periods. Shimura's conjecture was proved by Harris [**22**] under a technical hypothesis on the local components of the corresponding automorphic representation. This hypothesis was relaxed partly by Yoshida [**80**], who also used these period relations to prove a refined conjecture of Shimura ([**70**] Conj. 5.12, [**71**] Conj. 9.3) on the factorization of Petersson inner products into *fundamental periods* up to algebraic factors. In later papers [**24**] [**25**], Harris has considered the question of generalizing such period relations to the setting of unitary Shimura varieties. Specialized to the case of hermitian spaces of dimension two, these latter results provide more precise information about the fields of rationality of quadratic period ratios of quaternionic modular forms.

In this series of papers, we will study the corresponding *integrality* questions. The simplest interesting case is the period ratio

$$\frac{\langle f,f\rangle}{\langle g,g\rangle}$$

where f is a usual modular form of (even) weight $2k$ (for GL_2 over $\mathbb{Q}$) and trivial central character, and g is its lift to a Shimura curve corresponding to an indefinite quaternion algebra, also defined over $\mathbb{Q}$. The forms f and g here are assumed to be newforms and to be suitably integrally normalized. In this case, there is a very precise rationality result due to Harris and Kudla [**26**] which asserts that the ratio above lies in the field generated by the Hecke eigenvalues of f. As for the more refined integrality question, one knows the following:

(i) In the special case when the weight $2k$ equals 2 and f corresponds to an isogeny class of elliptic curves, it is shown in [**60**], §2.2.1 that the period ratio above equals (up to Eisenstein primes) an explicit product of Tamagawa numbers, which in turn are related to *level-lowering* congruences satisfied by the form f. This suggests that such period ratios contain rather deep arithmetic information. The proof in this case follows by combining *three* geometric ingredients:

- The work of Ribet on level-lowering [**64**] and its extension due to Ribet and Takahashi [**65**], which depend on a study of the geometry of Shimura curves, especially a description of their bad reduction and of the component groups of the Néron models of their Jacobians.

- The Tate conjecture for products of curves over number fields, which was proved by Faltings [**15**], and which implies that modular elliptic curves are equipped with a uniformization map $X \to E$, with X being a Shimura curve.
- A study of the *Manin constant* for the map $X \to E$, following [**52**], [**13**], [**1**].

(ii) In the more general case of weight $2k > 2$, such geometric arguments are not available. One of the main obstructions is that the Tate conjecture is unknown for products of Kuga–Sato varieties. Instead, one may try to use purely automorphic techniques. This is the strategy employed in [**60**], where we showed (using the theta correspondence and results from Iwasawa theory) that for f and g of arbitrary even weight, the ratio $\langle f, f\rangle/\langle g, g\rangle$ is integral outside of an explicit finite set of small primes, and further that it is always divisible by primes at which the form f satisfies certain level-lowering congruences. The converse divisibility and the more precise relation to Tamagawa numbers is also expected to hold in general, but seems harder to prove. This is one problem that we hope to eventually address by the methods of this paper.

Let us now elaborate a bit on the relation of this problem to the Tate conjecture. As described above, the case of weight two forms for GL_2 over $\mathbb{Q}$ is relatively simple since one knows by Faltings that the Tate conjecture holds for a product of curves. This implies that there exists an algebraic cycle on the product $X_1 \times X_2$, where X_1 and X_2 are modular and Shimura curves respectively, that at the level of cohomology, identifies the f and g-isotypic components of the "motives" $H^1(X_1)$ and $H^1(X_2)$ respectively. The rationality of the period ratio $\langle f, f\rangle/\langle g, g\rangle$ is then a simple consequence of the fact that such a cycle induces an isomorphism of the *Hodge–de Rham* structures [**23**] attached to f and g. For forms of higher weight, the Jacquet–Langlands correspondence can similarly be used to produce Tate classes on a product $W_1 \times W_2$ where W_1 and W_2 are Kuga–Sato varieties fibered over X_1 and X_2 respectively. However, we are far from understanding the Tate (or even Hodge) conjecture in this case. The case of Hilbert modular forms considered in this paper seems even harder: in the simplest setting, namely for forms of *parallel weight two* and trivial central character, the Tate conjecture predicts the existence of algebraic cycles on products of the form $X \times (X_1 \times X_2)$, where X, X_1 and X_2 are suitably chosen quaternionic Shimura varieties such that $\dim(X) = \dim(X_1) + \dim(X_2)$. Again, these cycles should induce isomorphisms of Hodge–de Rham structures $H^*(X)_\Pi \simeq H^*(X_1)_\Pi \otimes H^*(X_2)_\Pi$ that in turn should imply the predicted period relations up to rationality. (Here the subscript Π denotes the Π_f-isotypic component for a fixed automorphic representation $\Pi = \Pi_\infty \otimes \Pi_f$.) This point of view—at least the factorization of Hodge structures—occurs explicitly in the work of Oda ([**56**], [**57**]). It is worth remarking here that the Tate and Hodge conjectures are only expected to hold rationally in general and not integrally, and thus by themselves do not predict any statements about integrality of period ratios. Nevertheless, the discussion above on period ratios suggests that in the setting of arithmetic automorphic forms on Shimura varieties, such integral relations do hold and that their proofs lie much deeper than those of the corresponding rational relations.

With this background, we will outline the main results of this paper. Let F be a totally real number field with $[F : \mathbb{Q}] = d$, ring of integers $\mathcal{O}_F$, class number h_F and discriminant D_F. Let $\Pi = \otimes_v \Pi_v$ be an irreducible cuspidal automorphic representation of $\mathrm{GL}_2(\mathbb{A}_F)$ of weight $(\underline{k}, r) = (k_1, \ldots, k_d, r)$, conductor $\mathfrak{N}$ and central character ξ_Π. We assume that $k_1 \equiv k_2 \equiv \cdots \equiv k_d \equiv r \pmod 2$ and all the $k_i \geq 1$. Thus Π is the automorphic representation associated with a classical Hilbert modular form. (Note that we allow forms of partial or parallel weight one.)

For simplicity we will assume that at all finite places v where Π_v is ramified, it is either a special representation with square-free conductor (i.e., an unramified twist of the Steinberg representation) or a ramified principal series representation $\mathrm{Ind}(\chi_1 \otimes \chi_2)$ with χ_1 unramified and χ_2 ramified. We can thus factor the conductor $\mathfrak{N}$ of Π as

$$\mathfrak{N} = \mathfrak{N}_{\mathrm{s}} \cdot \mathfrak{N}_{\mathrm{ps}}$$

where $\mathfrak{N}_{\mathrm{s}}$ is the (square-free) product of the conductors at places where Π_v is special and $\mathfrak{N}_{\mathrm{ps}}$ is the product of the conductors at places where Π_v is ramified principal series.

Let K_Π be the number field generated by the Hecke eigenvalues of Π and $\mathcal{O}_{K_\Pi}$ the ring of integers of K_Π. We set $N_\Pi := \mathrm{N}\mathfrak{N}$, $k_\Pi := \max k_i$ and

$$N(\Pi) := 2 \cdot h_F \cdot D_F \cdot N_\Pi \cdot k_\Pi!, \qquad R := \mathcal{O}_{\overline{\mathbb{Q}}}[1/N(\Pi)].$$

Let Σ_F denote the set of all places of F and Σ_∞ and Σ_{fin} the subsets of infinite and finite places respectively. Let Σ_Π be the set of places v of F at which Π_v is discrete series. Thus, Σ_Π equals the union of $\Sigma_{\Pi,\infty}$ and $\Sigma_{\Pi,\mathrm{fin}}$, where

$$\begin{aligned}
\Sigma_{\Pi,\infty} &:= \Sigma_\Pi \cap \Sigma_\infty = \{v \in \Sigma_\infty : \ k_v \geq 2\}, \\
\Sigma_{\Pi,\mathrm{fin}} &:= \Sigma_\Pi \cap \Sigma_{\mathrm{fin}} = \{v \in \Sigma_{\mathrm{fin}} : \ \mathrm{ord}_v(\mathfrak{N}_s) > 0\}.
\end{aligned}$$

For any quaternion algebra B over F, let Σ_B denote the set of places of F at which B is ramified. Also set

$$\begin{aligned}
\Sigma_{B,\infty} &:= \Sigma_B \cap \Sigma_\infty, \\
\Sigma_{B,\mathrm{fin}} &:= \Sigma_B \cap \Sigma_{\mathrm{fin}}.
\end{aligned}$$

Henceforth we suppose that $\Sigma_B \subset \Sigma_\Pi$, so that by Jacquet–Langlands [**35**], Π transfers to an automorphic representation Π_B of $B^\times(\mathbb{A})$. To such a pair (B, Π), we will attach in §1.4 below a canonical quadratic period invariant

$$q_B(\Pi) \in \mathbb{C}^\times / R^\times.$$

This period invariant is essentially (i.e., up to some factors coming from normalizations of measures) equal to the Petersson inner product of a normalized eigenform f_B in Π_B. Here we use the assumption that $\mathfrak{N}_{\mathrm{s}}$ is square-free to first fix f_B up to a scalar. The scalar is then fixed by requiring that f_B correspond to an integrally normalized section of a suitable automorphic vector bundle on the Shimura variety associated with the algebraic group $B^\times$.

The goal of this paper and its sequels is to study the relations between the invariants $q_B(\Pi)$ for fixed Π as B varies over all quaternion algebras in Σ_Π. The following conjecture is a more precise version of [**61**] Conjecture 4.2 and provides an integral refinement of Shimura's conjecture on algebraic period relations. The reader may consult §4 of [**61**] for a discussion of the motivation behind this formulation. To state the conjecture, let $L(s, \Pi, \mathrm{ad})$ denote the adjoint (finite) L-function attached to Π and let $\Lambda(s, \Pi, \mathrm{ad})$ denote the corresponding completed L-function that includes the Γ-factors at the infinite places. Let us recall the following invariant of Π, which has played a crucial role in the study of congruences of modular forms (see [**29**], [**30**], [**77**]):

$$\Lambda(\Pi) := \Lambda(1, \Pi, \mathrm{ad}). \tag{I.1}$$

CONJECTURE A. *There exists a function*

$$c(\Pi) : \Sigma_\Pi \to \mathbb{C}^\times/R^\times, \quad v \mapsto c_v(\Pi),$$

such that:

(i) $c_v(\Pi)$ *lies in* R *(mod* $R^\times$*) if* v *is a finite place, and*
(ii) *for all* B *with* $\Sigma_B \subseteq \Sigma_\Pi$*, we have*

$$q_B(\Pi) = \frac{\Lambda(\Pi)}{\prod_{v\in\Sigma_B} c_v(\Pi)} \quad (\textit{in } \mathbb{C}^\times/R^\times).$$

REMARK 1. It is easy to see that, if it exists, the function $c(\Pi)$ is uniquely determined as long as $|\Sigma_\Pi| \geq 3$. Also, as the notation suggests, the invariants $c_v(\Pi)$ are *not* invariants of the local representation Π_v but rather are really invariants of the global representation Π.

REMARK 2. The conjecture should be viewed as describing *period relations* between the quadratic periods $q_B(\Pi)$ as B varies. Indeed, the number of B with $\Sigma_B \subseteq \Sigma_\Pi$ is $2^{|\Sigma_\Pi|-1}$ but the conjecture predicts that the corresponding invariants $q_B(\Pi)$ can all be described using only $|\Sigma_\Pi|+1$ invariants, namely the L-value $\Lambda(\Pi)$ and the additional invariants $c_v(\Pi)$, which are $|\Sigma_\Pi|$ in number.

REMARK 3. For $B = \mathrm{M}_2(F)$, the conjecture simply predicts that

$$q_{\mathrm{M}_2(F)}(\Pi) = \Lambda(\Pi) \quad \text{in } \mathbb{C}^\times/R^\times.$$

This piece of the conjecture is known to be true. Indeed, it follows from the fact that the integral normalization of f_B in the split case coincides with the q-expansion normalization (see [**11**], §5), combined with the well known relation between the Petersson inner product of a Whittaker normalized form $f \in \Pi$ and the value of the adjoint L-function at $s = 1$. (See Prop. 6.3.1 for instance.)

It is natural to ask for an independent description of the invariants $c_v(\Pi)$. Before discussing this, we recall the notion of *Eisenstein primes* for Π. To any finite place λ of K_Π, one can associate (by [**69**], [**9**], [**12**], [**67**], [**5**], [**76**], [**73**], [**37**]; see also [**4**]) an irreducible two dimensional Galois representation

$$\rho_{\Pi,\lambda} : \mathrm{Gal}(\overline{\mathbb{Q}}/F) \to \mathrm{GL}_2(K_{\Pi,\lambda})$$

that is characterized up to isomorphism by the requirement that

$$\operatorname{tr} \rho_{\Pi,\lambda}(\mathrm{Frob}_v) = a_v(\Pi)$$

for any finite place v of F that is prime to $\mathfrak{N}\cdot\mathrm{N}\lambda$, with $a_v(\Pi)$ being the eigenvalue of the Hecke operator T_v acting on a new-vector in Π_v. Choose a model for $\rho_{\Pi,\lambda}$ that takes values in $\mathrm{GL}_2(\mathcal{O}_{K_\Pi,\lambda})$ and denote by $\bar\rho_{\Pi,\lambda}$ the semisimplification of the mod λ reduction of $\rho_{\Pi,\lambda}$. The isomorphism class of $\bar\rho_{\Pi,\lambda}$ is independent of the choice of model of $\rho_{\Pi,\lambda}$. Let $\mathbf{F}_\lambda = \mathcal{O}_{K_\Pi}/\lambda$ be the residue field at λ. The prime λ is said to be Eisenstein for Π if

$$\bar\rho_{\Pi,\lambda} : \mathrm{Gal}(\overline{\mathbb{Q}}/F) \to \mathrm{GL}_2(\mathbf{F}_\lambda)$$

is (absolutely) reducible, and non-Eisenstein otherwise.

Let $N(\Pi)_{\mathrm{Eis}}$ be the product of the $\mathrm{N}\lambda$ as λ varies over all the Eisenstein primes for Π. (There are only finitely many such.) Let $\tilde{R}$ denote the ring

$$\tilde{R} := R[1/N(\Pi)_{\mathrm{Eis}}] = \mathcal{O}_{\overline{\mathbb{Q}}}[1/N(\Pi)N(\Pi)_{\mathrm{Eis}}].$$

The following conjecture characterizes the invariants $c_v(\Pi)$ for finite places v up to Eisenstein primes, relating them to *level-lowering congruences* for Π. (It is obviously conditional on the truth of Conjecture A. See also [**59**].)

CONJECTURE B. *Suppose that v belongs to $\Sigma_{\Pi,\mathrm{fin}}$. Then $c_v(\Pi)$ mod $\tilde{R}^\times$ has a representative in $K_\Pi \cap \tilde{R}$. Moreover, if λ is a non-Eisenstein prime for Π with $(\lambda, \mathrm{N}(\Pi)) = 1$, then $v_\lambda(c_v(\Pi))$ equals the largest integer n such that $\rho_{\Pi,\lambda}$ mod λ^n is unramified at v.*

At the infinite places v, one might hope to have similarly a description of the invariants $c_v(\Pi)$ purely in terms of the compatible system $\rho_{\Pi,\lambda}$ of two-dimensional Galois representations attached to Π. In principle, to such a system one should be able to attach a *motive* defined over F, and the $c_v(\Pi)$ should be related to periods of this motive taken with respect to suitable integral structures on the de Rham and Betti realizations. In practice, the only case in which one can make an unconditional definition is when Π satisfies the following conditions:

(a) Π is of *parallel* weight 2, that is $\underline{k} = (2, \dots, 2)$.
(b) Either $d (= [F : \mathbb{Q}])$ is odd or $\Sigma_{\Pi,\mathrm{fin}}$ is nonempty.

If Π satisfies both (a) and (b) above, it is known (using [**5**]) that one can associate to Π an abelian variety A over F (or more precisely, an isogeny class of abelian varieties) such that

- $\dim(A) = [K_\Pi : \mathbb{Q}]$;
- $\mathrm{End}_F(A) \otimes \mathbb{Q} \supset K_\Pi$;
- A has good reduction outside $\mathfrak{N}$;
- For any prime λ of K_Π lying over a rational prime ℓ, the representation $\rho_{\Pi,\lambda}$ is isomorphic to the representation of $\mathrm{Gal}(\overline{\mathbb{Q}}/F)$ on $H^1_{\mathrm{et}}(A_{\overline{\mathbb{Q}}}, \mathbb{Q}_\ell) \otimes_{K_\Pi \otimes \mathbb{Q}_\ell} K_{\Pi,\lambda}$.

In this isogeny class, we may pick an abelian variety A such that $\mathrm{End}_F(A) \supset \mathcal{O}_{K_\Pi}$. Then one can make a precise conjecture for $c_v(\Pi)$ for $v \in \Sigma_\infty$ in terms of the periods of A. Here, we will be content to state this conjecture in the case $K_\Pi = \mathbb{Q}$, namely when A is an elliptic curve over F. Let $\mathcal{A}$ denote the Néron model of A over $R_F := \mathcal{O}_F[1/N(\Pi)]$. Then $\mathcal{L} := H^0(\mathcal{A}, \Omega^1_{\mathcal{A}/R_F})$ is an invertible R_F-module. This module can be trivialized by picking a large enough number field $K \supseteq F$ and extending scalars to the ring $R_K := \mathcal{O}_K[1/N(\Pi)]$. Pick a generator ω for $\mathcal{L} \otimes_{R_F} R_K$ viewed as an R_K-module. Let v' be any archimedean place of K extending v, and denote by $\sigma_v : F \to \mathbb{R}$ the real embedding of F corresponding to v. The class of the integral

$$\frac{1}{(2\pi i)^2} \int_{A \otimes_{\sigma_v} \mathbb{C}} \omega_{v'} \wedge \bar{\omega}_{v'}$$

in $\mathbb{C}^\times / \tilde{R}^\times$ can be checked to be independent of the choices above.

CONJECTURE C. *Suppose that $K_\Pi = \mathbb{Q}$ so that A is an elliptic curve. Then*

$$c_v(\Pi) = \frac{1}{(2\pi i)^2} \int_{A \otimes_{\sigma_v} \mathbb{C}} \omega_{v'} \wedge \bar{\omega}_{v'} \quad \textit{in } \mathbb{C}^\times / \tilde{R}^\times.$$

REMARK 4. One expects that the invariants $c_v(\Pi)$ are transcendental for any infinite place v. Note that if A is the base change of an elliptic curve defined over a smaller totally real field F' (in which case Π is the base change of a Hilbert modular form for F'), then there are obvious algebraic relations between the $c_v(\Pi)$. It would be interesting to formulate a converse to this: namely, can one give a criterion for Π to be a base change purely in terms of the $\overline{\mathbb{Q}}$-algebra generated by the invariants $c_v(\Pi)$?

REMARK 5. It would also be interesting to formulate the conjectures above without inverting $N(\Pi)$. There are lots of obvious difficulties with primes that are small with respect to the weight as well as with integral models at primes of bad reduction. In **[32]**, we will extend the conjectures above in the case $F = \mathbb{Q}$ to include primes of bad reduction at which the local component of the automorphic representation Π is *ramified principal series*. The only Shimura varieties that occur then are Shimura curves and those associated with definite quaternion algebras over $\mathbb{Q}$. The geometric difficulties with primes of bad reduction can be dealt with in this case "by hand".

The goal of this first paper is to reformulate Conjecture A in terms of a new conjecture (Conjecture D below) on the arithmetic properties of a theta lift between quaternionic Shimura varieties. This reformulation has many advantages since the arithmetic of theta lifts can be studied via a range of automorphic techniques including the Rallis inner product formula and period integrals along tori. Moreover, the constructions involved seem to be useful in attacking several other related problems involving algebraic cycles. We will briefly discuss two such applications below.

Now we outline the main construction. Let B_1, B_2 and B be three quaternion algebras unramified outside of Σ_Π such that $B = B_1 \cdot B_2$ in the Brauer group of F. There is then, up to isometry, a unique skew-hermitian B-space $(V, \langle\cdot,\cdot\rangle)$ such that

$$\mathrm{GU}_B(V)^0 \simeq (B_1^\times \times B_2^\times)/F^\times.$$

Here $\mathrm{GU}_B(V)^0$ denotes the identity component of the group of quaternionic unitary similitudes of V. For computational purposes, we will need an explicit construction of such a space V. For this, we pick a CM extension E/F with

$$E = F + F\mathbf{i}, \quad \mathbf{i}^2 = u \in F^\times,$$

such that E embeds in B_1 and B_2. Fix embeddings

$$E \hookrightarrow B_1, \quad E \hookrightarrow B_2$$

and write

$$B_1 = E + E\mathbf{j}_1, \quad B_2 = E + E\mathbf{j}_2,$$

where $\mathbf{j}_1^2 = J_1$ and $\mathbf{j}_2^2 = J_2$ lie in F. Then there is an embedding of E in B such that

$$B = E + E\mathbf{j}, \quad \mathbf{j}^2 = J,$$

where $J = J_1 J_2$. Let $V = B_1 \otimes_E B_2$, viewed as a right E-vector space. In Chapter 2 below, we show that V can naturally be equipped with a right B-action extending the action of E, as well as a B-skew Hermitian form $\langle\cdot,\cdot\rangle$ such that the quaternionic unitary similitude group $\mathrm{GU}_B(V)^0$ admits the description above.

Let W be a one-dimensional B-space equipped with the standard B-hermitian form so that

$$\mathrm{GU}_B(W) = B^\times.$$

We wish to study the theta lift

$$\Theta : \mathcal{A}(\mathrm{GU}_B(W)) \longrightarrow \mathcal{A}(\mathrm{GU}_B(V)^0),$$

where $\mathcal{A}(G)$ denotes the space of cuspidal automorphic forms on $G(\mathbb{A})$. The pair $(\mathrm{U}_B(W), \mathrm{U}_B(V))$ is an example of a classical reductive dual pair. For our applications we need to work with the corresponding similitude groups. In order to construct the theta lift, one needs to first construct (local) splittings of the metaplectic cover over the subgroup

$$\{(g, h) \in \mathrm{GU}_B(V)^0 \times \mathrm{GU}_B(W) : \nu(g) = \nu(h)\},$$

that satisfy the product formula. (Here ν denotes the similitude character.) For quaternionic unitary similitude groups, this does not seem to be covered in the existing literature. This problem is handled in the appendices under the assumption that u, J_1 and J_2 are chosen such that for every finite place v of F, at least one of u, J_1, J_2 and J is locally a square. (See Remark 7 below.)

The splittings being chosen, the correspondence Θ above can be defined and studied. For any quaternion algebra B' with $\Sigma_{B'} \subseteq \Sigma_\Pi$, we let $\pi_{B'}$ denote the *unitary* representation associated with $\Pi_{B'}$. Thus

$$\pi_{B'} = \Pi_{B'} \otimes \|\nu_{B'}\|^{-r/2},$$

where $\nu_{B'}$ denotes the reduced norm on B'. In Chapter 4, we prove the following theorem regarding Θ (in the case $B \neq \mathrm{M}_2(F)$), which gives an explicit realization of the Jacquet–Langlands correspondence, extending the work of Shimizu [**68**].

THEOREM 1.

$$\Theta(\pi_B) = \pi_{B_1} \boxtimes \pi_{B_2}.$$

REMARK 6. Up to this point in the paper, we make no restrictions on F or Π. However from Chapter 5 onwards (and thus in the rest of the introduction), we assume for simplicity the following:

- $\mathfrak{N}$ is prime to $2\mathfrak{D}_{F/\mathbb{Q}}$, where $\mathfrak{D}_{F/\mathbb{Q}}$ denotes the different of $F/\mathbb{Q}$.

These assumptions simplify some of the local computations in Chapters 5 and 6, and could be relaxed with more work.

While Theorem 1 is an abstract representation theoretic statement, for our purposes we need to study a more explicit theta lift. The Weil representation used to define the theta lift above is realized on a certain Schwartz space $\mathcal{S}(\mathbb{X})$. In Chapter 5, we pick an explicit canonical Schwartz function $\varphi \in \mathcal{S}(\mathbb{X})$ with the property that $\theta_\varphi(f_B)$ is a scalar multiple of $f_{B_1} \boxtimes f_{B_2}$. Thus

$$\theta_\varphi(f_B) = \alpha(B_1, B_2) \cdot (f_{B_1} \times f_{B_2}),$$

for some scalar $\alpha(B_1, B_2) \in \mathbb{C}^\times$. The scalar $\alpha(B_1, B_2)$ depends not just on B_1 and B_2 but also on the other choices made above. However, we will omit these other dependencies in the notation.

That $\alpha(B_1, B_2)$ is nonzero follows from the following explicit version of the Rallis inner product formula, proved in Chapter 6. (The assumption $B \neq \mathrm{M}_2(F)$ in the statement below is made since the proof in the case $B = \mathrm{M}_2(F)$ would be somewhat different. See for instance §6 of [**16**]. Note that this case corresponds to the original setting of Shimuzu [**68**], and is not needed in this paper.)

THEOREM 2. *Suppose $B_1 \neq B_2$, or equivalently, $B \neq \mathrm{M}_2(F)$. Then*

$$|\alpha(B_1, B_2)|^2 \cdot \langle f_{B_1}, f_{B_1}\rangle \cdot \langle f_{B_2}, f_{B_2}\rangle = C \cdot \langle f, f\rangle \cdot \langle f_B, f_B\rangle,$$

where C is an explicit constant (see Thm. 6.3.2) and f is a Whittaker normalized form in Π (as in Remark 3).

The arithmetic properties of $\alpha(B_1, B_2)$ are of key importance. As such, the choice of measures needed to define the invariants $q_B(\Pi)$ requires us to work with a slight modification of $\alpha(B_1, B_2)$, denoted $\boldsymbol{\alpha}(B_1, B_2)$, as described in §7.2. We are especially interested in questions of integrality of $\boldsymbol{\alpha}(B_1, B_2)$ for which we may work one prime at a time. Thus we fix a prime ℓ not dividing $N(\Pi)$ and then choose all the data (for example, E, J_1, J_2, φ) to be suitably adapted to ℓ. The choices are described in detail in §7.1. Finally, we come to main conjecture of this paper, which is motivated by combining Theorem 2 with Conj. A.

CONJECTURE D. *Suppose that $B_1 \neq B_2$ and $\Sigma_{B_1} \cap \Sigma_{B_2} \cap \Sigma_\infty = \varnothing$, that is B_1 and B_2 have no infinite places of ramification in common. Then*

(i) *$\boldsymbol{\alpha}(B_1, B_2)$ lies in $\overline{\mathbb{Q}}^\times$.*
(ii) *$\boldsymbol{\alpha}(B_1, B_2)$ is integral at all primes above ℓ.*
(iii) *If in addition B_1 and B_2 have no finite places of ramification in common, then $\boldsymbol{\alpha}(B_1, B_2)$ is a unit at all primes above ℓ.*

While not immediately apparent, Conjecture D implies Conjecture A. Indeed, in §7.2, we show the following.

THEOREM 3. *Suppose that Conjecture D is true for all ℓ in some set of primes Ξ. Then Conjecture A holds with R replaced by $R[1/\ell : \ell \notin \Xi]$. Consequently, if Conjecture D is true for all $\ell \nmid N(\Pi)$, then Conjecture A is true.*

At this point, the reader may feel a bit underwhelmed since all we seem to have done is reformulate Conjecture A in terms of another conjecture that is not visibly easier. However, we believe that Conjecture D provides the correct perspective to attack these fine integrality questions about period ratios, for several reasons. Firstly, it does not require an a priori definition of the invariants c_v. Second, it fits into the philosophy that *theta lifts have excellent arithmetic properties* and is amenable to attack by automorphic methods of various kinds. Lastly, it is usually a very hard problem (in Iwasawa theory, say) to prove *divisibilities*; on the other hand, if a quantity is expected to be a unit, then this might be easier to show, for instance using congruences. Part (iii) of Conjecture D, which states that $\boldsymbol{\alpha}(B_1, B_2)$ is often a unit, has hidden in it a large number of divisibilities that would be very hard to show directly, but that might be more accessible when approached in this way. This is the approach taken in the sequels [**31**] and [**32**] where we study Conjecture D and give various applications to periods.

As mentioned earlier, the constructions discussed above also have concrete applications to problems about algebraic cycles. We mention two of these:

– In [**33**], we study the Bloch–Beilinson conjecture for Rankin–Selberg L-functions $L(f_E, \chi, s)$, where f is a modular form of weight k and χ is a Hecke character of an imaginary quadratic field E of infinity type $(k', 0)$ with $k' \geq k$. The simplest case is when $(k, k') = (2, 2)$. In this case we give an explicit construction of cycles corresponding to the vanishing of the L-function and prove a relation between the p-adic logarithms of such cycles and values of p-adic L-functions. (All previous constructions of cycles for such L-functions ([**19**], [**55**], [**3**]) only work in the case $k > k'$.) The key input from this paper is the embedding

$$\mathrm{GU}_B(V)^0 \to \mathrm{GU}_E(\mathbf{V}), \quad \mathbf{V} := \mathrm{Res}_{B/E}(V),$$

which provides a morphism of Shimura varieties that can be used to construct the relevant cycle.

– In [**34**] we consider the Tate conjecture for products $X_1 \times X_2$ where X_1 and X_2 are the Shimura varieties associated with two quaternion algebras B_1 and B_2 over a totally real field F that have identical ramification at the infinite places of F. As explained earlier, the Jacquet–Langlands correspondence gives rise to natural Tate classes on $X_1 \times X_2$ and the Tate conjecture predicts the existence of algebraic cycles on the product giving rise to these Tate classes. While we cannot as yet show the existence of such cycles, we are able to at least give an unconditional construction of the corresponding Hodge classes, and show moreover that these are compatible with all the Tate classes, i.e., are close to being *absolutely Hodge*. These Hodge classes are constructed not by comparing periods but rather by finding a morphism

$$X_1 \times X_2 \to X$$

into an auxiliary Shimura variety X and constructing Hodge classes on X that restrict nontrivially to $X_1 \times X_2$. The relation with the current paper is that $X_1 \times X_2$ and X may be viewed as the Shimura varieties associated with certain skew-hermitian B spaces, with $B = B_1 \cdot B_2$.

Finally, we give a brief outline of the contents of each chapter. In Chapter 1 we recall the theory of automorphic vector bundles on quaternionic Shimura varieties and define the canonical quadratic period invariants $q_B(\Pi)$. In Chapter 2, we give the key constructions involving quaternionic skew-hermitian forms. Chapter 3 discusses the general theory of the theta correspondence as well as the special case of quaternionic dual pairs, while Chapter 4 establishes the general form of the Rallis inner product formula in our situation and proves that the theta lift we are considering agrees with the Jacquet–Langlands correspondence. In Chapter 5, we pick explicit Schwartz functions, which are then used in Chapter 6 to compute the precise form of the Rallis inner product formula in our setting. In Chapter 7 we first discuss all the choices involved in formulating the main conjecture, Conjecture D above, and then show that it implies Conjecture A. Appendix A is strictly not necessary but is useful in motivating some constructions in Chapter 1. The results from Appendix B on metaplectic covers of symplectic *similitude* groups are used in the computations in Appendix C. Appendices C and D are invoked in Chapters 3, 4 and 5, and contain the construction of the relevant splittings, on which more is said in the remark below.

REMARK 7. The problem of constructing the required splittings and checking various compatibilities involving them turns out to be rather nontrivial and occupies the lengthy Appendices C and D. For *isometry groups*, these can be handled using the doubling method as in Kudla [**40**, §4]. This gives a collection of splittings (one for each place v)

$$s_{\mathrm{Kudla},v} : \mathrm{U}_B(V)(F_v) \times \mathrm{U}_B(W)(F_v) \to \mathbb{C}^{(1)}$$

that satisfy the *product formula*:

$$\prod_v s_{\mathrm{Kudla},v}(\gamma) = 1$$

for $\gamma \in \mathrm{U}_B(V)(F) \times \mathrm{U}_B(W)(F)$. The problem is really to extend these splittings to the groups

$$\{(g,h) \in \mathrm{GU}_B(V)^0(F_v) \times \mathrm{GU}_B(W)(F_v) : \nu(g) = \nu(h)\}$$

in such a way that they still satisfy the product formula. A similar problem for the dual pairs consisting of the unitary similitude groups of a hermitian E-space $\mathbf{V}$ and a skew-hermitian E-space $\mathbf{W}$ can be solved using the fact that $\mathbf{V} \otimes_E \mathbf{W}$ can be considered as a *skew-hermitian* E-space, and the group

$$\{(g,h) \in \mathrm{GU}_E(\mathbf{V}) \times \mathrm{GU}_E(\mathbf{W}) : \nu(g) = \nu(h)\}$$

(almost) embeds in $\mathrm{U}_E(\mathbf{V} \otimes_E \mathbf{W})$. This fails when working with B-spaces since B is non-commutative and the tensor product construction is not available. To circumvent this problem, we first construct by hand, splittings

$$s_v : \{(g,h) \in \mathrm{GU}_B(V)^0(F_v) \times \mathrm{GU}_B(W)(F_v) : \nu(g) = \nu(h)\} \to \mathbb{C}^{(1)}$$

in Appendix C and check that they satisfy several natural properties including the product formula (Proposition C.4.4). This suffices to construct the theta lift Θ. In order to prove the Rallis inner product formula, we need to check a further compatibility between our splittings s_v and the splittings $s_{\mathrm{Kudla},v}$, in the context of the doubling method. This is accomplished in Lemma D.4.2 in Appendix D.

Acknowledgments

During the preparation of this article, Ichino was partially supported by JSPS KAKENHI Grant Numbers 22740021, 26287003, and Prasanna was partially supported by NSF grants DMS 0854900, DMS 1160720, a grant from the Simons Foundation (305784) and by a Margaret and Herman Sokol Faculty Award from the University of Michigan.

CHAPTER 1

Quaternionic Shimura Varieties

1.1. Shimura varieties

1.1.1. Shimura varieties and canonical models. We recall quickly the general theory of Shimura varieties and their canonical models [**10**]. Let $\mathbb{S} := \mathrm{Res}_{\mathbb{C}/\mathbb{R}}\, \mathbf{G}_m$ denote the Deligne torus. There is an equivalence of categories

$$\mathbb{R}\text{-Hodge structures} \longleftrightarrow \mathbb{R}\text{-vector spaces with an algebraic action of } \mathbb{S},$$

described as follows. Suppose that V is an $\mathbb{R}$-vector space equipped with a pure Hodge structure of weight n. Thus we have a decomposition of $V_{\mathbb{C}} := V \otimes_{\mathbb{R}} \mathbb{C}$:

$$V_{\mathbb{C}} = \bigoplus_{p+q=n} V^{pq},$$

where $V^{pq} = \overline{V^{qp}}$. Define an action h of $\mathbb{C}^{\times}$ on $V_{\mathbb{C}}$ by

$$h(z)v = z^{-p}\bar{z}^{-q}v \quad \text{for} \quad v \in V^{pq}.$$

Since $h(z)$ commutes with complex conjugation, it is obtained by extension of scalars from an automorphism of V defined over $\mathbb{R}$. This gives a map on real points $h : \mathbb{S}(\mathbb{R}) = \mathbb{C}^{\times} \to \mathrm{GL}(V)(\mathbb{R})$, that comes from an algebraic map $\mathbb{S} \to \mathrm{GL}(V)$.

A Shimura datum is a pair (G, X) consisting of a reductive algebraic group G over $\mathbb{Q}$ and a $G(\mathbb{R})$-conjugacy class X of homomorphisms $h : \mathbb{S} \to G_{\mathbb{R}}$ satisfying the following conditions:

(i) For h in X, the Hodge structure on the Lie algebra $\mathfrak{g}$ of $G_{\mathbb{R}}$ given by $\mathrm{Ad} \circ h$ is of type $(0,0) + (-1,1) + (1,-1)$. (In particular, the restriction of such an h to $\mathbf{G}_{m,\mathbb{R}} \subset \mathbb{S}$ is central.)
(ii) For h in X, $\mathrm{Ad}\ h(i)$ is a Cartan involution on $G^{\mathrm{ad}}_{\mathbb{R}}$, where G^{ad} is the adjoint group of G.
(iii) G^{ad} has no factor defined over $\mathbb{Q}$ whose real points form a compact group.

These conditions imply that X has the natural structure of a disjoint union of Hermitian symmetric domains. The group $G(\mathbb{R})$ acts on X on the left by

$$(g \cdot h)(z) = g \cdot h(z) \cdot g^{-1}.$$

To agree with our geometric intuition, we will sometimes write τ_h (or simply τ) for h in X.

Let $\mathbb{A}$ and $\mathbb{A}_f$ denote respectively the ring of adèles and finite adèles of $\mathbb{Q}$. Let $\mathcal{K}$ be an open compact subgroup of $G(\mathbb{A}_f)$. The Shimura variety associated to $(G, X, \mathcal{K})$ is the quotient

$$\mathrm{Sh}_{\mathcal{K}}(G, X) = G(\mathbb{Q})\backslash X \times G(\mathbb{A}_f)/\mathcal{K}.$$

For $\mathcal{K}$ small enough, this has the natural structure of a smooth variety over $\mathbb{C}$. The inverse limit

$$\mathrm{Sh}(G,X) = \varprojlim_{\mathcal{K}} \mathrm{Sh}_{\mathcal{K}}(G,X)$$

is a pro-algebraic variety that has a canonical model over a number field $E(G,X)$, the reflex field of the Shimura datum (G,X). In particular, each $\mathrm{Sh}_{\mathcal{K}}(G,X)$ has a canonical model over $E(G,X)$.

We recall the definition of $E(G,X)$. This field is defined to be the field of definition of the conjugacy class of co-characters

$$\mu_h : \mathbf{G}_{m,\mathbb{C}} \to \mathbb{S}_{\mathbb{C}} \to G_{\mathbb{C}},$$

where the first map is $z \mapsto (z,1)$ and the second is the one induced by h. Let us say more precisely what this means. For any subfield k of $\mathbb{C}$, let $\mathcal{M}(k)$ denote the set of $G(k)$-conjugacy classes of homomorphisms $\mathbf{G}_{m,k} \to G_k$. Let $\overline{\mathbb{Q}}$ denote the algebraic closure of $\mathbb{Q}$ in $\mathbb{C}$. Then the inclusion $\overline{\mathbb{Q}} \hookrightarrow \mathbb{C}$ gives a bijection between $\mathcal{M}(\overline{\mathbb{Q}})$ and $\mathcal{M}(\mathbb{C})$, and thus a natural action of $\mathrm{Gal}(\overline{\mathbb{Q}}/\mathbb{Q})$ on $\mathcal{M}(\mathbb{C})$. The reflex field $E(G,X) \subset \mathbb{C}$ is then the fixed field of the subgroup

$$\{\sigma \in \mathrm{Gal}(\overline{\mathbb{Q}}/\mathbb{Q}) :\ \sigma M_X = M_X\}$$

where M_X is the conjugacy class of μ_h for any $h \in X$.

1.1.2. Automorphic vector bundles. We recall the basics of the theory of automorphic vector bundles following [**20**], [**21**], [**17**]. First, to any $\mu : \mathbf{G}_{m,\mathbb{C}} \to G_{\mathbb{C}}$ as above one can associate a filtration $\mathrm{Filt}(\mu)$ of $\mathrm{Rep}_{\mathbb{C}}(G_{\mathbb{C}})$. This is the functor which assigns to every complex representation (V,ρ) of $G_{\mathbb{C}}$ the filtered vector space $(V, F_\mu^{\cdot})$ where $F_\mu^{\cdot}$ is the filtration on V corresponding to $\rho \circ \mu$; that is, $F_\mu^p V = \oplus_{i \geq p} V_\mu^i$, where V_μ^i is the subspace of V on which $\mathbf{G}_m(\mathbb{C})$ acts via $z \mapsto z^i$. In particular, one obtains a filtration on $\mathfrak{g}_{\mathbb{C}}$ via the adjoint representation of $G(\mathbb{C})$. Let P_μ be the subgroup of $G_{\mathbb{C}}$ that preserves the filtration $F_\mu^{\cdot}$ in every representation (V,ρ). Then P_μ is a parabolic subgroup of $G_{\mathbb{C}}$ that contains the image of μ and has Lie algebra $F_\mu^0 \mathfrak{g}_{\mathbb{C}}$. The unipotent radical $R_u P_\mu$ of P_μ has Lie algebra $F_\mu^1 \mathfrak{g}_{\mathbb{C}}$ and is the subgroup that acts as the identity on $\mathrm{Gr}_\mu^{\cdot}(V)$ in every representation (V,ρ). The centralizer $Z(\mu)$ of μ in $G_{\mathbb{C}}$ is a Levi subgroup of P_μ, isomorphic to $P_\mu / R_u P_\mu$. Thus the composite map

$$\bar{\mu} : \mathbf{G}_{m,\mathbb{C}} \to P_\mu \to P_\mu / R_u P_\mu$$

is a central homomorphism. Then $\mathrm{Filt}(\mu)$ equals $\mathrm{Filt}(\mu')$ if and only if $P_\mu = P_{\mu'}$ and $\bar{\mu} = \bar{\mu'}$.

Let $\check{X}$ denote the compact dual Hermitian symmetric space to X. As a set, it may be defined as the set of filtrations of $\mathrm{Rep}_{\mathbb{C}}(G_{\mathbb{C}})$ that are $G(\mathbb{C})$-conjugate to $\mathrm{Filt}(\mu_h)$. Equivalently, it may be described as the set of equivalence classes $[(P,\mu)]$ of pairs where P is a parabolic in $G_{\mathbb{C}}$ and $\mu : \mathbf{G}_{m,\mathbb{C}} \to P$ is a co-character such that (P,μ) is $G(\mathbb{C})$-conjugate to (P_{μ_h}, μ_h) for some (and therefore every) $h \in X$. Here we say that (P,μ) is equivalent to (P',μ') if $P = P'$ and $\bar{\mu} = \bar{\mu'}$. Note that if (P,μ) is conjugate to (P,μ'), then $\bar{\mu} = \bar{\mu'}$. Indeed, if $g^{-1}(P,\mu)g = (P,\mu')$, then $g \in N_{G_{\mathbb{C}}}(P) = P$. Write $g = \ell u$, with $\ell \in Z(\mu)$ and $u \in R_u P$, we see that

$$\mu' = g^{-1} \mu g = u^{-1} \mu u,$$

so that $\bar{\mu'} = \bar{\mu}$ as claimed. Thus in a given conjugacy class of pairs (P,μ), the homomorphism $\bar{\mu}$ is determined entirely by P. Conversely, for any pair (P,μ) in the conjugacy class of (P_{μ_h}, μ_h), the parabolic P must equal P_μ so that μ determines P. It follows from this discussion that the natural map

$$G(\mathbb{C}) \times \check{X} \to \check{X}, \quad (g, [(P,\mu)]) \mapsto [g(P,\mu)g^{-1}]$$

makes $\check{X}$ into a homogeneous space for $G(\mathbb{C})$ and the choice of any basepoint $[(P,\mu)]$ gives a bijection $G(\mathbb{C})/P \simeq \check{X}$. Further, there is a unique way to make $\check{X}$ into a complex algebraic variety such that this map is an isomorphism of complex varieties for any choice of base point. Moreover, the map

$$\xi : M_X \to \check{X}, \quad \mu \mapsto [(P_\mu, \mu)]$$

is surjective and $\check{X}$ has the natural structure of a variety over $E(G,X)$ such that the map ξ is $\mathrm{Aut}(\mathbb{C}/E(G,X))$-equivariant. When we wish to emphasize the rational structure of $\check{X}$, we will write $\check{X}_{\mathbb{C}}$ instead of $\check{X}$.

There is a natural embedding (the Borel embedding)

$$\beta : X \hookrightarrow \check{X}, \quad h \mapsto [(P_h, \mu_h)],$$

where henceforth we write P_h for P_{μ_h}. Let $\check{\mathcal{V}}$ be a $G_{\mathbb{C}}$-vector bundle on $\check{X}$. The action of $G(\mathbb{C})$ on $\check{X}$ extends the $G(\mathbb{R})$ action on X. Thus $\mathcal{V} := \check{\mathcal{V}}|_X$ is a $G(\mathbb{R})$-vector bundle on X. For an open compact subgroup $\mathcal{K}$ of $G(\mathbb{A}_f)$, define

$$\mathcal{V}_{\mathcal{K}} = G(\mathbb{Q})\backslash \mathcal{V} \times G(\mathbb{A}_f)/\mathcal{K},$$

which we view as fibered over $\mathrm{Sh}_{\mathcal{K}}(G,X)$. In order that this define a vector bundle on $\mathrm{Sh}_{\mathcal{K}}(G,X)$, we need to assume that $\check{\mathcal{V}}$ satisfies the following condition:

$$\text{The action of } G_{\mathbb{C}} \text{ on } \check{\mathcal{V}} \text{ factors through } G^c_{\mathbb{C}}. \tag{1.1}$$

Here $G^c = G/Z_s$, where Z_s is the largest subtorus of the center Z_G of G that is split over $\mathbb{R}$ but that has no subtorus split over $\mathbb{Q}$. Assuming (1.1), for sufficiently small $\mathcal{K}$, $\mathcal{V}_{\mathcal{K}}$ is a vector bundle on $\mathrm{Sh}_{\mathcal{K}}(G,X)$. If $\check{\mathcal{V}}$ is defined over $E \supseteq E(G,X)$, then $\mathcal{V}_{\mathcal{K}}$ has a canonical model over E as well.

REMARK 1.1.1. The reader may keep in mind the following example which occurs in this paper. Let $G = \mathrm{Res}_{F/\mathbb{Q}} \mathrm{GL}_2$, with F a totally real field. Then $Z_G = \mathrm{Res}_{F/\mathbb{Q}} \mathbf{G}_m$ and $Z_s = \ker(\mathrm{N}_{F/\mathbb{Q}} : Z_G \to \mathbf{G}_m)$.

We now recall the relation between sections of the bundle $\mathcal{V}_{\mathcal{K}}$ and automorphic forms on $G(\mathbb{A})$. This requires the choice of a base point $h \in X$. Let K_h be the stabilizer in $G(\mathbb{R})$ of h. Let $\mathfrak{k}_h$ denote the Lie algebra of K_h and consider the decomposition of $\mathfrak{g}_{\mathbb{C}}$ with respect to the action of $\mathrm{Ad} \circ h$:

$$\mathfrak{g}_{\mathbb{C}} = \mathfrak{p}_h^+ \oplus \mathfrak{k}_{h,\mathbb{C}} \oplus \mathfrak{p}_h^-.$$

Here $\mathfrak{p}_h^+ = \mathfrak{g}_{\mathbb{C}}^{-1,1}$, $\mathfrak{p}_h^- = \mathfrak{g}_{\mathbb{C}}^{1,-1}$ and $\mathfrak{k}_{h,\mathbb{C}} = \mathfrak{g}_{\mathbb{C}}^{0,0}$ for the Hodge decomposition on $\mathfrak{g}_{\mathbb{C}}$ induced by $\mathrm{Ad} \circ h$. Thus $\mathfrak{p}_h^{\pm}$ correspond to the holomorphic and antiholomorphic tangent spaces of X at h. Then P_h is the parabolic subgroup of $G(\mathbb{C})$ with Lie algebra $\mathfrak{k}_{h,\mathbb{C}} \oplus \mathfrak{p}_h^-$. The choice of h gives identifications $X = G(\mathbb{R})/K_h$, $\check{X} = G(\mathbb{C})/P_h$ and the Borel embedding is given by the natural map

$$G(\mathbb{R})/K_h \hookrightarrow G(\mathbb{C})/P_h.$$

Let $\mathcal{V}_h$ denote the fiber of $\mathcal{V}$ at h; equivalently this is the fiber of the bundle $\check{\mathcal{V}}$ at $\beta(h) \in \check{X}$. This comes equipped with a natural action of K_h, denoted $\rho_{\mathcal{V}_h}$. Let ε_h denote the map

$$G(\mathbb{A}) \to \mathrm{Sh}_{\mathcal{K}}(G,X) = G(\mathbb{Q})\backslash X \times G(\mathbb{A}_f)/\mathcal{K}, \quad g = (g_\infty, g_f) \mapsto [(g_\infty(h), g_f)].$$

Then there is a canonical isomorphism

$$\varepsilon_h^*(\mathcal{V}_\mathcal{K}) \simeq G(\mathbb{A}) \times \mathcal{V}_h,$$

via which sections of $\mathcal{V}_\mathcal{K}$ can be identified with suitable functions from $G(\mathbb{A})$ into $\mathcal{V}_h$. This gives a canonical injective map

$$\mathrm{Lift}_h : \Gamma\left(\mathrm{Sh}_\mathcal{K}(G, X), \mathcal{V}\right) \to C^\infty(G(\mathbb{Q})\backslash G(\mathbb{A})/\mathcal{K}, \mathcal{V}_h)$$

whose image is the subspace $\mathcal{A}(G, \mathcal{K}, \mathcal{V}, h)$ consisting of $\varphi \in C^\infty(G(\mathbb{Q})\backslash G(\mathbb{A})/\mathcal{K}, \mathcal{V}_h)$ satisfying:

(i) $\varphi(gk) = \rho_{\mathcal{V}_h}(k)^{-1}\varphi(g)$, for $g \in G(\mathbb{A})$ and $k \in K_h$;
(ii) $Y \cdot \varphi = 0$ for all $Y \in \mathfrak{p}_h^-$;
(iii) φ is slowly increasing, K_h-finite and $Z(\mathfrak{g}_\mathbb{C})$-finite, where $Z(\mathfrak{g}_\mathbb{C})$ is the center of the universal enveloping algebra of $\mathfrak{g}_\mathbb{C}$.

Let us make explicit the map Lift_h. Fix some $\tau = \tau_h \in X$ and let s be a section of $\mathcal{V}_\mathcal{K}$. For any $g_f \in G(\mathbb{A}_f)$, there is a canonical identification

$$\mathcal{V}_\tau \simeq \mathcal{V}_{\mathcal{K},[\tau,g_f]}$$

where $[\tau, g_f]$ denotes the class of (τ, g_f) in $\mathrm{Sh}_\mathcal{K}(G, X)$. Let $g = (g_\infty, g_f) \in G(\mathbb{A}) = G(\mathbb{R}) \times G(\mathbb{A}_f)$. The section s gives an element $s([g_\infty\tau, g_f]) \in \mathcal{V}_{g_\infty\tau}$. However, the element g_∞ induces an isomorphism

$$t_{g_\infty} : \mathcal{V}_\tau \simeq \mathcal{V}_{g_\infty\tau}.$$

The map $\mathrm{Lift}_h(s) : G(\mathbb{A}) \to \mathcal{V}_\tau$ is then defined by sending

$$g \mapsto t_{g_\infty}^{-1} s([g_\infty\tau, g_f]).$$

REMARK 1.1.2. The subgroup P_h of $G_\mathbb{C}$ acts on the fiber $\check{\mathcal{V}}_{\beta(h)}$ at the point $\beta(h)$. This gives an equivalence of categories

$$G_\mathbb{C}\text{-vector bundles on } \check{X} \longleftrightarrow \text{complex representations of } P_h.$$

The functor in the opposite direction sends a representation (V, ρ) of P_h to the vector bundle

$$G_\mathbb{C} \times_\rho V = (G_\mathbb{C} \times V)/\{(gp, v) \sim (g, \rho(p)v),\ p \in P_h\},$$

which fibers over $G_\mathbb{C}/P_h$ in the obvious way. Sections of this vector bundle can be identified with functions

$$f : G(\mathbb{C}) \to V, \quad f(gp) = \rho(p)^{-1}f(g).$$

EXAMPLE 1.1.3. This example will serve to normalize our conventions. Let $G = \mathrm{GL}_{2,\mathbb{Q}}$ and X the $G(\mathbb{R})$-conjugacy class containing

$$h_0 : \mathbb{S} \to G_\mathbb{R}, \quad a + bi \mapsto \begin{pmatrix} a & b \\ -b & a \end{pmatrix}.$$

Identify X with $\mathfrak{h}^\pm$, the union of the upper and lower half planes; h_0 is identified with the point i. Then $E(G, X) = \mathbb{Q}$ and $\check{X} \simeq \mathbb{P}^1_\mathbb{Q} = G/P$, where P is the Borel subgroup (of upper triangular matrices) stabilizing ∞, for the standard action of G on $\mathbb{P}^1$. We will fix the isomorphism $\check{X} \simeq \mathbb{P}^1_\mathbb{Q}$ such that the map

$$\mathfrak{h}^\pm = X \overset{\beta}{\hookrightarrow} \check{X}_\mathbb{C}$$

is the identity map. For $k \equiv r \bmod 2$, let $\check{\mathcal{V}}_{k,r}$ be the homogeneous $G_{\mathbb{C}}$-bundle on $\check{X}_{\mathbb{C}}$ corresponding to the character

$$\chi_{k,r} : P_{\mathbb{C}} \to \mathbb{C}^{\times}, \quad \begin{pmatrix} a & * \\ & d \end{pmatrix} \mapsto d^k \det(\cdot)^{\frac{r-k}{2}}$$

of $P_{\mathbb{C}}$. Note that abstractly $\check{\mathcal{V}}_{k,r} \simeq \mathcal{O}(k)$ though the $G_{\mathbb{C}}$-action depends on r as well. For any $h \in X$, we write $\rho_{k,r}$ for the corresponding representation of K_h. The representation $\rho_{k,r}$ of $K_{h_0} = \mathbb{R}^{\times} \cdot \mathrm{SO}_2(\mathbb{R})$ is the character given by

$$z \cdot \kappa_\theta \mapsto z^r e^{ik\theta}, \quad \kappa_\theta = \begin{pmatrix} \cos\theta & \sin\theta \\ -\sin\theta & \cos\theta \end{pmatrix}.$$

For more general h, the character $\rho_{k,r}$ is given by composing the above character with the isomorphism $K_h \simeq K_{h_0}$ given by $x \mapsto \alpha^{-1} x \alpha$ for any $\alpha \in G(\mathbb{R})$ such that $\alpha h_0 \alpha^{-1} = h$. The corresponding automorphic line bundle $\mathcal{V}_{k,r,\mathcal{K}}$ is defined over $\mathbb{Q}$. Its dual $\mathcal{V}_{k,r,\mathcal{K}}^{\vee}$ is the usual bundle of modular forms of weight k and level $\mathcal{K}$. We can make this more explicit as follows.

Let $\mathcal{V} = \beta^*(\check{\mathcal{V}}_{k,r})$. The *connected* hermitian space $\mathfrak{h}^+$ carries a natural family of (polarized) elliptic curves, the fiber over $\tau \in \mathfrak{h}^+$ being the elliptic curve $A_\tau = \mathbb{C}/(\mathbb{Z}\tau + \mathbb{Z})$. Let $\underline{\omega}$ be the sheaf of relative one-forms; it is a line bundle on $\mathfrak{h}^+$ and there is a canonical isomorphism $\beta^* \check{\mathcal{V}}_{k,r}^{\vee}|_{\mathfrak{h}^+} = \mathcal{V}^{\vee} \simeq \underline{\omega}^k$. This gives a canonical trivialization $\mathrm{Triv}_h : \mathcal{V}_h^{\vee} \simeq \mathbb{C}$ for all $h \in \mathfrak{h}^+$, namely the map sending $dz^{\otimes k}$ to 1, where z is the coordinate on $\mathbb{C} = \mathrm{Lie}(A_\tau)$. Thus any section φ of $\mathcal{V}_{k,r,\mathcal{K}}^{\vee}$ on $\mathrm{Sh}_{\mathcal{K}}(G,X)$ gives rise (via $\mathrm{Triv}_h \circ \mathrm{Lift}_h$) to a function

$$\varphi_h : \mathrm{GL}_2(\mathbb{A}) \to \mathbb{C}, \quad h \in \mathfrak{h}^+,$$

such that $\varphi_h(g\kappa) = \rho_{k,r}^{\vee}(\kappa)^{-1}\varphi_h(g)$ for all $\kappa \in K_h$. In particular, for $z \cdot \kappa_\theta \in K_{h_0}$, we have

$$\varphi_{h_0}(g \cdot z \cdot \kappa_\theta) = \varphi_{h_0}(g) \cdot z^r \cdot e^{ik\theta}.$$

Finally, there is a unique modular form f of weight k on $\mathfrak{h}^+$ such that for all $h \in \mathfrak{h}^+$, we have

$$\varphi_h(g) = f(g_\infty(\tau_h)) j(g_\infty, \tau_h)^{-k} \det(g_\infty)^{\frac{r+k}{2}},$$

where $g = g_{\mathbb{Q}}(g_{\mathcal{K}} g_\infty)$ with $g_{\mathbb{Q}} \in G(\mathbb{Q})$, $g_{\mathcal{K}} \in \mathcal{K}$ and $g_\infty \in G(\mathbb{R})^+$. (Here $G(\mathbb{R})^+$ denotes the topological identity component of $G(\mathbb{R})$; moreover, as is the usual convention, we set $j(\gamma, z) = cz + d$ for $\gamma = \begin{bmatrix} a & b \\ c & d \end{bmatrix}$.)

EXAMPLE 1.1.4. Let $G = B^{\times}$, where B is a non-split indefinite quaternion algebra over $\mathbb{Q}$. Then $E(G,X) = \mathbb{Q}$ and $\check{X}$ is a form of $\mathbb{P}^1$; in fact it is a Severi–Brauer variety associated to the class of B in the Brauer group of $\mathbb{Q}$. The variety $\check{X}$ (over $\mathbb{C}$) carries the line bundles $\mathcal{O}(k)$ but only for k even do these descend to line bundles over $\mathbb{Q}$. Indeed, the canonical bundle on $\check{X}$ has degree -2, so $\mathcal{O}(-2)$ descends. On the other hand, $\mathcal{O}(1)$ does not descend since if it did, by Riemann–Roch it would admit a section whose zero locus is a rational point. Nevertheless, for any $\sigma \in \mathrm{Aut}(\mathbb{C}/\mathbb{Q})$, the line bundle $\mathcal{L} := \mathcal{O}(1)$ on $\check{X}_{\mathbb{C}}$ satisfies $\sigma^*\mathcal{L} \simeq \mathcal{L}$, so its *field of definition* is $\mathbb{Q}$.

1.1.3. Integral models. We assume in this section that the Shimura variety (G, X) is of abelian type. Let $\mathcal{O}$ denote the ring of integers of $E(G, X)$ and $\lambda \mid \ell$ a prime of $\mathcal{O}$. We assume that we are given a reductive group $\mathcal{G}_0$ over $\mathbb{Z}_{(\ell)}$ such that $\mathcal{G}_{\mathbb{Q}} = G$. Let $\mathcal{G} = \mathcal{G}_{0,\mathbb{Z}_\ell}$ and $\mathcal{K}_\ell = \mathcal{G}(\mathbb{Z}_\ell)$. Then $\mathcal{K}_\ell$ is a hyperspecial (maximal compact) subgroup of $G(\mathbb{Q}_\ell)$. Suppose that $\mathcal{K}$ is an open compact subgroup of $G(\mathbb{A}_f)$ of the form $\mathcal{K}_\ell \cdot \mathcal{K}^\ell$, with $\mathcal{K}_\ell$ as above and $\mathcal{K}^\ell$ a subgroup of $G(\mathbb{A}_f^\ell)$, where $\mathbb{A}_f^\ell$ denotes the ring of finite adeles with trivial component at ℓ. Then $\mathrm{Sh}_{\mathcal{K}}(G, X)$ admits a natural integral model $\mathcal{S}_{\mathcal{K},\lambda}(G, X)$ over $\mathcal{O}_{(\lambda)}$, the localization of $\mathcal{O}$ at the prime λ. More precisely, if one fixes $\mathcal{K}_\ell$ and allows $\mathcal{K}^\ell$ to vary, then Kisin [**38**] shows that the projective system $\varprojlim \mathrm{Sh}_{\mathcal{K}_\ell\mathcal{K}^\ell}(G, X)$ admits a *canonical model* $\mathcal{S}_{\mathcal{K}_\ell,\lambda}(G, X)$ over $\mathcal{O}_{(\lambda)}$, which is characterized by a certain extension property. We will also need integral models of automorphic vector bundles on $\mathrm{Sh}_{\mathcal{K}}(G, X)$. In the abelian case, these are constructed in the paper of Lovering [**49**], and we now summarize the relevant results.

Recall that the compact dual $\check{X}$ is naturally defined over $E(G, X)$. In addition, $\check{X}$ has a natural model $\check{\mathfrak{X}}$ over $\mathcal{O}_{(\lambda)}$ whose A-valued points for any $\mathcal{O}_{(\lambda)}$-algebra A are in bijection with equivalence classes of pairs (P, μ) consisting of a parabolic subgroup P of $\mathcal{G}_{0,A}$ and a cocharacter $\mu : \mathbf{G}_{m,A} \to P$, where $(P, \mu) \sim (P', \mu')$ if $P = P'$ and $\bar{\mu} = \bar{\mu}'$. The data needed to define integral models of automorphic vector bundles consists of the following:

— A finite extension L of $E(G, X)$ and a G_L-equivariant vector bundle $\check{\mathcal{V}}$ on $\check{X}_L$. The corresponding automorphic vector bundle $\mathcal{V}_{\mathcal{K}}$ on $\mathrm{Sh}_{\mathcal{K}}(G, X)$ has a canonical model over L.

— A prime λ of $\mathcal{O}_L$; we write λ for the induced prime of $\mathcal{O}$ as well.

— A $\mathcal{G}_0$-equivariant vector bundle $\check{\mathcal{V}}_\lambda$ on $\check{\mathfrak{X}}_{\mathcal{O}_{L,(\lambda)}}$, extending the G_L-equivariant vector bundle $\check{\mathcal{V}}$ on $\check{X}_L$.

To this data, one can associate (by the results of [**49**]) in a functorial way a vector bundle $\mathcal{V}_{\mathcal{K},\lambda}$ over $\mathcal{S}_{\mathcal{K},\lambda}(G, X) \otimes_{\mathcal{O}_{(\lambda)}} \mathcal{O}_{L,(\lambda)}$ which extends $\mathcal{V}_{\mathcal{K}}$. Likewise, if one fixes $\mathcal{K}_\ell$ and varies $\mathcal{K}^\ell$, one gets a vector bundle $\mathcal{V}_{\mathcal{K}_\ell,\lambda}$ over $\mathcal{S}_{\mathcal{K}_\ell,\lambda}(G, X) \otimes_{\mathcal{O}_{(\lambda)}} \mathcal{O}_{L,(\lambda)}$. If $f : \check{\mathcal{V}}_\lambda^1 \to \check{\mathcal{V}}_\lambda^2$ is a map of $\mathcal{G}_0$-equivariant vector bundles over $\check{\mathfrak{X}}_{\mathcal{O}_{L,(\lambda)}}$, there are natural associated maps $f_{\mathcal{K}} : \mathcal{V}_{\mathcal{K},\lambda}^1 \to \mathcal{V}_{\mathcal{K},\lambda}^2$ and $f_{\mathcal{K}_\ell} : \mathcal{V}_{\mathcal{K}_\ell,\lambda}^1 \to \mathcal{V}_{\mathcal{K}_\ell,\lambda}^2$.

1.1.3.1. *Models over* $\mathcal{O}_L[\frac{1}{N}]$. These are also constructed in [**49**] and we summarize the results. Suppose that we are given a reductive group $\mathcal{G}_0$ over $\mathbb{Z}[\frac{1}{N}]$ such that $\mathcal{G}_{0,\mathbb{Q}} = G$ and that $\mathcal{K}$ is of the form $\prod_\ell \mathcal{K}_\ell$, where $\mathcal{K}_\ell = \mathcal{G}_0(\mathbb{Z}_\ell)$ for all ℓ not dividing N, so that $\mathcal{K}_\ell$ is hyperspecial for such ℓ. Then the integral models of $\mathrm{Sh}_{\mathcal{K}}(G, X)$ for varying ℓ (not dividing N) patch together to give a canonical model $\mathcal{S}_{\mathcal{K},\mathcal{O}[\frac{1}{N}]}(G, X)$ over $\mathcal{O}[\frac{1}{N}]$.

The compact dual $\check{X}$ has a natural model $\check{\mathfrak{X}}$ over $\mathcal{O}[\frac{1}{N}]$ as well. Suppose that we are given moreover:

— A finite extension L of $E(G, X)$ and a G_L-equivariant vector bundle $\check{\mathcal{V}}$ on $\check{X}_L$.

— A $\mathcal{G}_0$-equivariant vector bundle $\check{\mathcal{V}}$ on $\check{\mathfrak{X}}_{\mathcal{O}_L[\frac{1}{N}]}$, extending the G_L-equivariant vector bundle $\check{\mathcal{V}}$ on $\check{X}_L$.

Then the integral models $\mathcal{V}_{\mathcal{K},\lambda}$ (as λ varies over the primes of $\mathcal{O}_L$ not dividing N) patch together to give an integral model $\mathcal{V}_{\mathcal{K},\mathcal{O}_L[\frac{1}{N}]}$ over $\mathcal{O}_L[\frac{1}{N}]$.

1.2. Automorphic vector bundles on quaternionic Shimura varieties

In this section, we review the connection between automorphic forms on the multiplicative group of a quaternion algebra over a totally real field and sections of automorphic vector bundles on the corresponding Shimura variety. We will also define canonical metrics on such bundles.

REMARK 1.2.1. Everything in this section goes through verbatim even in the case that the quaternion algebra B is *totally definite*, even though this does not strictly speaking give a Shimura variety in the sense of §1.1.

Let F be a totally real field and B a quaternion algebra over F. Let G_B denote the $\mathbb{Q}$-algebraic group $\mathrm{Res}_{F/\mathbb{Q}}(B^\times)$. Thus for any $\mathbb{Q}$-algebra R, the R-valued points of G_B are given by

$$G_B(R) = (B \otimes_{\mathbb{Q}} R)^\times.$$

Let Σ_B denote the set of places of F at which B is ramified.

We fix for the moment some choice of isomorphisms

$$B \otimes_{F,\sigma} \mathbb{R} \simeq \mathrm{M}_2(\mathbb{R}), \qquad \text{for } \sigma \in \Sigma_\infty \smallsetminus \Sigma_B; \tag{1.2}$$

$$B \otimes_{F,\sigma} \mathbb{R} \simeq \mathbb{H}, \qquad \text{for } \sigma \in \Sigma_{B,\infty}, \tag{1.3}$$

where $\mathbb{H}$ is the subalgebra

$$\left\{ \begin{pmatrix} \alpha & \beta \\ -\bar{\beta} & \bar{\alpha} \end{pmatrix} : \ \alpha, \beta \in \mathbb{C} \right\}$$

of $\mathrm{M}_2(\mathbb{C})$. (Later we will fix these isomorphisms more carefully.) The choice of isomorphisms above gives us identifications

$$G_B(\mathbb{R}) \simeq \prod_{\sigma \in \Sigma_\infty \smallsetminus \Sigma_B} \mathrm{GL}_2(\mathbb{R}) \times \prod_{\sigma \in \Sigma_{B,\infty}} \mathbb{H}^\times$$

and

$$G_B(\mathbb{C}) \simeq \prod_{\sigma \in \Sigma_\infty} \mathrm{GL}_2(\mathbb{C}).$$

Let X_B be the $G_B(\mathbb{R})$-conjugacy class of homomorphisms $\mathbb{S} \to G_{B,\mathbb{R}}$ containing

$$h_0 : \mathbb{S} \to G_{B,\mathbb{R}}, \quad h_0 := \prod_\sigma h_{0,\sigma}, \quad h_{0,\sigma}(z) = \begin{cases} z, & \text{if } \sigma \in \Sigma_\infty \smallsetminus \Sigma_B; \\ 1, & \text{if } \sigma \in \Sigma_{B,\infty}, \end{cases}$$

where we identify $\mathbb{C}$ with a subring of $\mathrm{M}_2(\mathbb{R})$ (see the remark below). Denote by $\check{X}_B$ the corresponding compact dual hermitian symmetric space. The choice of isomorphisms (1.2) and (1.3) above gives rise to an identification $\check{X}_B = (\mathbb{P}^1_{\mathbb{C}})^{d_B}$ and $X_B = (\mathfrak{h}^\pm)^{d_B}$, with d_B being the number of infinite places of F where B is split.

REMARK 1.2.2. (Choices) We embed $\mathbb{C}$ in $\mathrm{M}_2(\mathbb{R})$ by identifying $a+bi$ with the matrix

$$\begin{pmatrix} a & b \\ -b & a \end{pmatrix}.$$

In addition, we identify the homomorphism

$$\mathbb{S} \to \mathrm{GL}_{2,\mathbb{R}}, \quad a+bi \mapsto a+bi = \begin{pmatrix} a & b \\ -b & a \end{pmatrix}.$$

with the element $i \in \mathfrak{h}$. Note that this is *opposite* to the usual choice made by Shimura. Shimura would identify $i \in \mathfrak{h}$ with the map

$$a + bi \mapsto \begin{pmatrix} a & -b \\ b & a \end{pmatrix}.$$

1.2.1. Hermitian forms. For $\sigma \in \Sigma_\infty \smallsetminus \Sigma_B$, let $V_{\sigma,\mathbb{R}}$ denote the vector space $\mathbb{R}^2$ of column vectors viewed as a left $\mathrm{M}_2(\mathbb{R})$-module. Let $h : \mathbb{C}^\times = \mathbb{S}(\mathbb{R}) \to (B \otimes_{F,\sigma} \mathbb{R})^\times = \mathrm{GL}_2(\mathbb{R}) = \mathrm{GL}(V_{\sigma,\mathbb{R}})$ be any homomorphism that is $\mathrm{GL}_2(\mathbb{R})$-conjugate to $h_{0,\sigma}$. Then we can write

$$V_{\sigma,\mathbb{C}} = V_{\sigma,\mathbb{R}} \otimes_{\mathbb{R}} \mathbb{C} = V_{\sigma,h}^{-1,0} \oplus V_{\sigma,h}^{0,-1}, \tag{1.4}$$

where the decomposition on the right corresponds to the $\mathbb{C}$-subspaces on which $h(z) \otimes 1$ acts as $1 \otimes z$ and $1 \otimes \bar{z}$ respectively. The bilinear form

$$(x, y) \mapsto {}^t x \begin{pmatrix} 0 & -1 \\ 1 & 0 \end{pmatrix} y \tag{1.5}$$

on $V_{\sigma,\mathbb{R}}$ is almost $\mathrm{GL}_2(\mathbb{R})$-invariant:

$$(gx, gy) = \det(g) \cdot (x, y).$$

Further, it satisfies the following conditions:

(i) $(x, y) = -(y, x)$.
(ii) $(h(i)x, h(i)y) = (x, y)$.
(iii) The form $(x, h(i)y)$ is symmetric. (This follows formally from (i) and (ii).) Further, it is positive definite if h is $\mathrm{GL}_2(\mathbb{R})^+$-conjugate to h_0. (Otherwise it is negative definite.)

REMARK 1.2.3. Let τ be the unique point on the complex upper half plane fixed by K_h. The bilinear form above equals $\frac{1}{2\pi i}\lambda_\tau$ where λ_τ is the Weil pairing on $H_1(A_\tau)$ given in the ordered basis $\{\tau, 1\}$.

The composite map

$$V_{\sigma,\mathbb{R}} \to V_\sigma \otimes_{\mathbb{R}} \mathbb{C} \to V_{\sigma,h}^{-1,0}$$

is an $\mathbb{R}$-linear isomorphism; via this isomorphism one gets a skew-smmetric bilinear form on $V_{\sigma,h}^{-1,0}$, which is the *negative* of the imaginary part of a (necessarily unique) hermitian form H_h on $V_{\sigma,h}^{-1,0}$ defined by identifying $V_{\sigma,h}^{-1,0}$ with $V_{\sigma,\mathbb{R}}$ and setting

$$H_h(x, y) = (x, h(i)y) - i(x, y) = (x, iy) - i(x, y).$$

REMARK 1.2.4. The form H_h is linear in the first variable and conjugate linear in the second variable. If we denote the form (1.5) above by $\tilde{B}$, then H_h agrees with the form $2 \cdot \tilde{B}_{h(i)}$ of Appendix A, where:

$$\tilde{B}_{h(i)}(v, w) = \tilde{B}_{\mathbb{C}}(v, h(i)\bar{w}). \tag{1.6}$$

If h is $\mathrm{GL}_2(\mathbb{R})^+$-conjugate to h_0, the form H_h is positive definite on account of condition (iii) above. Note that

$$H_h(x, x) = (x, h(i)x). \tag{1.7}$$

The subgroup K_h preserves the decomposition (1.4) and the form H_h is K_h-invariant up to a scalar. In fact, for $\kappa \in K_h$, we have

$$H_h(\kappa x, \kappa y) = \det(\kappa) H_h(x, y).$$

Moreover, the natural action of $\mathrm{GL}_2(\mathbb{R})$ on $V_{\sigma,\mathbb{C}}$ takes $V_{\sigma,h}^{-1,0}$ isomorphically onto $V_{\sigma,g\cdot h}^{-1,0}$ (recall $g \cdot h = ghg^{-1}$) and we have

$$\begin{aligned} H_{g\cdot h}(gx, gy) = (gx, (gh(i)g^{-1})gy) - i(gx, gy) &= \det(g)\left[(x, h(i)y) - i(x,y)\right] \\ &= \det(g) H_h(x,y). \end{aligned}$$

We note also that $\det(V_{\sigma,\mathbb{C}})$ carries a natural bilinear form induced from the $\mathbb{C}$-linear extension of $(\cdot,\cdot)$. We equip $\det(V_{\sigma,\mathbb{C}})$ with the positive definite Hermitian form

$$H_{\det}(x,y) = (x, \bar{y}), \tag{1.8}$$

where the complex conjugation is with respect to the natural real structure coming from $\det(V_{\sigma,\mathbb{R}})$. This hermitian form satisfies

$$H_{\det}(gx, gy) = \det(g)^2 \cdot H_{\det}(x,y)$$

for all $g \in \mathrm{GL}_2(\mathbb{R})$. Equivalently, one can first extend $(\cdot,\cdot)$ to a $\mathbb{C}$-bilinear form on $V_{\mathbb{C}}$, convert it to a Hermitian form H on $V_{\mathbb{C}}$ by $H(x,y) = (x, \bar{y})$ and then let $H_{\det}$ be the induced hermitian form on $\det(V_{\mathbb{C}})$. Explicitly,

$$\begin{aligned} H_{\det}(v_1 \wedge v_2, w_1 \wedge w_2) &= \det \begin{bmatrix} H(v_1, w_1) & H(v_1, w_2) \\ H(v_2, w_1) & H(v_2, w_2) \end{bmatrix} \\ &= \det \begin{bmatrix} (v_1, \bar{w}_1) & (v_1, \bar{w}_2) \\ (v_2, \bar{w}_1) & (v_2, \bar{w}_2) \end{bmatrix}. \end{aligned}$$

For $\sigma \in \Sigma_{B,\infty}$, let $V_{\sigma,\mathbb{C}}$ denote the $\mathbb{C}$-vector space $\mathbb{C}^2$ of column vectors viewed as a left $\mathrm{M}_2(\mathbb{C})$-module. The form

$$(x,y) \mapsto {}^t x \begin{pmatrix} 0 & -1 \\ 1 & 0 \end{pmatrix} y$$

is almost $\mathrm{GL}_2(\mathbb{C})$-invariant:

$$(gx, gy) = \det(g) \cdot (x,y).$$

Let L be the $\mathbb{R}$-linear operator

$$L(x) := \begin{pmatrix} 0 & 1 \\ -1 & 0 \end{pmatrix} \bar{x}.$$

The operator L is the analog in this case of the operator $w \mapsto h(i)\bar{w}$ in (1.6) and the operator $x \mapsto h(i)x$ in (1.7) above. Define a hermitian form H on $V_{\sigma,\mathbb{C}}$ by

$$H(x,y) = (x, Ly) = {}^t x \bar{y}. \tag{1.9}$$

Note that L commutes with the left action of $\mathbb{H}$, hence the form H is $\mathbb{H}^\times$-invariant up to a scalar, which is also obvious from the formula above. More precisely, for $g \in \mathbb{H}^\times$, we have

$$H(gx, gy) = \nu(g) H(x,y)$$

where ν is the reduced norm. As before, we also get an induced Hermitian form $H_{\det}$ on $\det(V_{\sigma,\mathbb{C}})$, which satisfies

$$H_{\det}(gx, gy) = \nu(g)^2 H_{\det}(x,y)$$

for $g \in \mathbb{H}^\times$.

For any $\sigma \in \Sigma_\infty$, let $\rho_{\sigma,k,r}$ denote the representation

$$V_{\sigma,k,r} = \mathrm{Sym}^k(V_{\sigma,\mathbb{C}}) \otimes \det(V_{\sigma,\mathbb{C}})^{\frac{r-k}{2}}$$

of $\mathrm{GL}_2(\mathbb{C})$, viewed as a sub-representation of $V_{\sigma,\mathbb{C}}^{\otimes k} \otimes \det(V_{\sigma,\mathbb{C}})^{\frac{r-k}{2}}$. Note that the central character of $\rho_{\sigma,k,r}$ is $z \mapsto z^r$.

REMARK 1.2.5. For $\sigma \in \Sigma_{B,\infty}$ the hermitian form H of (1.9) induces a positive definite hermitian metric on $\mathrm{Sym}^k(V_{\sigma,\mathbb{C}})$ as follows. Let $V = V_{\sigma,\mathbb{C}}$. The symmetrization map

$$V^{\otimes k} \to V^{\otimes k}, \quad v_1 \otimes \cdots \otimes v_k \mapsto \frac{1}{k!} \sum_{\sigma \in \mathfrak{S}_k} v_{\sigma(1)} \otimes \cdots \otimes v_{\sigma(k)}$$

factors through $\mathrm{Sym}^k(V)$ and gives an injection

$$s : \mathrm{Sym}^k(V) \hookrightarrow V^{\otimes k} \tag{1.10}$$

which is a section of the natural map $V^{\otimes k} \to \mathrm{Sym}^k(V)$. The hermitian form H induces naturally a hermitian form on $V^{\otimes k}$ (the tensor product form) and then by restriction a hermitian form on $\mathrm{Sym}^k(V)$. It is easy to check that this gives a positive definite hermitian form on $\mathrm{Sym}^k(V)$. Combining this with the metric on $\det(V_{\sigma,\mathbb{C}})$, we get a positive definite hermitian metric on $V_{\sigma,k,r}$ for all $\sigma \in \Sigma_{B,\infty}$.

1.2.2. Hermitian metrics on automorphic vector bundles. Let $(\underline{k}, r)$ be a multi-index of integers with $\underline{k} = (k_\sigma)_{\sigma \in \Sigma_\infty}$ such that

$$k_\sigma \equiv r \pmod 2 \quad \text{for all } \sigma \in \Sigma_\infty.$$

We assume that $k_\sigma \geq 1$ if B is split at σ and that $k_\sigma \geq 0$ if B is ramified at σ.

Let $\rho_{\underline{k},r} = \otimes_\sigma \rho_{\sigma,k_\sigma,r}$ be the representation of $G_B(\mathbb{C})$ on

$$V_{\underline{k},r} = \bigotimes_\sigma V_{\sigma,k_\sigma,r}.$$

This gives rise to a $G_B(\mathbb{C})$-homogeneous vector bundle $\check{\mathcal{V}}_{\rho_{\underline{k},r}}$ on $\check{X}_B$:

$$\check{\mathcal{V}}_{\rho_{\underline{k},r}} = \check{X}_B \times V_{\underline{k},r},$$

where the $G_B(\mathbb{C})$ action is:

$$g \cdot (x, v) = (gx, gv).$$

By restriction one gets a $G_B(\mathbb{R})$-homogeneous vector bundle $\mathcal{V}_{\rho_{\underline{k},r}}$ on X_B. Further, the latter admits a unique $G_B(\mathbb{R})$-equivariant sub-bundle $\mathcal{V}_{\underline{k},r}$ corresponding to the K_h-subrepresentation $\rho_{\underline{k},r,h}$ on

$$\mathcal{V}_{\underline{k},r,h} = \bigotimes_{\sigma \in \Sigma_\infty \smallsetminus \Sigma_B} \left((V_{\sigma,h}^{-1,0})^{\otimes k_\sigma} \otimes \det(V_{\sigma,\mathbb{C}})^{\otimes \frac{r-k_\sigma}{2}} \right) \bigotimes_{\sigma \in \Sigma_{B,\infty}} V_{\sigma,k_\sigma,r}.$$

Here we consider $(V_{\sigma,h}^{-1,0})^{\otimes k_\sigma}$ as a subspace of $\mathrm{Sym}^{k_\sigma} V_{\sigma,\mathbb{C}}$ via the composite map $(V_{\sigma,h}^{-1,0})^{\otimes k_\sigma} \hookrightarrow V_{\sigma,\mathbb{C}}^{\otimes k_\sigma} \to \mathrm{Sym}^{k_\sigma} V_{\sigma,\mathbb{C}}$.

Let X_B^+ denote the connected component of X_B containing h_0. Note that for $h \in X_B^+$, the K_h-representation above carries a natural *positive definite* hermitian metric $\langle \cdot, \cdot \rangle_h$ obtained from the hermitian metrics in (1.7), (1.8) and (1.9) above. (See also Remark 1.2.5 above.) This gives a metric on $\mathcal{V}_{\underline{k},r}$ that is almost $G_B(\mathbb{R})^+$-equivariant; in fact, one has

$$\langle gx, gy \rangle_{g \cdot h} = \|\nu(g)\|^r \langle x, y \rangle_h$$

for $g \in G_B(\mathbb{R})^+$ and $x, y \in \mathcal{V}_{\underline{k},r,h}$, where $\|\cdot\|$ denotes the idelic norm. Now consider the vector bundle $\mathcal{V}_{\underline{k},r|X_B^+} \times G_B(\mathbb{A}_f)$ on $X_B^+ \times G_B(\mathbb{A}_f)$. We equip this with the hermitian metric that assigns to the fiber $\mathcal{V}_{\underline{k},r,h} \times \{g_f\}$ over $(h, g_f) \in X_B^+ \times G_B(\mathbb{A}_f)$ the metric defined above on $\mathcal{V}_{\underline{k},r,h}$ multiplied by the factor $\|\nu(g_f)\|^r$. Recall that

$$\mathrm{Sh}_{\mathcal{K}}(G_B, X_B) = G_B(\mathbb{Q})\backslash X_B \times G_B(\mathbb{A}_f)/\mathcal{K} = G_B(\mathbb{Q})^+\backslash X_B^+ \times G_B(\mathbb{A}_f)/\mathcal{K}$$

and

$$\mathcal{V}_{\underline{k},r,\mathcal{K}} = G_B(\mathbb{Q})\backslash \mathcal{V}_{\underline{k},r} \times G_B(\mathbb{A}_f)/\mathcal{K} = G_B(\mathbb{Q})^+\backslash \mathcal{V}_{\underline{k},r|X_B^+} \times G_B(\mathbb{A}_f)/\mathcal{K},$$

where $G_B(\mathbb{Q})^+ = G_B(\mathbb{R})^+ \cap G_B(\mathbb{Q})$.

PROPOSITION 1.2.6. *The metric on $\mathcal{V}_{\underline{k},r|X_B^+} \times G_B(\mathbb{A}_f)$ above descends to a (positive definite hermitian) metric on the vector bundle $\mathcal{V}_{\underline{k},r,\mathcal{K}}$ over $\mathrm{Sh}_{\mathcal{K}}(G_B, X_B)$.*

PROOF. Let (h, g_f) and (h', g_f') be two elements of $X_B^+ \times G_B(\mathbb{A}_f)$ whose classes in $\mathrm{Sh}_{\mathcal{K}}(G_B, X_B)$ are equal. Then there exist elements $\gamma \in G_B(\mathbb{Q})^+$ and $\kappa \in \mathcal{K}$ such that

$$(h', g_f') = \gamma(h, g_f)\kappa = (\gamma \cdot h, \gamma_f g_f \kappa).$$

Here γ_f is γ viewed as an element of $G_B(\mathbb{A}_f)$. We need to check that the bijection

$$\mathcal{V}_{\underline{k},r,h} \times \{g_f\} \to \mathcal{V}_{\underline{k},r,h'} \times \{g_f'\} = \mathcal{V}_{\underline{k},r,\gamma\cdot h} \times \{\gamma_f g_f \kappa\}$$

given by $(v, g_f) \mapsto (\gamma v, \gamma g_f \kappa)$ is metric preserving. But

$$\begin{aligned}\langle \gamma v_1, \gamma v_2\rangle_{\gamma\cdot h} \cdot \|\nu(\gamma_f g_f \kappa)\|^r &= \prod_{\sigma\in\Sigma_\infty} \sigma(\nu(\gamma))^r \cdot \langle v_1, v_2\rangle_h \cdot \|\nu(\gamma)_f\|^r \|\nu(g_f)\|^r \\ &= \langle v_1, v_2\rangle_h \cdot \|\nu(g_f)\|^r,\end{aligned}$$

using the product formula and the fact that $\|\nu(\kappa)\| = 1$. □

We will need to work with the dual vector bundle $\mathcal{V}_{\underline{k},r}^\vee$. This is motivated by observing that in the case of $\mathrm{GL}_2(\mathbb{Q})$, the bundle $\mathcal{V}_{\rho_{\underline{k},r}}$ corresponds to the relative homology of the universal elliptic curve and the sub-bundle $\mathcal{V}_{\underline{k},r}$ corresponds to its relative Lie algebra. The line bundle of usual modular forms corresponds to the bundle of relative differentials, which is why we need to replace $\mathcal{V}_{\underline{k},r}$ by its dual. We begin by making the following completely elementary remark, which we nevertheless state carefully to avoid any confusion.

REMARK 1.2.7. If ρ is a representation of a group G on a finite-dimensional complex vector space V, then $\rho^\vee$ is defined by

$$\rho^\vee(g)(L) = L \circ \rho(g^{-1})$$

for $L \in V^\vee = \mathrm{Hom}(V, \mathbb{C})$ and $g \in G$. Thus for the tautological pairing

$$(\cdot,\cdot) : V^\vee \times V \to \mathbb{C}, \quad (L, v) = L(v),$$

we have

$$(\rho^\vee(g^{-1})L, v) = (L, \rho(g)v).$$

Suppose V is equipped with a non-degenerate hermitian pairing $\langle\cdot,\cdot\rangle$ that is linear in the first variable and conjugate linear in the second variable, and such that

$$\langle gv, gw\rangle = \chi(g)\langle v, w\rangle$$

for some character $\chi : G \to \mathbb{C}^\times$. Since $\langle\cdot,\cdot\rangle$ is non-degenerate, it induces a conjugate linear isomorphism

$$V \simeq V^\vee, \quad w \mapsto L_w, \quad L_w(v) = \langle v, w\rangle.$$

Composing the inverse of this isomorphism with the canonical isomorphism $V \simeq V^{\vee\vee}$ gives a conjugate linear isomorphism $V^\vee \simeq (V^\vee)^\vee$, which one may view as a hermitian form on $V^\vee$. (There is some choice here, but it is fixed by requiring the form to be linear in the first variable.) Explicitly this isomorphism sends L_w to the linear functional $\mathrm{eval}_w \in (V^\vee)^\vee$, so that for any $L \in V^\vee$, we have

$$\langle L, L_w \rangle = L(w).$$

Note that

$$gL_w(v) = L_w(g^{-1}v) = \langle g^{-1}v, w\rangle = \chi(g)^{-1}\langle v, gw\rangle = \chi(g)^{-1}L_{gw}(v),$$

so that $gL_w = \chi(g)^{-1}L_{gw}$. For any $L \in V^\vee$, we have

$$\begin{aligned}\langle gL, gL_w\rangle &= \langle gL, \chi(g)^{-1}L_{gw}\rangle = \overline{\chi(g)}^{-1}\langle gL, L_{gw}\rangle \\ &= \overline{\chi(g)}^{-1}(gL)(gw) = \overline{\chi(g)}^{-1}L(w) \\ &= \overline{\chi(g)}^{-1}\langle L, L_w\rangle,\end{aligned}$$

so for any $L_1, L_2 \in V^\vee$, we have $\langle gL_1, gL_2\rangle = \overline{\chi(g)}^{-1}\langle L_1, L_2\rangle$.

From the remark above, it is clear that for $x, y \in \mathcal{V}^\vee_{\underline{k},r,h}$ and $g \in G_B(\mathbb{R})^+$, we have

$$\langle gx, gy\rangle_{gh} = \|\nu(g)\|^{-r}\langle x, y\rangle_h.$$

Thus we take on $\mathcal{V}^\vee_{\underline{k},r}{}_{|X_B^+} \times G_B(\mathbb{A}_f)$ the metric which on $\mathcal{V}^\vee_{\underline{k},r,h} \times \{g_f\}$ is $\|\nu(g_f)\|^{-r}$ times the induced metric on $\mathcal{V}^\vee_{\underline{k},r,h}$. This descends to a positive definite hermitian metric $\langle\!\langle \cdot, \cdot \rangle\!\rangle$ on $\mathcal{V}^\vee_{\underline{k},r,\mathcal{K}}$. (See Prop. 1.2.6 above.)

1.2.3. Petersson norms and scalar valued forms.

DEFINITION 1.2.8. A holomorphic automorphic form of weight $(\underline{k}, r)$ and level $\mathcal{K}$ on G_B is a *holomorphic* section s of the bundle $\mathcal{V}^\vee_{\underline{k},r,\mathcal{K}}$ on $\mathrm{Sh}_{\mathcal{K}}(G_B, X_B)$. Let $\tilde{\mathcal{K}} \supseteq \mathcal{K}$ be any open compact subgroup of $G_B(\mathbb{A}_f)$ such that $\langle\!\langle s(x), s(x)\rangle\!\rangle$ descends to a function on $\mathrm{Sh}_{\tilde{\mathcal{K}}}(G_B, X_B)$. Then the Petersson norm of the section s (normalized with respect to $\tilde{\mathcal{K}}$) is defined to be the integral

$$\langle\!\langle s, s\rangle\!\rangle_{\tilde{\mathcal{K}}} := \int_{\mathrm{Sh}_{\tilde{\mathcal{K}}}(G_B, X_B)} \langle\!\langle s(x), s(x)\rangle\!\rangle \, d\mu_x$$

where $d\mu_x$ is the measure on $\mathrm{Sh}_{\tilde{\mathcal{K}}}(G_B, X_B)$ defined in §6.1.2.

REMARK 1.2.9. Definition 1.2.8 above has the advantage that it does not depend on any choice of base point. In practice though, one usually needs to pick a base point to make any computation at all, and so we shall now discuss the translation between these two points of view.

Pick a base point $h \in X_B^+$. Via the isomorphism Lift_h, the space of holomorphic automorphic forms s as above is identified with the space of functions $\mathcal{A}(G_B, \mathcal{K}, \mathcal{V}^\vee_{\underline{k},r}, h)$. An element

$$F : G_B(\mathbb{Q})\backslash G_B(\mathbb{A})/\mathcal{K} \to \mathcal{V}^\vee_{\underline{k},r,h}$$

in $\mathcal{A}(G_B, \mathcal{K}, \mathcal{V}^\vee_{\underline{k},r}, h)$ satisfies in particular the condition

$$F(g\kappa_h) = \rho^\vee_{\underline{k},r,h}(\kappa_h)^{-1}F(g), \quad \text{for all } \kappa_h \in K_h. \tag{1.11}$$

Henceforth we will fix a character ξ of $F^\times\backslash\mathbb{A}_F^\times$ which satisfies

$$\xi(z\cdot z_\infty)=\mathrm{N}(z_\infty)^r\cdot\xi(z)$$

for $z\in\mathbb{A}_F^\times$, $z_\infty\in\mathbb{A}_{F,\infty}^\times$, and assume that the section s satisfies the following invariance under the center $Z_{G_B}(\mathbb{A}_f)=\mathbb{A}_{F,f}^\times$:

$$s(x\cdot\alpha)=\xi(\alpha)\cdot s(x). \tag{1.12}$$

This enables us to take $\tilde{\mathcal{K}}$ containing the maximal open compact subgroup of $Z_{G_B}(\mathbb{A}_f)$, and implies that the corresponding function F above satisfies the following invariance property: for $\alpha\in\mathbb{A}_F^\times=Z_{G_B}(\mathbb{A})$, we have

$$F(g\cdot\alpha)=\xi(\alpha)\cdot F(g)$$

and

$$\langle F(g\cdot\alpha),F(g\cdot\alpha)\rangle=\|\alpha\|^{2r}\cdot\langle F(g),F(g)\rangle.$$

PROPOSITION 1.2.10. *Suppose* $\mathrm{Lift}_h(s)=F$. *Let* $\mathcal{K}_0$ *denote any maximal compact subgroup of* $G_B(\mathbb{A}_f)$ *containing* $\tilde{\mathcal{K}}$. *Then*

$$\langle\langle s,s\rangle\rangle_{\tilde{\mathcal{K}}}=2^{|\Sigma_\infty\smallsetminus\Sigma_B|}\cdot h_F\cdot[\mathcal{K}_0:\tilde{\mathcal{K}}]\cdot\langle F,F\rangle_h,$$

where

$$\langle F,F\rangle_h=\int_{[G_B]}\langle F(g),F(g)\rangle_h\cdot\|\nu(g)\|^{-r}\,dg.$$

Here and henceforth we write $[G_B]$ for $G_B(\mathbb{Q})Z_{G_B}(\mathbb{A})\backslash G_B(\mathbb{A})$. Also, dg denotes the *standard measure* on $[G_B]$ which is defined in §6.1.2.

PROOF. Recall that if $g=(g_\infty,g_f)$, we have

$$F(g)=g_\infty^{-1}s[(g_\infty\cdot h,g_f)],$$

where we view $s[(g_\infty\cdot h,g_f)]$ as an element in $\mathcal{V}^\vee_{\underline{k},r,g_\infty\cdot h}$. Let

$$G_B(\mathbb{A})^+:=\{g\in G_B(\mathbb{A}):\ g=(g_\infty,g_f),\ g_\infty\in G_B(\mathbb{R})^+\}.$$

Now for $g\in G_B(\mathbb{A})^+$, we have

$$\begin{aligned}\langle F(g),F(g)\rangle_h&=\|\nu(g_\infty)\|^r\langle s[(g_\infty\cdot h,g_f)],s[(g_\infty\cdot h,g_f)]\rangle_{g_\infty\cdot h}\\&=\|\nu(g_\infty)\|^r\|\nu(g_f)\|^r\cdot\langle\langle s[(g_\infty\cdot h,g_f)],s[(g_\infty\cdot h,g_f)]\rangle\rangle\\&=\|\nu(g)\|^r\langle\langle s[(g_\infty\cdot h,g_f)],s[(g_\infty\cdot h,g_f)]\rangle\rangle.\end{aligned}$$

Since F and s are left $G_B(\mathbb{Q})$-invariant and $G_B(\mathbb{A})=G_B(\mathbb{Q})G_B(\mathbb{A})^+$, the equality above holds for all $g\in G_B(\mathbb{A})$. The proposition follows from this and the comparison of measures in Lemma 6.1.3. □

Next, we simplify further to scalar valued forms. For $\kappa=(z_\sigma e^{i\theta_\sigma})_{\sigma\in\Sigma_\infty}\in(\mathbb{C}^\times)^d$, let κ_h be the element of $K_h\subset G_B(\mathbb{R})$ defined by:

$$\kappa_{h,\sigma}=\begin{cases}h_\sigma(z_\sigma e^{i\theta_\sigma}), & \text{if }\sigma\in\Sigma_\infty\smallsetminus\Sigma_B;\\ h'_\sigma(z_\sigma e^{i\theta_\sigma}), & \text{if }\sigma\in\Sigma_{B,\infty},\end{cases}$$

where for $\sigma\in\Sigma_{B,\infty}$, we let h'_σ be some embedding of $\mathbb{C}^\times$ in $(B\otimes_{F,\sigma}\mathbb{R})^\times$. (The discussion below thus depends on the choice of such an h'_σ for all $\sigma\in\Sigma_{B,\infty}$; for the moment the exact choice is unimportant but later in Chapter 7, we shall make the choice more precise.) The equation (1.11) can be rewritten as

$$F(g\kappa_h) = \prod_{\sigma\in\Sigma_\infty} z_\sigma^r \cdot \prod_{\sigma\in\Sigma_\infty\setminus\Sigma_B} e^{ik_\sigma\theta_\sigma} \cdot \bigotimes_{\sigma\in\Sigma_{B,\infty}} \rho^\vee_{\sigma,k_\sigma,r}(h'_\sigma(e^{-i\theta_\sigma}))F(g).$$

For $\sigma \in \Sigma_\infty \setminus \Sigma_B$, let v_{σ,k_σ} be any nonzero vector in the one-dimensional $\mathbb{C}$-vector space

$$(V_{\sigma,h}^{-1,0})^{\otimes k_\sigma} \otimes \det(V_{\sigma,\mathbb{C}})^{\otimes \frac{r-k_\sigma}{2}},$$

so that

$$\rho_{\sigma,k_\sigma,r}(\kappa_{h,\sigma}) \cdot v_{\sigma,k_\sigma} = z_\sigma^r e^{ik_\sigma\theta_\sigma} \cdot v_{\sigma,k_\sigma}. \tag{1.13}$$

Note that v_{σ,k_σ} may also be characterized as the unique vector (up to scaling) in $V_{\sigma,k_\sigma,r}$ that satisfies (1.13) for all $\kappa \in (\mathbb{C}^\times)^d$. For $\sigma \in \Sigma_{B,\infty}$, we likewise let $v_{\sigma,k_\sigma} \in V_{\sigma,k_\sigma,r}$ be any nonzero vector such that the condition (1.13) is satisfied for all $\kappa \in (\mathbb{C}^\times)^d$. Such a vector is again well-defined up to scaling.

Set $v_{\underline{k}} = \otimes_{\sigma\in\Sigma_\infty} v_{\sigma,k_\sigma} \in \mathcal{V}_{\underline{k},r,h}$. Define

$$\phi_F(g) = (F(g), v_{\underline{k}}).$$

Then $\phi_F(g)$ satisfies

$$\phi_F(g\kappa_h) = \prod_{\sigma\in\Sigma_\infty} z_\sigma^r e^{ik_\sigma\theta_\sigma} \cdot \phi_F(g) \tag{1.14}$$

and

$$\phi_F(\alpha g) = \xi(\alpha)\phi_F(g), \quad \text{for } \alpha \in Z_{G_B}(\mathbb{A}) = \mathbb{A}_F^\times. \tag{1.15}$$

PROPOSITION 1.2.11. *The map $F \mapsto \phi_F$ is injective.*

PROOF. This follows immediately from the fact that $\mathcal{V}_{\underline{k},r,h}$ is irreducible as a module over $G = \prod_{\sigma\in\Sigma_{B,\infty}} (B \otimes_{F,\sigma} \mathbb{R})^\times$. Indeed, given any $w \in \mathcal{V}_{\underline{k},r,h}$, there exist elements $\kappa_i \in G$ and $\alpha_i \in \mathbb{C}$ such that

$$w = \sum_i \alpha_i \rho(\kappa_i) v_{\underline{k}},$$

where ρ denotes the natural action of G on $\mathcal{V}_{\underline{k},r,h}$. Then

$$\begin{aligned}(F(g), w) &= \sum_i \alpha_i (F(g), \rho(\kappa_i) v_{\underline{k}}) = \sum_i \alpha_i (\rho^\vee(\kappa_i)^{-1} F(g), v_{\underline{k}}) \\ &= \sum_i \alpha_i (F(g\kappa_i), v_{\underline{k}}) = \sum_i \alpha_i \phi_F(g\kappa_i).\end{aligned}$$

Thus if ϕ_F is identically zero, then so is F. □

We will now compare $\langle F, F\rangle$ to $\langle \phi_F, \phi_F\rangle$, where

$$\langle \phi_F, \phi_F \rangle = \int_{[G_B]} \phi_F(g)\overline{\phi_F(g)} \cdot \|\nu(g)\|^{-r}\, dg.$$

We use the following well known lemma.

LEMMA 1.2.12. *Let K be a compact Lie group and V a (finite dimensional) irreducible complex representation of K. Let $\langle\cdot,\cdot\rangle$ be a nonzero K-invariant hermitian form on V (such a form is unique up to scalar multiples) and denote also by $\langle\cdot,\cdot\rangle$ the induced hermitian form on $V^\vee$. Then for all $v_1, v_2 \in V$ and $L_1, L_2 \in V^\vee$, we have*

$$\int_K (\rho^\vee(k)L_1, v_1)\overline{(\rho^\vee(k)L_2, v_2)}\, dk = \frac{1}{\dim(V)} \cdot \langle v_1, v_2\rangle\langle L_1, L_2\rangle,$$

where dk is Haar measure normalized to have total volume 1.

REMARK 1.2.13. It is immediate to check that if the form $\langle\cdot,\cdot\rangle$ on V is scaled by $\alpha \in \mathbb{C}^\times$, then the form $\langle\cdot,\cdot\rangle$ on $V^\vee$ is scaled by $\bar{\alpha}^{-1}$, so the right hand side is independent of the choice of $\langle\cdot,\cdot\rangle$.

PROPOSITION 1.2.14.

$$\langle F, F\rangle_h = \frac{\operatorname{rank} \mathcal{V}_{\underline{k},r}}{\langle v_{\underline{k}}, v_{\underline{k}}\rangle_h} \cdot \langle \phi_F, \phi_F\rangle.$$

PROOF. Let K_h^0 denote the maximal compact subgroup of K_h. Since $\mathcal{V}_{\underline{k},r,h}$ is an irreducible representation of K_h^0, using Lemma 1.2.12 we get

$$\begin{aligned}
\langle \phi_F, \phi_F\rangle &= \int_{[G_B]} \phi_F(g)\overline{\phi_F(g)} \cdot \|\nu(g)\|^{-r}\, dg \\
&= \int_{K_h^0}\int_{[G_B]} \phi_F(g)\overline{\phi_F(g)} \cdot \|\nu(g)\|^{-r}\, dg\, d\kappa \\
&= \int_{K_h^0}\int_{[G_B]} \phi_F(g\kappa^{-1})\overline{\phi_F(g\kappa^{-1})} \cdot \|\nu(g\kappa^{-1})\|^{-r}\, dg\, d\kappa \\
&= \int_{K_h^0}\int_{[G_B]} (F(g\kappa^{-1}), v_{\underline{k}})\overline{(F(g\kappa^{-1}), v_{\underline{k}})} \cdot \|\nu(g)\|^{-r}\, dg\, d\kappa \\
&= \int_{[G_B]}\int_{K_h^0} (\rho^\vee(\kappa)F(g), v_{\underline{k}})\overline{(\rho^\vee(\kappa)F(g), v_{\underline{k}})} \cdot \|\nu(g)\|^{-r}\, d\kappa\, dg \\
&= \frac{1}{\operatorname{rank} \mathcal{V}_{\underline{k},r}} \langle v_{\underline{k}}, v_{\underline{k}}\rangle_h \int_{[G_B]} \langle F(g), F(g)\rangle_h \|\nu(g)\|^{-r}\, dg.
\end{aligned}$$

□

1.3. Rational and integral structures

As in the introduction, let $\Pi = \otimes_v \Pi_v$ be an irreducible cuspidal automorphic representation of $\mathrm{GL}_2(\mathbb{A}_F)$ corresponding to a Hilbert modular form of weight $(\underline{k}, r)$, character ξ_Π and conductor $\mathfrak{N} = \mathfrak{N}_{\mathrm{s}} \cdot \mathfrak{N}_{\mathrm{ps}}$. Thus the character ξ_Π satisfies

$$\xi_\Pi(z \cdot z_\infty) = \mathrm{N}(z_\infty)^r \cdot \xi_\Pi(z)$$

for $z \in \mathbb{A}_F^\times$ and $z_\infty \in \mathbb{A}_{F,\infty}^\times$. We also let $\pi = \otimes_v \pi_v$ denote the corresponding *unitary* representation:

$$\pi := \Pi \otimes \|\det(\cdot)\|^{-r/2}.$$

Recall that Σ_Π denotes the set of all places v of F at which Π_v is discrete series. Let B be any quaternion algebra over F such that $\Sigma_B \subseteq \Sigma_\Pi$, where Σ_B denotes the set of places v of F where B is ramified. By the Jacquet–Langlands correspondence, there exists a unique irreducible (cuspidal) automorphic representation

$\Pi_B \simeq \otimes_v \Pi_{B,v}$ of $G_B(\mathbb{A})$ such that $\Pi_{B,v} \simeq \Pi_v$ for all $v \notin \Sigma_B$. Let $\underline{k}_B = (k_{B,\sigma})_{\sigma\in\Sigma_\infty}$ be defined by:

$$k_{B,\sigma} = \begin{cases} k_\sigma, & \text{if } B \text{ is split at } \sigma, \\ k_\sigma - 2, & \text{if } B \text{ is ramified at } \sigma. \end{cases} \tag{1.16}$$

We say then that Π_B has *weight* $(\underline{k}_B, r)$ at infinity.

Choose a maximal order $\mathcal{O}_B$ in B. Recall that we have assumed that the conductor $\mathfrak{N}$ of Π satisfies

$$\mathfrak{N} = \mathfrak{N}_{\mathrm{s}} \cdot \mathfrak{N}_{\mathrm{ps}}$$

where $\mathfrak{N}_{\mathrm{s}}$ is divisible exactly by those primes at which Π_v is discrete series and $\mathfrak{N}_{\mathrm{ps}}$ is divisible exactly by those primes at which Π_v is ramified principal series. Let $\mathfrak{d}_B$ be the (finite part of the) discriminant of B, so that $\mathfrak{d}_B$ divides $\mathfrak{N}_{\mathrm{s}}$. Then there is a unique integral ideal $\mathfrak{N}_B$ in $\mathcal{O}_F$ such that

$$\mathfrak{N} = \mathfrak{N}_B \cdot \mathfrak{d}_B,$$

and we may choose and fix an Eichler order $\mathcal{O}_B(\mathfrak{N}_B)$ in $\mathcal{O}_B$ of level $\mathfrak{N}_B$. We will also fix an *orientation* of this order at the places dividing $\mathfrak{N}_{\mathrm{ps}}$. By this, we mean a ring homomorphism

$$o : \mathcal{O}_B(\mathfrak{N}_B) \to \mathcal{O}_F/\mathfrak{N}_{\mathrm{ps}}.$$

This choice determines an open compact subgroup $\mathcal{K} = \prod_\ell \mathcal{K}_\ell$ of $G_B(\mathbb{A}_f)$, namely $\mathcal{K}_\ell = \prod_{v|\ell} \mathcal{K}_v$ where for any finite place v of F, we have

$$\mathcal{K}_v = \ker\left[o_v : (\mathcal{O}_B(\mathfrak{N}_B) \otimes_{\mathcal{O}_F} \mathcal{O}_{F,v})^\times \to (\mathcal{O}_{F,v}/\mathfrak{N}_{\mathrm{ps}}\mathcal{O}_{F,v})^\times\right].$$

Here o_v is the natural map induced by the orientation o. For all rational primes ℓ such that $(\ell, N(\Pi)) = 1$, the subgroup $\mathcal{K}_\ell$ is a hyperspecial maximal compact subgroup of $G_B(\mathbb{Q}_\ell)$.

Now, we will assume that B is not *totally definite*, relegating the case of totally definite B to Remark 1.3.3 at the end of this section. Let ℓ be such that $(\ell, N(\Pi)) = 1$. Then for each prime λ of $E(G_B, X_B)$ dividing such an ℓ, one has (see §1.1.3) an associated canonical integral model $\mathcal{S}_{\mathcal{K},\lambda} = \mathcal{S}_{\mathcal{K},\lambda}(G_B, X_B)$ of $\mathrm{Sh}_{\mathcal{K}}(G_B, X_B)$ defined over $\mathcal{O}_{E(G_B,X_B),(\lambda)}$.

We will now fix more carefully the isomorphism

$$\phi_B : B \otimes_{\mathbb{Q}} \mathbb{R} \simeq \prod_{\sigma\in\Sigma_\infty\smallsetminus\Sigma_B} \mathrm{M}_2(\mathbb{R}) \times \prod_{\sigma\in\Sigma_{B,\infty}} \mathbb{H}. \tag{1.17}$$

Note that the vector bundles previously denoted by $\mathcal{V}_{\rho_{\underline{k}_B},r,\mathcal{K}}$ and $\mathcal{V}_{\underline{k}_B,r,\mathcal{K}}$ actually depend on the choice of ϕ_B. In this section alone, we will be pedantic and write $\mathcal{V}^{\phi_B}_{\rho_{\underline{k}_B},r,\mathcal{K}}$ and $\mathcal{V}^{\phi_B}_{\underline{k}_B,r,\mathcal{K}}$ to indicate the dependence on ϕ_B. Let $L \subset \overline{\mathbb{Q}}$ be a number field that contains the images of all embeddings σ of F in $\overline{\mathbb{Q}}$ and such that for all such σ, we have

$$B \otimes_{F,\sigma} L \simeq \mathrm{M}_2(L).$$

Then L contains $E(G_B, X_B)$. We pick the isomorphism ϕ_B above such that B maps into $\prod_{\sigma\in\Sigma_\infty} \mathrm{M}_2(L)$. This data defines an L-rational structure ([**20**], [**53**]) on the automorphic vector bundle $\mathcal{V}^{\phi_B}_{\rho_{\underline{k}_B},r,\mathcal{K}}$ on $\mathrm{Sh}_{\mathcal{K}}(G_B, X_B)$ associated to the $G_B(\mathbb{R})$-homogeneous vector bundles $\mathcal{V}^{\phi_B}_{\rho_{\underline{k}_B},r}$ as well as the sub-bundles $\mathcal{V}^{\phi_B}_{\underline{k}_B,r,\mathcal{K}}$. To define integral models of these vector bundles, we first pick a rational prime ℓ prime to $N(\Pi)$ and insist that the isomorphism ϕ_B satisfy

$$\phi_B(\mathcal{O}_B) \subset \prod_\sigma \mathrm{M}_2(\mathcal{O}_{L,(\ell)}), \tag{1.18}$$

so that ϕ_B gives an isomorphism

$$\mathcal{O}_B \otimes_{\mathbb{Z}} \mathcal{O}_{L,(\ell)} \simeq \prod_\sigma \mathcal{O}_B \otimes_{\mathcal{O}_F,\sigma} \mathcal{O}_{L,(\ell)} \simeq \prod_\sigma \mathrm{M}_2(\mathcal{O}_{L,(\ell)}). \tag{1.19}$$

For later use, we will need the following lemma.

LEMMA 1.3.1. *Let ϕ_B and ϕ'_B be two isomorphisms satisfying the requirements above. Then there exists an element*

$$t \in \left(\prod_{\sigma \in \Sigma_\infty \smallsetminus \Sigma_B} \mathrm{GL}_2(\mathbb{R}) \times \prod_{\sigma \in \Sigma_{B,\infty}} \mathbb{H}^\times \right) \cap \prod_\sigma \mathrm{GL}_2(\mathcal{O}_{L,(\ell)})$$

such that ϕ_B and ϕ'_B are conjugate by t.

PROOF. We can work one embedding at a time. First suppose $\sigma \in \Sigma_\infty \smallsetminus \Sigma_B$. Then we have a diagram as below for ϕ_B and a similar one for ϕ'_B,

$$\begin{array}{ccc} B \otimes_{F,\sigma} \mathbb{R} & \xrightarrow[\simeq]{\phi_B} & \mathrm{M}_2(\mathbb{R}) \\ \uparrow & & \uparrow \\ \mathcal{O}_B \otimes_{\mathcal{O}_F,\sigma} \mathcal{O}_{L_0,(\ell)} & \xrightarrow[\simeq]{\phi_B} & \mathrm{M}_2(\mathcal{O}_{L_0,(\ell)}) \\ \downarrow & & \downarrow \\ \mathcal{O}_B \otimes_{\mathcal{O}_F,\sigma} \mathcal{O}_{L,(\ell)} & \xrightarrow[\simeq]{\phi_B} & \mathrm{M}_2(\mathcal{O}_{L,(\ell)}) \end{array}$$

where $L_0 := L \cap \mathbb{R}$. The map in the middle is an isomorphism of Azumaya algebras over the semi-local ring $\mathcal{O}_{L_0,(\ell)}$, so the Skolem–Noether theorem applies ([**39**] Chap. III, 5.2.3) and there exists $t \in \mathrm{GL}_2(\mathcal{O}_{L_0,(\ell)})$ such that ϕ_B and ϕ'_B are conjugate by t. The proof in the case $\sigma \in \Sigma_{B,\infty}$ proceeds along similar lines. We first note that $\mathcal{O}_B \otimes_{\mathcal{O}_F,\sigma} \mathcal{O}_{L_0,(\ell)}$ is free of rank four over $\mathcal{O}_{L_0,(\ell)}$ and pick a basis $b_1, \ldots, b_4$ for this $\mathcal{O}_{L_0,(\ell)}$-module. Then $(b_1, \ldots, b_4)$ is a basis both for $B \otimes_{F,\sigma} \mathbb{R}$ as an $\mathbb{R}$-vector space and for $\mathcal{O}_B \otimes_{\mathcal{O}_{F,\sigma}} \mathcal{O}_{L,(\ell)}$ as an $\mathcal{O}_{L,(\ell)}$-module. From this it follows that

$$(B \otimes_{F,\sigma} \mathbb{R}) \cap (\mathcal{O}_B \otimes_{\mathcal{O}_F,\sigma} \mathcal{O}_{L,(\ell)}) = \mathcal{O}_B \otimes_{\mathcal{O}_F,\sigma} \mathcal{O}_{L_0,(\ell)}.$$

Thus there is a diagram as above:

$$\begin{array}{ccc} B \otimes_{F,\sigma} \mathbb{R} & \xrightarrow[\simeq]{\phi_B} & \mathbb{H} \\ \uparrow & & \uparrow \\ \mathcal{O}_B \otimes_{\mathcal{O}_F,\sigma} \mathcal{O}_{L_0,(\ell)} & \xrightarrow[\simeq]{\phi_B} & \mathrm{M}_2(\mathcal{O}_{L,(\ell)}) \cap \mathbb{H} \\ \downarrow & & \downarrow \\ \mathcal{O}_B \otimes_{\mathcal{O}_F,\sigma} \mathcal{O}_{L,(\ell)} & \xrightarrow[\simeq]{\phi_B} & \mathrm{M}_2(\mathcal{O}_{L,(\ell)}) \end{array}$$

Moreover, $A := \mathcal{O}_B \otimes_{\mathcal{O}_F,\sigma} \mathcal{O}_{L_0,(\ell)}$ is an Azumaya algebra over $\mathcal{O}_{L_0,(\ell)}$ (by [**39**] Chap. III, 5.1.1) since for every maximal ideal $\mathfrak{m}$ of $\mathcal{O}_{L_0,(\ell)}$, the algebra $A/\mathfrak{m}A \simeq \mathrm{M}_2(k_\mathfrak{m})$ is central simple over $k_{\mathfrak{m}} := \mathcal{O}_{L_0,(\ell)}/\mathfrak{m}$. Thus the map in the middle of the diagram is an isomorphism of Azumaya algebras over $\mathcal{O}_{L_0,(\ell)}$, and the Skolem–Noether theorem applies as before. □

We now return to our discussion of integral models. Using the discussion in §1.1.3, we will see that the choice of Eichler order $\mathcal{O}_B(\mathfrak{N}_B)$ and the isomorphism ϕ_B defines for all primes λ' of L with $\lambda' \mid \lambda \mid \ell$ natural integral models for the bundles $\mathcal{V}^{\phi_B}_{\rho_{\underline{k}_B},r,\mathcal{K}}$ and $\mathcal{V}^{\phi_B}_{\underline{k}_B,r,\mathcal{K}}$ over $\mathcal{S}_{\mathcal{K},\lambda} \otimes_{\mathcal{O}_{E(G_B,X_B),(\lambda)}} \mathcal{O}_{L,(\lambda')}$. Indeed, the choice of the Eichler order $\mathcal{O}_B(\mathfrak{N}_B)$ determines a reductive group $\mathcal{G}_0$ over $\mathbb{Z}_{(\ell)}$ such that $\mathcal{G}_{0,\mathbb{Q}} = G_B$; namely for any $\mathbb{Z}_{(\ell)}$-algebra A, we have

$$\mathcal{G}_0(A) = (\mathcal{O}_B(\mathfrak{N}_B) \otimes A)^\times.$$

Further, the map ϕ_B induces an isomorphism

$$\mathcal{G}_0 \otimes \mathcal{O}_{L,(\ell)} \simeq \prod_{\sigma\in\Sigma_\infty} \mathrm{GL}_{2/\mathcal{O}_{L,(\ell)}}. \tag{1.20}$$

This gives an integral model over $\mathcal{O}_{L,(\ell)}$ for the compact dual symmetric space and the vector bundle $\check{\mathcal{V}}_{\underline{k}_B,r}$. Via the identification (1.20) above, the integral model $\check{\mathfrak{X}}_{\mathcal{O}_{L,(\ell)}}$ for the compact dual is simply the conjugacy class of the parabolic subgroup

$$\mathcal{P} := \prod_{\sigma\in\Sigma_\infty\smallsetminus\Sigma_B} \mathcal{B} \times \prod_{\sigma\in\Sigma_{B,\infty}} \mathrm{GL}_{2/\mathcal{O}_{L,(\ell)}}$$

of $\mathcal{G}_0 \otimes \mathcal{O}_{L,(\ell)}$, where

$$\mathcal{B} = \left\{\begin{pmatrix} * & * \\ 0 & * \end{pmatrix}\right\} \subset \mathrm{GL}_{2/\mathcal{O}_{L,(\ell)}}.$$

Thus $\check{\mathfrak{X}}_{\mathcal{O}_{L,(\ell)}}$ is isomorphic to $\prod_{\sigma\in\Sigma_\infty\smallsetminus\Sigma_B} \mathbb{P}^1_{\mathcal{O}_{L,(\ell)}}$, the isomorphism depending on the choice of ϕ_B. Let $\mathcal{L}_\sigma = \mathcal{O}^2_{L,(\ell)}$ with the obvious left action of $\mathrm{GL}_2(\mathcal{O}_{L,(\ell)})$. Then the integral models of the vector bundles $\check{\mathcal{V}}_{\rho_{\underline{k}_B},r}$ and $\check{\mathcal{V}}_{\underline{k}_B,r}$ over $\check{\mathfrak{X}}_{\mathcal{O}_{L,(\ell)}}$ are the vector bundles $\check{\mathcal{V}}^{\phi_B}_{\rho_{\underline{k}_B},r,\mathcal{O}_{L,(\ell)}}$ and $\check{\mathcal{V}}^{\phi_B}_{\underline{k}_B,r,\mathcal{O}_{L,(\ell)}}$ corresponding respectively to the representations

$$\bigotimes_{\sigma\in\Sigma_\infty} \mathrm{Sym}^{k_{B,\sigma}}(\mathcal{L}_\sigma) \otimes \det(\mathcal{L}_\sigma)^{\frac{r-k_{B,\sigma}}{2}}$$

and

$$\prod_{\sigma\in\Sigma_\infty\smallsetminus\Sigma_B} \chi_{k_{B,\sigma},r} \cdot \bigotimes_{\sigma\in\Sigma_{B,\infty}} \mathrm{Sym}^{k_{B,\sigma}}(\mathcal{L}_\sigma) \otimes \det(\mathcal{L}_\sigma)^{\frac{r-k_{B,\sigma}}{2}}$$

of $\mathcal{P}$. (Recall that $\chi_{k,r}$ has been defined in Eg. 1.1.3.) Finally we apply the discussion in §1.1.3 to get corresponding canonical integral models of $\mathcal{V}^{\phi_B}_{\rho_{\underline{k}_B},r,\mathcal{K}}$ and $\mathcal{V}^{\phi_B}_{\underline{k}_B,r,\mathcal{K}}$ (and their duals) over $\mathcal{S}_{\mathcal{K},\lambda} \otimes_{\mathcal{O}_{E(G_B,X_B),(\lambda)}} \mathcal{O}_{L,(\lambda')}$.

For all finite places v of F at which B is split, we will fix an isomorphism

$$i_v : B \otimes F_v \simeq \mathrm{M}_2(F_v) \tag{1.21}$$

such that for all but finitely many v, we have

$$i_v : \mathcal{O}_B(\mathfrak{N}_B) \otimes \mathcal{O}_{F,v} \simeq \mathrm{M}_2(\mathcal{O}_{F,v}). \tag{1.22}$$

Let Δ be a large enough finite set of places of F such that

- Δ contains all the infinite places, and all the finite places v at which Π_v is ramified.

• For all $v \notin \Delta$, the condition (1.22) holds.

For all finite places v not in Δ, we get (using i_v) an identification

$$\mathcal{K}_v \simeq \mathrm{GL}_2(\mathcal{O}_{F,v}), \quad \mathcal{H}'_v \simeq \mathcal{H}_v \tag{1.23}$$

where $\mathcal{H}'_v$ and $\mathcal{H}_v$ denote the spherical Hecke algebras on $B_v^\times$ and $\mathrm{GL}_2(F_v)$ constructed using the maximal compact subgroups $\mathcal{K}_v$ and $\mathrm{GL}_2(\mathcal{O}_{F,v})$ respectively. Let

$$\mathcal{H}'_\Delta = \bigotimes_{v\notin\Delta} \mathcal{H}'_v, \quad \mathcal{H}_\Delta = \bigotimes_{v\notin\Delta} \mathcal{H}_v.$$

Note that $\mathcal{H}'_\Delta$ acts naturally on the space of sections of $(\mathcal{V}^{\phi_B}_{\underline{k}_B,r,\mathcal{K}})^\vee$ and we have an identification $\mathcal{H}'_\Delta \simeq \mathcal{H}_\Delta$. Also $\mathcal{H}_\Delta$ acts on $\otimes'_{v\notin\Delta}\Pi_v$. Let $\varphi = \otimes_{v\notin\Delta}\varphi_v$ be a new-vector in the space $\otimes'_{v\notin\Delta}\Pi_v$, so that φ is an eigenvector for the action of $\mathcal{H}_\Delta$. Let Λ_Π denote the corresponding character of $\mathcal{H}_\Delta$.

PROPOSITION 1.3.2. *There exists up to scaling a unique non-zero section s_B of the bundle $(\mathcal{V}^{\phi_B}_{\underline{k}_B,r,\mathcal{K}})^\vee$ which satisfies the following conditions:*

(i) *s is an eigenvector for the action of $\mathcal{H}'_\Delta$ and $\mathcal{H}'_\Delta$ acts on it by Λ_Π, via the identification $\mathcal{H}'_\Delta \simeq \mathcal{H}_\Delta$ above.*
(ii) *s satisfies (1.12) for $\xi = \xi_\Pi$.*

PROOF. Let s be any section of $(\mathcal{V}^{\phi_B}_{\underline{k}_B,r,\mathcal{K}})^\vee$. Pick some point $h \in X_B$. Let $F_{s,h} = \mathrm{Lift}_h(s)$ and set $\phi_{s,h} = \phi_{F_{s,h}}$, notations as in the previous section. By strong multiplicity one, the assignment $s \mapsto \phi_{s,h}$ gives a bijection of the space of sections of $(\mathcal{V}^{\phi_B}_{\underline{k}_B,r,\mathcal{K}})^\vee$ on which $\mathcal{H}'_\Delta$ acts by Λ_Π with the space of functions

$$\phi : G_B(\mathbb{Q})\backslash G_B(\mathbb{A})/\mathcal{K} \to \mathbb{C}$$

that satisfy (1.14) and (1.15) and on which $\mathcal{H}'_\Delta$ acts by Λ_Π. By the Jacquet–Langlands correspondence and the uniqueness of newforms [**6**], this latter space is one-dimensional, generated by a nonzero element ϕ. If s_B is such that $\phi_{s_B,h} = \phi$, then s_B is our required section. □

Let us enlarge L if necessary so that $E_\Pi \subset L$ where E_Π is the field generated by the Hecke eigenvalues of Π. By [**22**] Prop. 2.2.4, the section s_B of Prop. 1.3.2 can be chosen to be L-rational. Further, for $\lambda' \mid \lambda \mid \ell$ as above, the integral model of $(\mathcal{V}^{\phi_B}_{\underline{k}_B,r,\mathcal{K}})^\vee$ over $\mathcal{S}_{\mathcal{K},\lambda} \otimes_{\mathcal{O}_{E(G_B,X_B),(\lambda)}} \mathcal{O}_{L,(\lambda')}$ defines an $\mathcal{O}_{L,(\lambda')}$-lattice $\mathcal{M}_{\lambda'}$ in

$$H^0(\mathrm{Sh}_{\mathcal{K}}(G_B,X_B)_{/L}, (\mathcal{V}^{\phi_B}_{\underline{k}_B,r,\mathcal{K}})^\vee).$$

Fixing ℓ, choose s_B (by suitably scaling) such that for all $\lambda' \mid \lambda \mid \ell$, it is a generator for the rank one $\mathcal{O}_{L,(\lambda')}$-lattice $\mathcal{M}_{\lambda'} \cap Ls_B$. We will say that the section s_B is ℓ-normalized.

REMARK 1.3.3. In this remark we deal with the case of totally definite B. Pick ϕ_B satisfying (1.18), (1.19) above with an appropriate choice of L. Then $X_B = \{h_0\}$, and sections s of $(\mathcal{V}^{\phi_B}_{\underline{k}_B,r,\mathcal{K}})^\vee$ are identified with functions

$$F : G_B(\mathbb{Q})\backslash G_B(\mathbb{A})/\mathcal{K} \to \mathcal{V}^\vee_{\underline{k}_B,r} = \bigotimes_{\sigma\in\Sigma_\infty} V^\vee_{\sigma,k_{B,\sigma},r}$$

satisfying the appropriate invariance property under the right action of $G_B(\mathbb{R})$. Now $\mathcal{V}_{\underline{k}_B,r}$ admits a natural L-rational structure as well as a natural $\mathcal{O}_{L,(\ell)}$-submodule:

$$\mathcal{V}_{\underline{k}_B,r} \supset \mathcal{V}_{\underline{k}_B,r}(L) = \bigotimes_{\sigma\in\Sigma_\infty} \mathrm{Sym}^{k_{B,\sigma}} L^2 \otimes \det(L^2)^{\frac{r-k_{B,\sigma}}{2}}$$

$$\supset \mathcal{V}_{\underline{k}_B,r}(\mathcal{O}_{L,(\ell)}) = \bigotimes_{\sigma\in\Sigma_\infty} \mathrm{Sym}^{k_{B,\sigma}} \mathcal{O}_{L,(\ell)}^2 \otimes \det(\mathcal{O}_{L,(\ell)}^2)^{\frac{r-k_{B,\sigma}}{2}}.$$

The same holds for $\mathcal{V}^\vee_{\underline{k}_B,r}$ as well; this gives an L-rational structure and an $\mathcal{O}_{L,(\ell)}$-integral structure on the space of sections of $(\mathcal{V}^{\phi_B}_{\underline{k}_B,r,\mathcal{K}})^\vee$, namely we take sections s which on $G_B(\mathbb{A}_f)$ take values in $\mathcal{V}^\vee_{\underline{k}_B,r}(L)$ and $\mathcal{V}^\vee_{\underline{k}_B,r}(\mathcal{O}_{L,(\ell)})$ respectively. We pick isomorphisms i_v as above and Prop. 1.3.2 continues to hold. Finally, we pick s_B to be ℓ-normalized with respect to the integral structure provided by $\mathcal{V}^\vee_{\underline{k}_B,r}(\mathcal{O}_{L,(\ell)})$.

Finally, for the convenience of the reader, we summarize the discussion in this section in the following proposition which makes clear all the choices involved.

PROPOSITION 1.3.4. *Suppose that the following data has been chosen:*

(i) *A maximal order* $\mathcal{O}_B$ *in* B.
(ii) *An Eichler order* $\mathcal{O}_B(\mathfrak{N}_B) \subset \mathcal{O}_B$.
(iii) *An orientation* $o : \mathcal{O}_B(\mathfrak{N}_B) \to \mathcal{O}_F/\mathfrak{N}_{\mathrm{ps}}$.
(iv) *A number field* $L \subset \overline{\mathbb{Q}}$ *containing* $\sigma(F)$ *for all embeddings* $\sigma : F \hookrightarrow \overline{\mathbb{Q}} \subset \mathbb{C}$, *and containing the field generated by the Hecke eigenvalues of* Π.
(v) *An isomorphism*

$$\phi_B : B \otimes \mathbb{R} \simeq \prod_{\sigma\in\Sigma_\infty\smallsetminus\Sigma_B} \mathrm{M}_2(\mathbb{R}) \times \prod_{\sigma\in\Sigma_{B,\infty}} \mathbb{H},$$

such that

$$\phi_B(\mathcal{O}_B) \subset \prod_\sigma \mathrm{M}_2(\mathcal{O}_{L,(\ell)}).$$

(vi) *For all finite places* v *of* F *at which* B *is split, an isomorphism*

$$i_v : B \otimes F_v \simeq \mathrm{M}_2(F_v)$$

such that for almost all v, i_v *induces an isomorphism*

$$\mathcal{O}_B(\mathfrak{N}_B) \otimes \mathcal{O}_{F,v} \simeq \mathrm{M}_2(\mathcal{O}_{F,v}).$$

The data of (i) *to* (vi) *determines an open compact subgroup* $\mathcal{K}$ *of* $G_B(\mathbb{A}_f)$ *and a canonical section* s_B *of the vector bundle* $(\mathcal{V}^{\phi_B}_{\underline{k}_B,r,\mathcal{K}})^\vee$ *that satisfies the conditions* (i) *and* (ii) *of Prop.* 1.3.2. *Moreover,* s_B *can be chosen such that it is well defined up to multiplication by an* ℓ*-adic unit.*

In the next section, we shall define a canonical quadratic period invariant using the section s_B. The key point will be to show that this period invariant is (up to ℓ-adic units) independent of the choices made in the proposition above.

1.4. Canonical quadratic period invariants

We can now define the canonical quadratic period invariants attached to Π and state the main conjecture relating these invariants. Let B be a quaternion algebra such that $\Sigma_B \subseteq \Sigma_\Pi$. As in the introduction, let R be the ring $\mathcal{O}_{\overline{\mathbb{Q}}}[1/N(\Pi)]$. For any rational prime ℓ prime to $N(\Pi)$, we define an invariant $q_B(\Pi,\ell) \in \mathbb{C}^\times/R^\times_{(\ell)}$

as follows. Let $\tilde{\mathcal{K}} \supseteq \mathcal{K}$ be the open compact subgroup of $G_B(\mathbb{A}_f)$ defined by $\tilde{\mathcal{K}} = \prod_v \tilde{\mathcal{K}}_v$ with

$$\tilde{\mathcal{K}}_v = (\mathcal{O}_B(\mathfrak{N}_B) \otimes_{\mathcal{O}_F} \mathcal{O}_{F,v})^\times.$$

Choosing a section s_B as above that is ℓ-normalized, define

$$q_B(\Pi, \ell) := \langle\!\langle s_B, s_B \rangle\!\rangle_{\tilde{\mathcal{K}}} \in \mathbb{C}^\times / R_{(\ell)}^\times,$$

to be the Petersson norm of the section s_B as in Defn. 1.2.8.

PROPOSITION 1.4.1. *The invariant $q_B(\Pi, \ell)$ is well defined, in that as an element of $\mathbb{C}^\times/R_{(\ell)}^\times$, it does not depend on the choices of the number field L, the pair $(\mathcal{O}_B(\mathfrak{N}_B), o)$ consisting of the Eichler order $\mathcal{O}_B(\mathfrak{N}_B)$ and the orientation $o : \mathcal{O}_B(\mathfrak{N}_B) \to \mathcal{O}_F/\mathfrak{N}_{\mathrm{ps}}$, the isomorphism* (1.17) *satisfying* (1.18), (1.19) *above and the collection of isomorphisms* (1.21).

PROOF. We will give the argument in the case when B is not totally definite. In the case of a totally definite B, a similar (but simpler) argument can be given which we leave to the reader.

Independence of the choice of L is clear since we can always replace L by a larger field without changing the choice of s_B. Implicitly in the arguments below we may need to make such a field extension and we do this without comment. Let us first show that fixed choices of other data, there is no dependence on the choice of isomorphisms (1.21). Indeed, for all but finitely many v, the isomorphisms i_v must satisfy (1.22). Let $\{i'_v\}$ be a different set of choices. Then for all but finitely many v, the isomorphisms i_v and i'_v must differ by conjugation by an element of $\mathcal{K}_v$. For such v, the identifications $\mathcal{H}'_v \simeq \mathcal{H}_v$ given by i_v and i'_v are the same. This implies that the same choice of s_B can be used if $\{i_v\}$ is replaced by $\{i'_v\}$ and the norm $\langle\!\langle s_B, s_B \rangle\!\rangle_{\tilde{\mathcal{K}}}$ is unchanged.

Next let us look at the dependence on the choice of isomorphism ϕ_B in (1.17), for fixed choices of other data. Let ϕ'_B be a different choice of isomorphism satisfying (1.18). By Lemma 1.3.1, the isomorphisms ϕ_B and ϕ'_B differ by conjugation by an element

$$t \in \prod_\sigma \mathrm{GL}_2(\mathcal{O}_{L,(\ell)}) \cap \left(\prod_{\sigma \in \Sigma_\infty \smallsetminus \Sigma_B} \mathrm{GL}_2(\mathbb{R}) \times \prod_{\sigma \in \Sigma_{B,\infty}} \mathbb{H}^\times \right).$$

Then there is a natural morphism of integral models

$$\check{\mathcal{V}}^{\phi_B}_{\mathcal{O}_{L,(\ell)}} \simeq \check{\mathcal{V}}^{\phi'_B}_{\mathcal{O}_{L,(\ell)}}$$

which is just given by the (left) action of t on the fibers. This induces an isomorphism between the integral models of the corresponding automorphic vector bundles that is also given by the action of t on the fibers. (Keep in mind that the $G(\mathbb{Q})$-actions on the fibers are different, and differ by conjugation by t, so the left action of t on the fibers is indeed a map of bundles.) Thus if s_B is an ℓ-normalized section of $(\mathcal{V}^{\phi_B}_{\underline{k}_B,r,\mathcal{K}})^\vee$, then $s'_B := t \cdot s_B$ is an ℓ-normalized section of $(\mathcal{V}^{\phi'_B}_{\underline{k}_B,r,\mathcal{K}})^\vee$. Then the inner products $\langle\!\langle s_B, s_B \rangle\!\rangle_{\tilde{\mathcal{K}}}$ and $\langle\!\langle s'_B, s'_B \rangle\!\rangle_{\tilde{\mathcal{K}}}$ differ by a power of $\|\nu(t)\|$, which is a unit at ℓ.

Finally, we consider dependence on the choice of the pair $(\mathcal{O}_B(\mathfrak{N}_B), o)$. Let $(\mathcal{O}_B(\mathfrak{N}_B)', o')$ be another such pair and let ϕ'_B (respectively i'_v) denote our choices of isomorphism (1.17) satisfying (1.18) (respectively the isomorphisms (1.21) satisfying (1.22) for all but finitely many v). Let us suppose first that the pair

$(\mathcal{O}_B(\mathfrak{N}_B)', o')$ is conjugate to $(\mathcal{O}_B(\mathfrak{N}_B), o)$ by an element in $B^\times$, say $\mathcal{O}_B(\mathfrak{N}_B)' = b^{-1}\mathcal{O}_B(\mathfrak{N}_B)b$ and $o'(x) = o(bxb^{-1})$. By what we have shown so far we may assume that

$$\phi'_B(x) = \phi_B(bxb^{-1}), \quad i'_v(x) = i_v(bxb^{-1}).$$

The open compact subgroup $\mathcal{K}'$ of $G_B(\mathbb{A}_f)$ determined by the pair $(\mathcal{O}_B(\mathfrak{N}_B)', o')$ satisfies $\mathcal{K}' = b^{-1}\mathcal{K}b$. Let us write $b = b_\infty \cdot b_f$ where b_∞ and b_f denote the infinite and finite components of b respectively, viewed as elements in $G_B(\mathbb{A})$. There is a natural isomorphism of Shimura varieties

$$\begin{aligned}\mathrm{Sh}_{\mathcal{K}}(G_B, X_B) = G_B(\mathbb{Q})\backslash X_B \times G_B(\mathbb{A}_f)/\mathcal{K} &\overset{\xi_b}{\simeq} G_B(\mathbb{Q})\backslash X_B \times G_B(\mathbb{A}_f)/\mathcal{K}' \\ &= \mathrm{Sh}_{\mathcal{K}'}(G_B, X_B),\end{aligned}$$

given by

$$(h, g_f) \mapsto (h, g_f b_f).$$

Further, there is a natural isomorphism

$$\begin{aligned}(\mathcal{V}^{\phi_B}_{\underline{k}_B,r,\mathcal{K}})^\vee = G_B(\mathbb{Q})\backslash(\mathcal{V}^{\phi_B}_{\underline{k}_B,r})^\vee \times G_B(\mathbb{A}_f)/\mathcal{K} &\overset{\tilde{\xi}_b}{\simeq} G_B(\mathbb{Q})\backslash(\mathcal{V}^{\phi'_B}_{\underline{k}_B,r})^\vee \times G_B(\mathbb{A}_f)/\mathcal{K}' \\ &= (\mathcal{V}^{\phi'_B}_{\underline{k}_B,r,\mathcal{K}'})^\vee,\end{aligned}$$

covering ξ_b, given by

$$(v, g_f) \mapsto (\phi_B(b) \cdot v, g_f b_f). \tag{1.24}$$

Note that if γ is an element in $G_B(\mathbb{Q})$ then

$$\begin{aligned}\gamma \cdot (v, g_f) = (\phi_B(\gamma) \cdot v, \gamma g_f) &\mapsto (\phi_B(b)\phi_B(\gamma) \cdot v, \gamma g_f b_f) \\ &= (\phi'_B(\gamma)\phi_B(b) \cdot v, \gamma g_f b_f) = \gamma \cdot (\phi_B(b) \cdot v, g_f b_f),\end{aligned}$$

so that the assignment in (1.24) does descend to equivalence classes for the $G_B(\mathbb{Q})$-action. Note also that $\tilde{\xi}_B$ is the map on automorphic vector bundles corresponding to a morphism of vector bundles that extend to the integral models, since these integral models are defined using the triples $(\mathcal{O}_B(\mathfrak{N}_B), o, \phi_B)$ and $(\mathcal{O}_B(\mathfrak{N}_B)', o', \phi'_B)$ respectively. Thus $\tilde{\xi}_b$ is an isomorphism at the level of integral models, and so we may assume that $s'_B = \tilde{\xi}_b(s_B)$. But then we see from the definition of the metrics on the vector bundles $(\mathcal{V}^{\phi_B}_{\underline{k}_B,r,\mathcal{K}})^\vee$ and $(\mathcal{V}^{\phi'_B}_{\underline{k}_B,r,\mathcal{K}'})^\vee$ and the product formula that

$$\langle\!\langle s'_B, s'_B \rangle\!\rangle_{\tilde{\mathcal{K}}'} = \|\nu(b_\infty)\|^{-r} \cdot \|\nu(b_f)\|^{-r} \cdot \langle\!\langle s_B, s_B \rangle\!\rangle_{\tilde{\mathcal{K}}} = \langle\!\langle s_B, s_B \rangle\!\rangle_{\tilde{\mathcal{K}}}.$$

In general, it may not be true that the pairs $(\mathcal{O}_B(\mathfrak{N}_B), o)$ and $(\mathcal{O}_B(\mathfrak{N}_B)', o')$ are conjugate by an element of $B^\times$. Nevertheless, we can always find an element $\beta_f \in B^\times(\mathbb{A}_f)$ such that

$$\mathcal{O}_B(\mathfrak{N}_B)' = \beta_f^{-1}\mathcal{O}_B(\mathfrak{N}_B)\beta_f, \quad o'(x) = o(\beta_f x \beta_f^{-1}).$$

Let b be an element of $B^\times$ approximating $\beta_f = (\beta_v)$ at ℓ so that

$$\mathcal{O}_B(\mathfrak{N}_B)' \otimes \mathbb{Z}_{(\ell)} = (b^{-1}\mathcal{O}_B(\mathfrak{N}_B)b) \otimes \mathbb{Z}_{(\ell)}.$$

Then we may assume that

$$\phi'_B(x) = \phi_B(bxb^{-1}), \quad i'_v(x) = i_v(\beta_v x \beta_v^{-1}).$$

The open compact subgroup $\mathcal{K}'$ satisfies $\mathcal{K}' = \beta_f^{-1}\mathcal{K}\beta_f$. We now run through the same argument as above, defining

$$\xi_b[(h, g_f)] = [(h, g_f\beta_f)], \quad \tilde{\xi}_b[(v, g_f)] = [(\phi_B(b) \cdot v, g_f\beta_f)].$$

The result follows from observing that $\|\nu(b_\infty)\| \cdot \|\nu(\beta_f)\|$, while not necesarily 1, is still an element in $R_{(\ell)}^\times$. $\square$

Finally, we define an invariant $q_B(\Pi) \in \mathbb{C}^\times/R^\times$ such that the class of $q_B(\Pi)$ in $\mathbb{C}^\times/R_{(\ell)}^\times$ equals $q_B(\Pi, \ell)$. Let us pick the isomorphism ϕ_B, the number field L and the maximal order $\mathcal{O}_B$ in B such that

$$\phi_B(\mathcal{O}_B) \subset \prod_\sigma \mathrm{M}_2(\mathcal{O}_L).$$

Choose a pair $(\mathcal{O}_B(\mathfrak{N}_B), o)$ consisting of an Eichler order and an orientation. The constructions in §1.3 can be copied verbatim to give integral models for everything in sight over $\mathcal{O}_L[\frac{1}{N(\Pi)}]$. (See §1.1.3.1.) By enlarging L if need be, we can pick a section s_B that is ℓ-normalized at *all* rational primes that are prime to $N(\Pi)$. Then we set

$$q_B(\Pi) := \langle\langle s_B, s_B \rangle\rangle_{\tilde{\mathcal{K}}}$$

for such a choice of s_B. This is an element of $\mathbb{C}^\times$ that maps to $q_B(\Pi, \ell)$ under the natural map $\mathbb{C}^\times \to \mathbb{C}^\times/R_{(\ell)}^\times$ for all ℓ such that $(\ell, N(\Pi)) = 1$. Since the map

$$\mathbb{C}^\times/R^\times \to \prod_{(\ell,N(\Pi))=1} \mathbb{C}^\times/R_{(\ell)}^\times$$

is injective, the class of $q_B(\Pi)$ in $\mathbb{C}^\times/R^\times$ is well defined. This defines the invariants needed in the formulation of Conjecture A of the introduction.

CHAPTER 2

Unitary and Quaternionic Unitary Groups

In §2.1, we review the general theory of hermitian and skew-hermitian forms over local fields and number fields. In §2.2, we describe the construction of a certain skew-hermitian space over a quaternion algebra (over a number field), which plays an important role throughout the paper, while in §2.3 we review the connection between this construction and the failure of the Hasse principle for quaternionic skew-hermitian forms.

2.1. Hermitian and skew-hermitian spaces

2.1.1. Hermitian spaces. Let F be a field of characteristic zero and E a quadratic extension of F, possibly split. Let $\mathbf{V}$ be a right E-vector space of dimension n (i.e., a free E-module of rank n), equipped with a Hermitian form

$$(\cdot,\cdot):\mathbf{V}\times\mathbf{V}\to E.$$

Such a form is linear in one variable and antilinear in the other, and we fix any one convention at this point. For example, if $(\cdot,\cdot)$ is antilinear in the first variable and linear in the second, then:

$$(v\alpha, v'\beta)=\alpha^{\rho}(v,v')\beta,\quad (v,v')=(v',v)^{\rho},$$

where ρ denotes the nontrivial involution of E/F.

To such a $\mathbf{V}$, one associates the following invariants: $\dim(\mathbf{V})=n$ and $\mathrm{disc}(\mathbf{V})\in F^{\times}/\mathrm{N}_{E/F}E^{\times}$, where

$$\mathrm{disc}(\mathbf{V})=(-1)^{n(n-1)/2}\det\left((v_i,v_j)\right),$$

with $\{v_i\}$ an E-basis for $\mathbf{V}$. Since $(\cdot,\cdot)$ is Hermitian, $\mathrm{disc}(\mathbf{V})$ lies in $F^{\times}$ and its class in $F^{\times}/\mathrm{N}_{E/F}E^{\times}$ is independent of the choice of basis.

Let $\mathrm{GU}(\mathbf{V})$ denote the unitary similitude group of $\mathbf{V}$. (Occasionally, we will write $\mathrm{GU}_E(\mathbf{V})$ for clarity.) This is an algebraic group over F such that for any F-algebra R, we have

$$\mathrm{GU}(\mathbf{V})(R):=\{g\in\mathrm{GL}(\mathbf{V}\otimes R):\ (gv,gv')=\nu(g)(v,v')\text{ for all }v,v',\,\nu(g)\in R^{\times}\}.$$

If $E=F\times F$, then $\mathrm{GU}(\mathbf{V})\simeq\mathrm{GL}_n\times\mathrm{GL}_1$. If E is a field, the various possibilities for $\mathrm{GU}(\mathbf{V})$ are discussed below.

2.1.1.1. *p-adic local fields.* Let F be p-adic. As a Hermitian space, $\mathbf{V}$ is determined up to isomorphism by its dimension and discriminant. Further, given any choice of dimension and discriminant, there is a space $\mathbf{V}$ with these as its invariants. If $\dim(\mathbf{V})$ is odd, the group $\mathrm{GU}(\mathbf{V})$ is (up to isomorphism) independent of $\mathrm{disc}(\mathbf{V})$ and is quasi-split. If $\dim(\mathbf{V})$ is even, there are two posibilities for $\mathrm{GU}(\mathbf{V})$ up to isomorphism and $\mathrm{GU}(\mathbf{V})$ is quasi-split if and only if $\mathrm{disc}(\mathbf{V})=1$.

2.1.1.2. *Archimedean fields.* Let $F = \mathbb{R}$ and $E = \mathbb{C}$. Then the form $(\cdot, \cdot)$ can be put into the diagonal form $(1, \ldots, 1, -1, \ldots, -1)$ which is called the signature of $\mathbf{V}$; we say $\mathbf{V}$ is of type (p, q) if the number of 1s is p and the number of -1s is q. Hermitian spaces are classified up to isomorphism by their signature (which determines both the dimension and discrminant) and we write $\mathrm{GU}(p, q)$ for the associated group. The only isomorphisms between these groups are $\mathrm{GU}(p, q) \simeq \mathrm{GU}(q, p)$.

2.1.1.3. *Number fields.* Let E/F be a quadratic extension of number fields. If $\mathbf{V}$ is a Hermitian E-space, then for each place v of F, one gets a local space $\mathbf{V}_v$ which is a Hermitian space for E_v/F_v and such that for almost all v, the discriminant of $\mathbf{V}_v$ is 1. The Hasse principle says that $\mathbf{V}$ is determined up to isomorphism by this collection of local spaces. Conversely, suppose we are given for each place v a local space $\mathbf{V}_v$ (of some fixed dimension n) such that almost all of the local discriminants are equal to 1. The collection of local discriminants gives an element of $\mathbb{A}_F^\times / \mathrm{N}_{E/F} \mathbb{A}_E^\times$. Such a collection of local spaces comes from a global space if and only if this element lies in the image of $F^\times$, i.e., is trivial in the quotient $\mathbb{A}_F^\times / F^\times \mathrm{N}_{E/F} \mathbb{A}_E^\times$, which has order 2.

2.1.2. Skew-Hermitian spaces. Let E/F be a quadratic extension as in the beginning of the previous section. Skew-hermitian E-spaces are defined similarly to hermitian spaces but with the condition

$$(v, v') = -(v', v)^\rho.$$

We can go back and forth between hermitian and skew-hermitian spaces simply by multiplying the form by an element in $E^\times$ of trace zero. Indeed, pick a trace zero element $\mathbf{i} \in E^\times$. If $(\cdot, \cdot)$ is skew-hermitian form on $\mathbf{V}$, the product $(\cdot, \cdot)' := \mathbf{i} \cdot (\cdot, \cdot)$ is hermitian. The group $\mathrm{GU}(\mathbf{V})$ is the same for both $(\cdot, \cdot)$ and $(\cdot, \cdot)'$. Thus the classification of skew-hermitian forms (and the corresponding groups) can be deduced from the hermitian case.

2.1.3. Quaternionic hermitian spaces. Let F be a field and B a quaternion algebra over F. Let $a \mapsto a^*$ denote the main involution on B. A B-Hermitian space is a right B-space V equipped with a B-valued form

$$\langle \cdot, \cdot \rangle : V \times V \to B$$

satisfying

$$\langle v\alpha, v'\beta \rangle = \alpha^* \langle v, v' \rangle \beta, \quad \langle v, v' \rangle = \langle v', v \rangle^*,$$

for $v, v' \in V$ and $\alpha, \beta \in B$.

Let $\mathrm{GU}(V)$ denote the unitary similitude group of V. (Sometimes, we write $\mathrm{GU}_B(V)$ for clarity.) This is an algebraic group over F such that for any F-algebra R, we have

$$\mathrm{GU}(V)(R) := \{g \in \mathrm{GL}(V \otimes R) : \ \langle gv, gv' \rangle = \nu(g) \langle v, v' \rangle \text{ for all } v, v', \, \nu(g) \in R^\times\}.$$

If B is split, there is a unique such space V of any given dimension n over B. The corresponding group $\mathrm{GU}(V)$ is identified with $\mathrm{GSp}(2n)$. If B is nonsplit, the classification of such spaces over p-adic fields and number fields is recalled below.

2.1.3.1. *p-adic fields.* If F is a p-adic field, there is a unique such space of any given dimension, up to isometry. The corresponding group is the unique nontrivial inner form of $\mathrm{GSp}(2n)$.

2.1.3.2. *Archimedean fields.* If $F = \mathbb{R}$, such spaces are classified by dimension and signature. If the signature is of type (p, q), the associated group is denoted $\mathrm{GSp}(p, q)$. The only isomorphisms between these are $\mathrm{GSp}(p, q) \simeq \mathrm{GSp}(q, p)$.

2.1.3.3. *Global fields.* The Hasse principle holds in this case, so a global B-hermitian space is determined up to isometry by the collection of corresponding local B_v-Hermitian spaces. Conversely, given any collection of B_v-hermitian spaces, there is a (unique) B-Hermitian space that gives rise to this local collection up to isometry.

2.1.4. Quaternionic skew-hermitian spaces. These are defined similarly to B-hermitian spaces but with the condition

$$\langle v, v' \rangle = -\langle v', v \rangle^*.$$

To such a space V is associated the invariant $\det(V) \in F^\times/(F^\times)^2$ as follows. Pick a B-basis $\{v_i\}$ for V and set

$$\det(V) = \nu_B(\langle v_i, v_j \rangle).$$

Here ν_B denotes the reduced norm. (Often, we will omit the subscript B when the choice of quaternion algebra is clear.) The group $\mathrm{GU}(V)$ is defined similarly as above. It is however not connected as an algebraic group. We now recall the classification of such spaces and the associated groups. Note that if B is split, we can associate to V a quadratic space $V^\dagger$ over F of dimension $2n$ (where $n = \dim_B(V)$) and $\mathrm{GU}(V) \simeq \mathrm{GO}(V^\dagger)$.

2.1.4.1. *p-adic fields.* Let F be p-adic. If B is split, V is determined by $\dim(V)$, $\det(V)$ and the Hasse invariant of $V^\dagger$. If B is nonsplit, V is determined by $\dim(V)$ and $\det(V)$.

2.1.4.2. *Archimedean fields.* If $F = \mathbb{R}$ and B is split, V is determined by the signature of $V^\dagger$. The group $\mathrm{GU}(V)$ is isomorphic to $\mathrm{GO}(p, q)$ where (p, q) is the signature. If B is nonsplit, V is determined just by $n = \dim_B(V)$. The group $\mathrm{GU}(V)$ is isomorphic to $\mathrm{GO}^*(2n)$. If $F = \mathbb{C}$, then B must be split and there is a unique skew-hermitian space of any given dimension. Then $\mathrm{GU}(V) \simeq \mathrm{GO}(2n, \mathbb{C})$.

2.1.4.3. *Global fields.* Let F be a number field. If B is split, then the classification reduces to that for quadratic spaces via the assignment $V \mapsto V^\dagger$. In this case, the Hasse principle holds. If B is nonsplit, then the Hasse principle does not hold. Let Σ_B be the set of places v where B is ramified and let $s = |\Sigma_B|$. The space V gives rise to a collection of local spaces and up to isometry there are exactly 2^{s-2} global B-skew-hermitian spaces that give rise to the same set of local spaces. Conversely a collection of local B_v-skew-hermitian spaces V_v arises from a global B-skew-hermitian space V if and only if there exists a global element $d \in F^\times$ such that $\det(V_v) = d$ in $F_v^\times/(F_v^\times)^2$ for all v and for almost all v, the Hasse invariant of $V_v^\dagger$ is trivial.

2.2. The key constructions

In this section, we assume that B_1 and B_2 are two quaternion algebras over a number field F and E/F is a quadratic extension that embeds in both B_1 and B_2. We will fix embeddings $E \hookrightarrow B_1$ and $E \hookrightarrow B_2$. Via these embeddings, B_1 and B_2 are hermitian spaces over E. Let τ_i and ν_i be respectively the reduced trace and norm on B_i. We think of B_1 and B_2 as right E-vector spaces, the Hermitian form being described below. Write

$$B_1 = E + E\mathbf{j}_1 = E + \mathbf{j}_1 E, \quad B_2 = E + E\mathbf{j}_2 = E + \mathbf{j}_2 E,$$

where $\tau_1(\mathbf{j}_1) = \tau_2(\mathbf{j}_2) = 0$. We write pr_i for the projection $B_i \to E$ onto the "first coordinate" and $*_i$ for the main involution on B_i. Then B_i is a right Hermitian E-space, the form being given by:

$$(x, y)_i = \mathrm{pr}_i(x^{*_i} y).$$

If $x = a + \mathbf{j}_i b$, $y = c + \mathbf{j}_i d$, then

$$(x, y)_i = (a + \mathbf{j}_i b, c + \mathbf{j}_i d)_i = a^\rho c - J_i b^\rho d,$$

where $J_i := -\nu_i(\mathbf{j}_i) = \mathbf{j}_i^2$. This form satisfies the relations

$$(x\alpha, y\beta)_i = \alpha^\rho (x, y)_i \beta, \quad \text{for} \quad \alpha, \beta \in E,$$

and

$$(x, y)_i = (y, x)_i^\rho.$$

We note that $B_i^\times$ acts naturally on B_i by left multiplication, and this action is E-linear. Further,

$$(\boldsymbol{\alpha} x, \boldsymbol{\alpha} y)_i = \nu_i(\boldsymbol{\alpha})(x, y)_i \tag{2.1}$$

for all $\boldsymbol{\alpha} \in B_i$. Thus $B_i^\times$ embeds naturally in $\mathrm{GU}_E(B_i)$. In fact,

$$F^\times \backslash (B_i^\times \times E^\times) \simeq \mathrm{GU}_E(B_i),$$

where $E^\times$ acts on B_i by right multiplication, and we think of $F^\times$ as a subgroup of $B_i^\times \times E^\times$ via $\lambda \mapsto (\lambda^{-1}, \lambda)$.

Consider the (right) E-vector space

$$V := B_1 \otimes_E B_2.$$

REMARK 2.2.1. If $x \in B_1$, $y \in B_2$, $\alpha \in E$, then by definition,

$$(x \otimes y)\alpha = x\alpha \otimes y = x \otimes y\alpha.$$

The E-vector space V is equipped with a natural Hermitian form given by the tensor product $(\cdot,\cdot)_1 \otimes (\cdot,\cdot)_2$. We fix a nonzero element $\mathbf{i} \in E$ of trace 0, and define $(\cdot,\cdot)$ on V by

$$(\cdot,\cdot) := \mathbf{i} \cdot (\cdot,\cdot)_1 \otimes (\cdot,\cdot)_2. \tag{2.2}$$

Clearly, $(\cdot,\cdot)$ satisfies

$$(x\alpha, y\beta) = \alpha^\rho (x, y)\beta, \quad \text{for} \quad \alpha, \beta \in E,$$

$$(x, y) = -(y, x)^\rho.$$

Thus $(\cdot,\cdot)$ is an E-skew-Hermitian form on V.

It will be useful to write down the form $(\cdot,\cdot)$ explicitly in terms of coordinates with respect to a suitable E-basis. We pick the following (orthogonal) basis:

$$\mathbf{e}_1 := 1 \otimes 1, \qquad \mathbf{e}_2 := \mathbf{j}_1 \otimes 1, \qquad \mathbf{e}_3 := 1 \otimes \mathbf{j}_2, \qquad \mathbf{e}_4 := \mathbf{j}_1 \otimes \mathbf{j}_2. \tag{2.3}$$

In this basis,

$$\begin{aligned}(\mathbf{e}_1 a + \mathbf{e}_2 b + \mathbf{e}_3 c + \mathbf{e}_4 d, \mathbf{e}_1 a' + \mathbf{e}_2 b' + \mathbf{e}_3 c' + \mathbf{e}_4 d')& \\ = \mathbf{i} \cdot \left[a^\rho a' - J_1 b^\rho b' - J_2 c^\rho c' + J_1 J_2 d^\rho d'\right].&\end{aligned}$$

There is a natural map

$$\mathrm{GU}_E(B_1) \times \mathrm{GU}_E(B_2) \to \mathrm{GU}_E(V),$$

given by the actions of $\mathrm{GU}_E(B_1)$ and $\mathrm{GU}_E(B_2)$ on the first and second component respectively of $V = B_1 \otimes_E B_2$. The kernel of this map is

$$\mathrm{Z} := \left\{([\lambda_1, \alpha_1], [\lambda_2, \alpha_2]) : \lambda_i \in F^\times, \alpha_i \in E^\times, \lambda_1\lambda_2\alpha_1\alpha_2 = 1\right\}.$$

Let $B := B_1 \cdot B_2$ be the product in the Brauer group over F. Then E embeds in B as well, and we will fix an embedding $E \to B$. We may write

$$B = E + E\mathbf{j}$$

where $\tau(\mathbf{j}) = 0$ and $J := -\nu(\mathbf{j}) = \mathbf{j}^2$ satisfies

$$J = J_1 J_2.$$

Here τ and ν are respectively the reduced trace and reduced norm on B. We define a right action of B on V (extending the right E-action on V) by setting

$$(1 \otimes 1) \cdot \mathbf{j} = \mathbf{j}_1 \otimes \mathbf{j}_2 \tag{2.4}$$
$$(\mathbf{j}_1 \otimes 1) \cdot \mathbf{j} = J_1(1 \otimes \mathbf{j}_2) \tag{2.5}$$
$$(1 \otimes \mathbf{j}_2) \cdot \mathbf{j} = J_2(\mathbf{j}_1 \otimes 1) \tag{2.6}$$
$$(\mathbf{j}_1 \otimes \mathbf{j}_2) \cdot \mathbf{j} = J_1 J_2(1 \otimes 1) \tag{2.7}$$

and requiring the right action by $\mathbf{j}$ on V to be conjugate E-linear. (It is straightforward to check that this gives an action.) Then V is a free rank-2 right B-module. For example, a basis for V as a right B-module is given either by

$$\{1 \otimes 1, \mathbf{j}_1 \otimes 1\}$$

or

$$\{1 \otimes 1, 1 \otimes \mathbf{j}_2\}.$$

Further, one checks that (equivalently)

$$(x\mathbf{j}, y) = (y\mathbf{j}, x) \tag{2.8}$$
$$(x\mathbf{j}, y\mathbf{j})^\rho = -J(x, y) \tag{2.9}$$

for all $x, y \in V$.

We will now show that there is a B-valued skew-Hermitian form $\langle \cdot, \cdot \rangle$ on V such that

$$\mathrm{pr} \circ \langle \cdot, \cdot \rangle = (\cdot, \cdot).$$

Indeed, define

$$\langle x, y \rangle = (x, y) - \frac{1}{J} \cdot \mathbf{j} \cdot (x\mathbf{j}, y). \tag{2.10}$$

It may be checked (using (2.8) and (2.9)) that

$$\langle x\boldsymbol{\alpha}, y\boldsymbol{\beta} \rangle = \boldsymbol{\alpha}^* \langle x, y \rangle \boldsymbol{\beta}, \tag{2.11}$$
$$\langle x, y \rangle = -\langle y, x \rangle^*, \tag{2.12}$$

for all $\boldsymbol{\alpha}, \boldsymbol{\beta} \in B$. For future reference, we write down the matrix of inner products $\langle \mathbf{e}_i, \mathbf{e}_j \rangle$.

$\langle\cdot,\cdot\rangle$	$\mathbf{e}_1$	$\mathbf{e}_2$	$\mathbf{e}_3$	$\mathbf{e}_4$
$\mathbf{e}_1$	$\mathbf{i}$	0	0	$\mathbf{ij}$
$\mathbf{e}_2$	0	$-J_1\mathbf{i}$	$-\mathbf{ij}$	0
$\mathbf{e}_3$	0	$-\mathbf{ij}$	$-J_2\mathbf{i}$	0
$\mathbf{e}_4$	$\mathbf{ij}$	0	0	$J\mathbf{i}$

From the table, we see that $\det(V) = \nu(-J_1u) = 1$ in $F^\times/(F^\times)^2$.

Notice that $B_1^\times$ and $B_2^\times$ act on V by left multiplication on the first and second factor of $V = B_1 \otimes_E B_2$ respectively. These actions are (right) E-linear and in fact (right) B-linear as is easily checked using (2.4) through (2.7). Further, it follows from (2.1), (2.2), and (2.10) that

$$\langle \boldsymbol{\alpha}_i \cdot x, \boldsymbol{\alpha}_i \cdot y\rangle = \nu_i(\boldsymbol{\alpha}_i)\langle x, y\rangle \tag{2.13}$$

for $\boldsymbol{\alpha}_i \in B_i^\times$. Clearly, the actions of $B_1^\times$ and $B_2^\times$ commute, hence one gets an embedding

$$F^\times\backslash(B_1^\times \times B_2^\times) \hookrightarrow \mathrm{GU}_B(V), \tag{2.14}$$

the quaternionic unitary group of the B-skew-Hermitian form $\langle\cdot,\cdot\rangle$. (Here we think of $F^\times$ as embedded antidiagonally in $B_1^\times \times B_2^\times$ via $\lambda \mapsto (\lambda^{-1}, \lambda)$.) Then (2.14) gives an isomorphism

$$F^\times\backslash(B_1^\times \times B_2^\times) \simeq \mathrm{GU}_B(V)^0,$$

where $\mathrm{GU}_B(V)^0$ denotes the identity component of $\mathrm{GU}_B(V)$. Further, one has a commutative diagram

$$\begin{array}{ccc} F^\times\backslash(B_1^\times \times B_2^\times) & \xrightarrow{\ \simeq\ } & \mathrm{GU}_B(V)^0 \\ \big\downarrow & & \big\downarrow \\ \mathrm{Z}\backslash(\mathrm{GU}_E(B_1) \times \mathrm{GU}_E(B_2)) & \longrightarrow & \mathrm{GU}_E(V) \end{array}$$

where the vertical map

$$F^\times\backslash(B_1^\times \times B_2^\times) \hookrightarrow \mathrm{Z}\backslash(\mathrm{GU}_E(B_1) \times \mathrm{GU}_E(B_2))$$

is

$$[b_1, b_2] \mapsto [[(b_1, 1)], [(b_2, 1)]]\,,$$

and the vertical map $\mathrm{GU}_B(V)^0 \hookrightarrow \mathrm{GU}_E(V)$ is just the natural inclusion.

Let $\mathbb{V} = \mathrm{Res}_{E/F}(V)$, that is $\mathbb{V}$ is just V thought of as an F-space, with non-degenerate symplectic form

$$\langle\!\langle v_1, v_2\rangle\!\rangle := \frac{1}{2}\,\mathrm{tr}_{E/F}(v_1, v_2).$$

Let

$$\mathbb{X} = F\mathbf{e}_1 \oplus F\mathbf{e}_2 \oplus F\mathbf{e}_3 \oplus F\mathbf{e}_4 \subset \mathbb{V}.$$

Since $\mathbb{X}$ is maximal isotropic for $\langle\!\langle\cdot,\cdot\rangle\!\rangle$, there exists a unique maximal isotropic subspace $\mathbb{Y}$ in $\mathbb{V}$, such that $\mathbb{V} = \mathbb{X} \oplus \mathbb{Y}$. Let $(\mathbf{e}_1^*, \ldots, \mathbf{e}_4^*)$ be an F-basis for $\mathbb{Y}$ that is dual to $(\mathbf{e}_1, \ldots, \mathbf{e}_4)$. We can identify this basis precisely: letting $\mathbf{i}^2 = u \in F^\times$, we have

$$\mathbf{e}_1^* = \frac{1}{u}\mathbf{e}_1\mathbf{i}, \qquad \mathbf{e}_2^* = -\frac{1}{J_1u}\mathbf{e}_2\mathbf{i}, \qquad \mathbf{e}_3^* = -\frac{1}{J_2u}\mathbf{e}_3\mathbf{i}, \qquad \mathbf{e}_4^* = \frac{1}{Ju}\mathbf{e}_4\mathbf{i}. \tag{2.15}$$

2.2.1. The unitary group $\mathrm{U}_E(V)$ at infinite places. This section will not be relevant in this paper. We simply record for future use the isomorphism class of the unitary group $\mathrm{U}_E(V)$ at the infinite places v assuming that $F_v = \mathbb{R}$ and $E_v = \mathbb{C}$. The skew-hermitian form is given by

$$\mathbf{i} \cdot [a^\rho a' - J_1 b^\rho b' - J_2 c^\rho c' + J d^\rho d']$$

Thus we have the following table which summarizes the relation between the ramification of B_1 and B_2 at v and the isomorphism class of $\mathrm{U}_E(V)$.

B_1, B_2	J_1, J_2	$\mathrm{U}_E(V)$
split, split	$J_1 > 0, J_2 > 0$	$\mathrm{U}(2,2)$
ramified, split	$J_1 < 0, J_2 > 0$	$\mathrm{U}(2,2)$
split, ramified	$J_1 > 0, J_2 < 0$	$\mathrm{U}(2,2)$
ramified, ramified	$J_1 < 0, J_2 < 0$	$\mathrm{U}(4,0)$

2.3. The failure of the Hasse principle

The constructions above illustrate the failure of the Hasse principle for skew-hermitian B-spaces. Indeed, let us fix a B and consider pairs (B_1, B_2) such that $\Sigma_{B_1} \cap \Sigma_{B_2} = \Sigma_0$, where Σ_0 is some fixed set of places not intersecting Σ_B. Let E/F be a quadratic extension that is nonsplit at all the places in $\Sigma_B \cup \Sigma_0$. Then E embeds in B,B_1,B_2, so the constructions from the previous section apply. The various spaces V obtained by taking different choices of B_1 and B_2 are all locally isometric, since all of them have $\det(V) = 1$ and the Hasse invariant of $V_v^\dagger$ is independent of V for $v \notin \Sigma_B$. Since interchanging B_1 and B_2 gives an isometric global space, the number of different global spaces obtained in this way (up to isometry) is exactly 2^{s-2}, where $s = |\Sigma_B|$.

Conversely, suppose that we are given a quaternion *division* algebra B and a collection of local B_v-skew-hermitian spaces V_v such that $\det(V_v) = 1$ for all v and the Hasse invariant of $V_v^\dagger$ (for $v \notin \Sigma_B$) is 1 for all but finitely many v. Then there are up to isometry 2^{s-2} different global skew-hermitian spaces that give rise to this collection of local spaces, and all of them may be obtained by the construction above, by suitably choosing B_1, B_2 and $\mathbf{i}$.

CHAPTER 3

Weil Representations

In this chapter, we recall some basic properties of metaplectic groups and Weil representations. The material in this chapter is standard, but we include it for the convenience of the reader. We also consider splittings over some quaternionic unitary similitude groups, but the actual construction will be deferred to Appendix C.

3.1. Preliminaries

3.1.1. Weil indices. Let F be a local field of characteristic not 2 and fix a non-trivial additive character ψ of F. For a non-degenerate symmetric F-bilinear form q, we let $\gamma_F(\psi \circ q) \in \mu_8$ denote the Weil index associated to the character of second degree $x \mapsto \psi(q(x,x))$ (see [**75**], [**63**, Appendix]). When $q(x,y) = xy$ for $x, y \in F$, we write $\gamma_F(\psi) := \gamma_F(\psi \circ q)$. Put

$$\gamma_F(a, \psi) := \frac{\gamma_F(a\psi)}{\gamma_F(\psi)}$$

for $a \in F^\times$, where $(a\psi)(x) := \psi(ax)$ for $x \in F$. Then we have

$$\begin{aligned}
\gamma_F(ab^2, \psi) &= \gamma_F(a, \psi), \\
\gamma_F(ab, \psi) &= \gamma_F(a, \psi) \cdot \gamma_F(b, \psi) \cdot (a, b)_F, \\
\gamma_F(a, b\psi) &= \gamma_F(a, \psi) \cdot (a, b)_F, \\
\gamma_F(a, \psi)^2 &= (-1, a)_F, \\
\gamma_F(a, \psi)^4 &= 1, \\
\gamma_F(\psi)^2 &= \gamma_F(-1, \psi)^{-1}, \\
\gamma_F(\psi)^8 &= 1
\end{aligned}$$

for $a, b \in F^\times$ (see [**63**, p. 367]). Here $(\cdot, \cdot)_F$ is the quadratic Hilbert symbol of F.

Let q be a non-degenerate symmetric F-bilinear form. Let $\det q \in F^\times/(F^\times)^2$ and $h_F(q) \in \{\pm 1\}$ denote the determinant and the Hasse invariant of q. For example, when

$$q(x, y) = a_1 x_1 y_1 + \cdots + a_m x_m y_m$$

for $x = (x_1, \ldots, x_m)$, $y = (y_1, \ldots, y_m) \in F^m$, then

$$\det q = \prod_{1 \le i \le m} a_i, \quad h_F(q) = \prod_{1 \le i < j \le m} (a_i, a_j)_F.$$

Moreover, we have

$$\gamma_F(\psi \circ q) = \gamma_F(\psi)^m \cdot \gamma_F(\det q, \psi) \cdot h_F(q) \tag{3.1}$$

(see [**63**, pp. 367–368]).

3.1.2. Leray invariants. Let $\mathbb{V}$ be a $2n$-dimensional F-vector space with a non-degenerate symplectic form $\langle\langle\cdot,\cdot\rangle\rangle : \mathbb{V}\times\mathbb{V}\to F$. For maximal isotropic subspaces $\mathbb{Y}$, $\mathbb{Y}'$, $\mathbb{Y}''$ of $\mathbb{V}$, the Leray invariant $q(\mathbb{Y},\mathbb{Y}',\mathbb{Y}'')$ is a non-degenerate symmetric F-bilinear form defined as follows. (See also Definitions 2.4 and 2.10 of [**63**].)

Suppose first that $\mathbb{Y}$, $\mathbb{Y}'$, $\mathbb{Y}''$ are pairwise transverse. Let $P_{\mathbb{Y}}$ be the maximal parabolic subgroup of $\mathrm{Sp}(\mathbb{V})$ stabilizing $\mathbb{Y}$ and let $N_{\mathbb{Y}}$ be its unipotent radical. By Lemma 2.3 of [**63**], there exists a unique $g\in N_{\mathbb{Y}}$ such that $\mathbb{Y}'g=\mathbb{Y}''$. We write

$$g=\begin{pmatrix}1 & b\\ & 1\end{pmatrix},\quad b\in \mathrm{Hom}(\mathbb{Y}',\mathbb{Y})$$

with respect to the complete polarization $\mathbb{V}=\mathbb{Y}'+\mathbb{Y}$. Then $q=q(\mathbb{Y},\mathbb{Y}',\mathbb{Y}'')$ is a symmetric bilinear form on $\mathbb{Y}'$ defined by

$$q(x',y'):=\langle\langle x',y'b\rangle\rangle.$$

In general, we consider a symplectic space $\mathbb{V}_{\mathbb{R}}:=\mathbb{R}^{\perp}/\mathbb{R}$, where

$$\mathbb{R}:=(\mathbb{Y}\cap\mathbb{Y}')+(\mathbb{Y}'\cap\mathbb{Y}'')+(\mathbb{Y}''\cap\mathbb{Y}),$$

and maximal isotropic subspaces

$$\mathbb{Y}_{\mathbb{R}}:=(\mathbb{Y}\cap\mathbb{R}^{\perp})/\mathbb{R},\quad \mathbb{Y}'_{\mathbb{R}}:=(\mathbb{Y}'\cap\mathbb{R}^{\perp})/\mathbb{R},\quad \mathbb{Y}''_{\mathbb{R}}:=(\mathbb{Y}''\cap\mathbb{R}^{\perp})/\mathbb{R}$$

of $\mathbb{V}_{\mathbb{R}}$. By Lemma 2.9 of [**63**], $\mathbb{Y}_{\mathbb{R}}$, $\mathbb{Y}'_{\mathbb{R}}$, $\mathbb{Y}''_{\mathbb{R}}$ are pairwise transverse. We put

$$q(\mathbb{Y},\mathbb{Y}',\mathbb{Y}''):=q(\mathbb{Y}_{\mathbb{R}},\mathbb{Y}'_{\mathbb{R}},\mathbb{Y}''_{\mathbb{R}}).$$

By Theorem 2.11 of [**63**], we have

$$q(\mathbb{Y}g,\mathbb{Y}'g,\mathbb{Y}''g)=q(\mathbb{Y},\mathbb{Y}',\mathbb{Y}'')$$

for $g\in\mathrm{Sp}(\mathbb{V})$.

3.2. Weil representation for metaplectic groups

3.2.1. Heisenberg group, Heisenberg representations. Let F be a local field of characteristic not 2. For simplicity, we assume that F is non-archimedean.

Let $\mathbb{V}$ be a finite dimensional F-vector space equipped with a non-degenerate symplectic form

$$\langle\langle\cdot,\cdot\rangle\rangle:\mathbb{V}\times\mathbb{V}\longrightarrow F.$$

The Heisenberg group $H(\mathbb{V})$ is defined by

$$H(\mathbb{V}):=\mathbb{V}\oplus F$$

as a set, with group law

$$(v_1,z_1)\cdot(v_2,z_2)=\left(v_1+v_2,z_1+z_2+\frac{1}{2}\langle\langle v_1,v_2\rangle\rangle\right).$$

The center of $H(\mathbb{V})$ is F.

Let ψ be a nontrivial additive character of F. Then by the Stone–von Neumann theorem, $H(\mathbb{V})$ admits a unique (up to isomorphism) irreducible representation ρ_ψ on which F acts via ψ. This representation can be realized as follows. Fix a complete polarization

$$\mathbb{V}=\mathbb{X}\oplus\mathbb{Y},$$

i.e., $\mathbb{X}$ and $\mathbb{Y}$ are maximal isotropic subspaces of $\mathbb{V}$. We construct a character $\psi_{\mathbb{Y}}$ of $H(\mathbb{Y}) = \mathbb{Y} \oplus F$ by setting

$$\psi_{\mathbb{Y}}(y, z) = \psi(z).$$

Define

$$S_{\mathbb{Y}} := \mathrm{Ind}_{H(\mathbb{Y})}^{H(\mathbb{V})} \psi_{\mathbb{Y}}.$$

i.e. $S_{\mathbb{Y}}$ is the space of functions $f : H(\mathbb{V}) \to \mathbb{C}$ satisfying

(i) $f(\tilde{y}\tilde{v}) = \psi_{\mathbb{Y}}(\tilde{y}) f(\tilde{v})$ for $\tilde{y} \in H(\mathbb{Y})$, $\tilde{v} \in H(\mathbb{V})$.

(ii) f is smooth i.e. there exists an open compact subgroup (a lattice !) L in $\mathbb{V}$ such that

$$f(\tilde{v}\ell) = f(\tilde{v}) \quad \text{for all } \ell \in L \subset \mathbb{V} \subset H(\mathbb{V}).$$

Then $H(\mathbb{V})$ acts on $S_{\mathbb{Y}}$ on the right naturally. We can identify $S_{\mathbb{Y}}$ with $\mathcal{S}(\mathbb{X})$, the Schwartz space of locally constant functions with compact support on $\mathbb{X}$, via the restriction of functions to $\mathbb{X}$.

3.2.2. Metaplectic group. Let $\mathrm{Sp}(\mathbb{V})$ be the symplectic group of $\mathbb{V}$. Following Weil, we let $\mathrm{Sp}(\mathbb{V})$ act on $\mathbb{V}$ on the right. Recall that $\mathrm{Sp}(\mathbb{V})$ acts on $H(\mathbb{V})$ by

$$(v, z)^g := (vg, z).$$

Let $\widetilde{\mathrm{Sp}}(\mathbb{V})$ be the unique non-trivial 2-fold central extension of $\mathrm{Sp}(\mathbb{V})$. The metaplectic group $\mathrm{Mp}(\mathbb{V})$ is a central extension

$$1 \longrightarrow \mathbb{C}^1 \longrightarrow \mathrm{Mp}(\mathbb{V}) \longrightarrow \mathrm{Sp}(\mathbb{V}) \longrightarrow 1$$

defined by

$$\mathrm{Mp}(\mathbb{V}) := \widetilde{\mathrm{Sp}}(\mathbb{V}) \times_{\{\pm 1\}} \mathbb{C}^1.$$

LEMMA 3.2.1. *Any automorphism of* $\mathrm{Mp}(\mathbb{V})$ *which lifts the identity map of* $\mathrm{Sp}(\mathbb{V})$ *and which restricts to the identity map of* $\mathbb{C}^1$ *must be the identity map of* $\mathrm{Mp}(\mathbb{V})$.

PROOF. Let $p : \mathrm{Mp}(\mathbb{V}) \to \mathrm{Sp}(\mathbb{V})$ be the projection. Let $f : \mathrm{Mp}(\mathbb{V}) \to \mathrm{Mp}(\mathbb{V})$ be such an automorphism. Since $p(f(g)) \cdot p(g)^{-1} = 1$ for $g \in \mathrm{Mp}(\mathbb{V})$, there exists a character $\kappa : \mathrm{Mp}(\mathbb{V}) \to \mathbb{C}^1$ such that $f(g) \cdot g^{-1} = \kappa(g)$. Since $f(z) = z$ for $z \in \mathbb{C}^1$, κ is trivial on $\mathbb{C}^1$, and hence induces a character of $\mathrm{Sp}(\mathbb{V})$. Since $[\mathrm{Sp}(\mathbb{V}), \mathrm{Sp}(\mathbb{V})] = \mathrm{Sp}(\mathbb{V})$, this character must be trivial. □

One can realize $\mathrm{Mp}(\mathbb{V})$ explicitly as follows. Put

$$z_{\mathbb{Y}}(g_1, g_2) = \gamma_F(\frac{1}{2}\psi \circ q(\mathbb{Y}, \mathbb{Y}g_2^{-1}, \mathbb{Y}g_1))$$

for $g_1, g_2 \in \mathrm{Sp}(\mathbb{V})$. By Theorem 4.1 of [**63**], $z_{\mathbb{Y}}$ is a 2-cocycle valued in 8th roots of unity (called the Leray cocycle), and the group

$$\mathrm{Mp}(\mathbb{V})_{\mathbb{Y}} := \mathrm{Sp}(\mathbb{V}) \times \mathbb{C}^1$$

with group law

$$(g_1, z_1) \cdot (g_2, z_2) = (g_1 g_2, z_1 z_2 \cdot z_{\mathbb{Y}}(g_1, g_2))$$

is isomorphic to $\mathrm{Mp}(\mathbb{V})$. Moreover, by Lemma 3.2.1, this isomorphism is canonical. If there is no confusion, we identify $\mathrm{Mp}(\mathbb{V})_{\mathbb{Y}}$ with $\mathrm{Mp}(\mathbb{V})$. We also remark that there is a 2-cocycle valued in $\{\pm 1\}$ realizing $\mathrm{Mp}(\mathbb{V})$ (called the Ranga Rao cocycle; see [**63**, Theorem 5.3]), but the Leray cocycle is more convenient for our purposes (see e.g. Lemma B.1.1 below).

Let $\mathbb{V} = \mathbb{X}' \oplus \mathbb{Y}'$ be another complete polarization.

LEMMA 3.2.2. *We have*

$$z_{\mathbb{Y}'}(g_1, g_2) = \lambda(g_1 g_2)\lambda(g_1)^{-1}\lambda(g_2)^{-1} \cdot z_{\mathbb{Y}}(g_1, g_2),$$

where $\lambda : \mathrm{Sp}(\mathbb{V}) \to \mathbb{C}^1$ *is given by*

$$\lambda(g) := \gamma_F(\frac{1}{2}\psi \circ q(\mathbb{Y}, \mathbb{Y}'g^{-1}, \mathbb{Y}')) \cdot \gamma_F(\frac{1}{2}\psi \circ q(\mathbb{Y}, \mathbb{Y}', \mathbb{Y}g)).$$

In particular, the bijection

$$\begin{aligned} \mathrm{Mp}(\mathbb{V})_{\mathbb{Y}} &\longrightarrow \mathrm{Mp}(\mathbb{V})_{\mathbb{Y}'} \\ (g, z) &\longmapsto (g, z \cdot \lambda(g)) \end{aligned}$$

is an isomorphism.

PROOF. See Lemma 4.2 of [**40**]. □

Suppose that $\mathbb{V} = \mathbb{V}_1 \oplus \mathbb{V}_2$, where each $\mathbb{V}_i$ is a non-degenerate symplectic subspace. One can lift the natural embedding $\mathrm{Sp}(\mathbb{V}_1) \times \mathrm{Sp}(\mathbb{V}_2) \hookrightarrow \mathrm{Sp}(\mathbb{V})$ to a homomorphism

$$\mathrm{Mp}(\mathbb{V}_1) \times \mathrm{Mp}(\mathbb{V}_2) \longrightarrow \mathrm{Mp}(\mathbb{V}).$$

If $\mathbb{V}_i = \mathbb{X}_i \oplus \mathbb{Y}_i$ is a complete polarization and

$$\mathbb{X} = \mathbb{X}_1 \oplus \mathbb{X}_2, \quad \mathbb{Y} = \mathbb{Y}_1 \oplus \mathbb{Y}_2,$$

then this homomorphism is given by

$$\begin{aligned} \mathrm{Mp}(\mathbb{V}_1)_{\mathbb{Y}_1} \times \mathrm{Mp}(\mathbb{V}_2)_{\mathbb{Y}_2} &\longrightarrow \mathrm{Mp}(\mathbb{V})_{\mathbb{Y}}, \\ ((g_1, z_1), (g_2, z_2)) &\longmapsto (g_1 g_2, z_1 z_2) \end{aligned}$$

i.e., we have

$$z_{\mathbb{Y}_1}(g_1, g_1') \cdot z_{\mathbb{Y}_2}(g_2, g_2') = z_{\mathbb{Y}}(g_1 g_2, g_1' g_2')$$

for $g_i, g_i' \in \mathrm{Sp}(\mathbb{V}_i)$ (see Theorem 4.1 of [**63**]).

Let L be a self-dual lattice in $\mathbb{V}$ and let K be the stabiliser of L in $\mathrm{Sp}(\mathbb{V})$. If the *residual* characteristic of F is not 2, then there exists a splitting

$$\begin{array}{ccc} & & \mathrm{Mp}(\mathbb{V}) \\ & \nearrow & \downarrow \\ K & \longrightarrow & \mathrm{Sp}(\mathbb{V}) \end{array}$$

Moreover, if the residue field of F has at least four elements, then $[K, K] = K$ (see Lemma 11.1 of [**54**]), and hence such a splitting is unique. In the next section, we shall describe this splitting by using the Schrödinger model.

3.2.3. Weil representation, Schrödinger model. Recall that ρ_ψ is the unique (up to isomorphism) irreducible smooth representation of $H(\mathbb{V})$ with central character ψ. Let S be the underlying space of ρ_ψ. The Weil representation ω_ψ of $\mathrm{Mp}(\mathbb{V})$ on S is a smooth representation characterized by the following properties:

- $\rho_\psi(h^g) = \omega_\psi(g)^{-1}\rho_\psi(h)\omega_\psi(g)$ for all $h \in H(\mathbb{V})$ and $g \in \mathrm{Mp}(\mathbb{V})$.
- $\omega_\psi(z) = z \cdot \mathrm{id}_S$ for all $z \in \mathbb{C}^1$.

One can realize ω_ψ explicitly as follows. We regard $\mathbb{V} = F^{2n}$ as the space of row vectors. Fix a complete polarization $\mathbb{V} = \mathbb{X} \oplus \mathbb{Y}$. Choose a basis $\mathbf{e}_1, \dots, \mathbf{e}_n, \mathbf{e}_1^*, \dots, \mathbf{e}_n^*$ of $\mathbb{V}$ such that

$$\mathbb{X} = F\mathbf{e}_1 + \cdots + F\mathbf{e}_n, \quad \mathbb{Y} = F\mathbf{e}_1^* + \cdots + F\mathbf{e}_n^*, \quad \langle\langle \mathbf{e}_i, \mathbf{e}_j^* \rangle\rangle = \delta_{ij}.$$

Then we have

$$\mathrm{Sp}(\mathbb{V}) = \left\{ g \in \mathrm{GL}_{2n}(F) \,\middle|\, g \begin{pmatrix} & \mathbf{1}_n \\ -\mathbf{1}_n & \end{pmatrix} {}^t g = \begin{pmatrix} & \mathbf{1}_n \\ -\mathbf{1}_n & \end{pmatrix} \right\}.$$

The Weil representation ω_ψ of $\mathrm{Mp}(\mathbb{V})_\mathbb{Y}$ on the Schwartz space $\mathcal{S}(\mathbb{X})$ is given as follows:

$$\omega_\psi \left(\begin{pmatrix} a & \\ & {}^t a^{-1} \end{pmatrix}, z \right) \varphi(x) = z \cdot |\det a|^{1/2} \cdot \varphi(xa)$$

$$\omega_\psi \left(\begin{pmatrix} \mathbf{1}_n & b \\ & \mathbf{1}_n \end{pmatrix}, z \right) \varphi(x) = z \cdot \psi \left(\frac{1}{2} x b {}^t x \right) \cdot \varphi(x)$$

$$\omega_\psi \left(\begin{pmatrix} & \mathbf{1}_n \\ -\mathbf{1}_n & \end{pmatrix}, z \right) \varphi(x) = z \cdot \int_{F^n} \varphi(y) \psi(x {}^t y) \, dy.$$

for $\varphi \in \mathcal{S}(\mathbb{X})$, $x \in \mathbb{X} \cong F^n$, $a \in \mathrm{GL}(\mathbb{X}) \cong \mathrm{GL}_n(F)$, $b \in \mathrm{Hom}(\mathbb{X}, \mathbb{Y}) \cong \mathrm{M}_n(F)$ with ${}^t b = b$, and $z \in \mathbb{C}^1$. More generally, for $(g, z) \in \mathrm{Mp}(\mathbb{V})_\mathbb{Y}$ with $g = \left(\begin{smallmatrix} a & b \\ c & d \end{smallmatrix} \right) \in \mathrm{Sp}(\mathbb{V})$, we have

$$\begin{aligned} &\omega_\psi(g, z) \varphi(x) \\ &= z \cdot \int_{F^n / \ker(c)} \psi \left(\frac{1}{2} (xa, xb) + (xb, yc) + \frac{1}{2} (yc, yd) \right) \varphi(xa + yc) \, d\mu_g(y), \end{aligned}$$

where $(x, y) = x {}^t y$ for row vectors $x, y \in F^n$, and the measure $d\mu_g(y)$ on $F^n / \ker(c)$ is normalized so that this operator is unitary (see [**42**, Proposition 2.3]). In particular, if $\det c \neq 0$, then $\omega_\psi(g, 1)\varphi(x)$ is equal to

$$\begin{aligned} &\int_{F^n} \psi \left(\frac{1}{2} (xa, xb) + (xb, yc) + \frac{1}{2} (yc, yd) \right) \varphi(xa + yc) \, d\mu_g(y) \\ &= |\det c|^{-1} \cdot \psi \left(\frac{1}{2} (xa, xb) \right) \\ &\quad \times \int_{F^n} \psi \left((xb, y - xa) + \frac{1}{2} (y - xa, yc^{-1} d - xac^{-1} d) \right) \varphi(y) \, d\mu_g(y) \\ &= |\det c|^{-1} \cdot \psi \left(\frac{1}{2} (xa, x(ac^{-1} d - b)) \right) \\ &\quad \times \int_{F^n} \psi \left((y, x(b - ac^{-1} d)) + \frac{1}{2} (y, yc^{-1} d) \right) \varphi(y) \, d\mu_g(y) \\ &= |\det c|^{-1} \cdot \psi \left(\frac{1}{2} (xa, x {}^t c^{-1}) \right) \cdot \int_{F^n} \psi \left(-(y, x {}^t c^{-1}) + \frac{1}{2} (y, yc^{-1} d) \right) \varphi(y) \, d\mu_g(y) \\ &= |\det c|^{-1/2} \cdot \psi \left(\frac{1}{2} (xac^{-1}, x) \right) \cdot \int_{F^n} \psi \left(-(x {}^t c^{-1}, y) + \frac{1}{2} (yc^{-1} d, y) \right) \varphi(y) \, dy, \end{aligned}$$

where dy is the self-dual Haar measure on F^n with respect to the pairing $\psi \circ (\cdot, \cdot)$.

If the residual characteristic of F is not 2, let K be the stabilizer of the self-dual lattice

$$\mathfrak{o}\mathbf{e}_1 + \cdots + \mathfrak{o}\mathbf{e}_n + \mathfrak{o}\mathbf{e}_1^* + \cdots + \mathfrak{o}\mathbf{e}_n^*.$$

Then the splitting $K \to \mathrm{Mp}(\mathbb{V})_{\mathbb{Y}}$ is given as follows. Assume that ψ is of order zero. Let $\varphi^0 \in \mathcal{S}(\mathbb{X})$ be the characteristic function of $\mathfrak{o}\mathbf{e}_1 + \cdots + \mathfrak{o}\mathbf{e}_n \cong \mathfrak{o}^n$. Since the residual characteristic of F is not 2, we see that

$$\omega_\psi(k,1)\varphi^0 = \varphi^0$$

for

$$k = \begin{pmatrix} a & \\ & {}^t a^{-1} \end{pmatrix}, \begin{pmatrix} \mathbf{1}_n & b \\ & \mathbf{1}_n \end{pmatrix} \text{ and } \begin{pmatrix} & \mathbf{1}_n \\ -\mathbf{1}_n & \end{pmatrix},$$

where $a \in \mathrm{GL}_n(\mathfrak{o})$ and $b \in \mathrm{M}_n(\mathfrak{o})$. Since these elements generate K, there exists a function $s_{\mathbb{Y}} : K \to \mathbb{C}^1$ such that

$$\omega_\psi(k,1)\varphi^0 = s_{\mathbb{Y}}(k)^{-1} \cdot \varphi^0 \quad \text{for all } k \in K.$$

Thus we obtain

$$z_{\mathbb{Y}}(k_1,k_2) = s_{\mathbb{Y}}(k_1k_2)s_{\mathbb{Y}}(k_1)^{-1}s_{\mathbb{Y}}(k_2)^{-1}$$

for $k_1, k_2 \in K$, so that the map $k \mapsto (k, s_{\mathbb{Y}}(k))$ is the desired splitting.

3.2.4. Change of models. Suppose $\mathbb{V} = \mathbb{X}' \oplus \mathbb{Y}'$ is another polarization of $\mathbb{V}$. Then likewise the representation ρ_ψ can be realized on $S_{\mathbb{Y}'} \simeq \mathcal{S}(\mathbb{X}')$. We will need an explicit isomorphism between these representations of $H(\mathbb{V})$. At the level of the induced representations, this is given by the map

$$S_{\mathbb{Y}'} \to S_{\mathbb{Y}}, \quad f' \mapsto f$$

$$f(\tilde{v}) = \int_{\mathbb{Y}\cap\mathbb{Y}'\backslash\mathbb{Y}} f'((y,0)\cdot\tilde{v}) \cdot \psi_{\mathbb{Y}}(y,0)^{-1}\,dy = \int_{\mathbb{Y}\cap\mathbb{Y}'\backslash\mathbb{Y}} f'((y,0)\cdot\tilde{v})\,dy.$$

For now, we will take any Haar measure on $\mathbb{Y}$ to define this. We will fix this more carefully later. Let us now write down this isomorphism in terms of Schwartz spaces.

LEMMA 3.2.3. *Suppose that $\varphi \in \mathcal{S}(\mathbb{X})$ and $\varphi' \in \mathcal{S}(\mathbb{X}')$ correspond to $f \in S_{\mathbb{Y}}$ and $f' \in S_{\mathbb{Y}'}$ respectively. Then we have*

$$\varphi(x) = \int_{\mathbb{Y}\cap\mathbb{Y}'\backslash\mathbb{Y}} \psi\left(\frac{1}{2}\langle\langle x',y'\rangle\rangle - \frac{1}{2}\langle\langle x,y\rangle\rangle\right)\varphi'(x')\,dy, \tag{3.2}$$

where $x' = x'(x,y) \in \mathbb{X}'$ and $y' = y'(x,y) \in \mathbb{Y}'$ are given by $x' + y' = x + y \in \mathbb{V}$.

PROOF. Let $\varphi' \in \mathcal{S}(\mathbb{X}')$. Let $(x'+y',z) \in H(\mathbb{V})$. We write this as:

$$(x'+y',z) = \left(y', z - \frac{1}{2}\langle\langle y',x'\rangle\rangle\right)\cdot(x',0).$$

Thus if $f' \in S_{\mathbb{Y}'}$ corresponds to φ', then

$$f'(x'+y',z) = \psi\left(z - \frac{1}{2}\langle\langle y',x'\rangle\rangle\right)\cdot\varphi'(x').$$

We can rewrite this as: (with $v = x' + y'$)

$$f'(v,z) = \psi\left(z - \frac{1}{2}\langle\langle v,x'\rangle\rangle\right)\cdot\varphi'(x').$$

Thus f' corresponds to $f \in S_{\mathbb{Y}}$ given by

$$\begin{aligned} f(x+y,z) &= \int_{\mathbb{Y}\cap\mathbb{Y}'\backslash\mathbb{Y}} f'((y_0,0)\cdot(x+y,z))\,dy_0 \\ &= \int_{\mathbb{Y}\cap\mathbb{Y}'\backslash\mathbb{Y}} f'\left(x+y+y_0, z+\frac{1}{2}\langle\langle y_0,x\rangle\rangle\right)dy_0. \end{aligned}$$

Thus

$$\begin{aligned} \varphi(x) &= \int_{\mathbb{Y}\cap\mathbb{Y}'\backslash\mathbb{Y}} f'\left(x+y_0, \frac{1}{2}\langle\langle y_0,x\rangle\rangle\right)dy_0 \\ &= \int_{\mathbb{Y}\cap\mathbb{Y}'\backslash\mathbb{Y}} \psi\left(\frac{1}{2}\langle\langle y_0,x\rangle\rangle\right) f'(x+y_0,0)\,dy_0 \\ &= \int_{\mathbb{Y}\cap\mathbb{Y}'\backslash\mathbb{Y}} \psi\left(-\frac{1}{2}\langle\langle x+y_0,x'\rangle\rangle+\frac{1}{2}\langle\langle y_0,x\rangle\rangle\right)\varphi'(x')\,dy_0. \end{aligned}$$ □

Thus we obtain an $H(\mathbb{V})$-equivariant isomorphism $\mathcal{S}(\mathbb{X}') \cong \mathcal{S}(\mathbb{X})$ defined by the partial Fourier transform (3.2). Using the characterization of the Weil representation of $\mathrm{Mp}(\mathbb{V})$, one sees that this isomorphism is also $\mathrm{Mp}(\mathbb{V})$-equivariant.

The isomorphism $\mathcal{S}(\mathbb{X}') \cong \mathcal{S}(\mathbb{X})$ is in fact a partial Fourier transform. To see this, we assume for simplicity that there exists a basis $\mathbf{e}_1,\dots,\mathbf{e}_n,\mathbf{e}_1^*,\dots,\mathbf{e}_n^*$ of $\mathbb{V}$ such that

$$\begin{aligned} \mathbb{X} &= F\mathbf{e}_1+\cdots+F\mathbf{e}_n, & \mathbb{X}' &= F\mathbf{e}_1+\cdots+F\mathbf{e}_k+F\mathbf{e}_{k+1}^*+\cdots+F\mathbf{e}_n^*, \\ \mathbb{Y} &= F\mathbf{e}_1^*+\cdots+F\mathbf{e}_n^*, & \mathbb{Y}' &= F\mathbf{e}_1^*+\cdots+F\mathbf{e}_k^*+F\mathbf{e}_{k+1}+\cdots+F\mathbf{e}_n, \end{aligned}$$

and $\langle\langle \mathbf{e}_i,\mathbf{e}_j^*\rangle\rangle = \delta_{ij}$, where $k = \dim(\mathbb{Y}\cap\mathbb{Y}')$. In particular, we have $\mathbb{Y}\cap\mathbb{Y}' = F\mathbf{e}_1^*+\cdots+F\mathbf{e}_k^*$. Let $\varphi \in \mathcal{S}(\mathbb{X})$ and $\varphi' \in \mathcal{S}(\mathbb{X}')$ be as in (3.2). We also regard φ' as a function on F^n via the basis $\mathbf{e}_1,\dots,\mathbf{e}_k,\mathbf{e}_{k+1}^*,\dots,\mathbf{e}_n^*$. Write $x+y=x'+y'$ with $x\in\mathbb{X}$, $y\in\mathbb{Y}$, $x'\in\mathbb{X}'$, $y'\in\mathbb{Y}'$. If we write

$$x = x_1\mathbf{e}_1+\cdots+x_n\mathbf{e}_n, \quad y = y_1\mathbf{e}_1^*+\cdots+y_n\mathbf{e}_n^*$$

with $x_i, y_j \in F$, then

$$\begin{aligned} x' &= x_1\mathbf{e}_1+\cdots+x_k\mathbf{e}_k+y_{k+1}\mathbf{e}_{k+1}^*+\cdots+y_n\mathbf{e}_n^*, \\ y' &= y_1\mathbf{e}_1^*+\cdots+y_k\mathbf{e}_k^*+x_{k+1}\mathbf{e}_{k+1}+\cdots+x_n\mathbf{e}_n, \end{aligned}$$

and

$$\begin{aligned} \langle\langle x,y\rangle\rangle &= x_1y_1+\cdots+x_ny_n, \\ \langle\langle x',y'\rangle\rangle &= x_1y_1+\cdots+x_ky_k-x_{k+1}y_{k+1}-\cdots-x_ny_n. \end{aligned}$$

Hence we have

$$\begin{aligned} \varphi(x) &= \int_{\mathbb{Y}\cap\mathbb{Y}'\backslash\mathbb{Y}} \psi\left(\frac{1}{2}\left(\langle\langle x',y'\rangle\rangle-\langle\langle x,y\rangle\rangle\right)\right)\varphi'(x')\,dy \\ &= \int_{F^{n-k}} \psi(-x_{k+1}y_{k+1}-\cdots-x_ny_n)\,\varphi'(x_1,\dots,x_k,y_{k+1},\dots,y_n)\,dy_{k+1}\cdots dy_n. \end{aligned}$$

3.2.5. Over global fields. In this section, let F be a number field with ring of adeles $\mathbb{A}$. Let $\mathbb{V}$ be a symplectic space over F. The global metaplectic group $\mathrm{Mp}(\mathbb{V})_{\mathbb{A}}$ is defined as follows.

Fix a lattice L in $\mathbb{V}$. For each finite place v, let K_v be the stabilizer of L_v in $\mathrm{Sp}(\mathbb{V}_v)$. For almost all v, L_v is self-dual and there exists a unique splitting $K_v \hookrightarrow \mathrm{Mp}(\mathbb{V}_v)$, in which case we identify K_v with its image in $\mathrm{Mp}(\mathbb{V}_v)$.

For a finite set S of places of F including all archimedean places, we define a central extension

$$1 \longrightarrow \mathbb{C}^1 \longrightarrow \mathrm{Mp}(\mathbb{V})_S \longrightarrow \prod_{v \in S} \mathrm{Sp}(\mathbb{V}_v) \longrightarrow 1$$

by

$$\mathrm{Mp}(\mathbb{V})_S := \left(\prod_{v \in S} \mathrm{Mp}(\mathbb{V}_v) \right) \Big/ \left\{ (z_v) \in \prod_{v \in S} \mathbb{C}^1 \mid \prod_{v \in S} z_v = 1 \right\}.$$

Put $K^S := \prod_{v \notin S} K_v$. If $S \subset S'$ are sufficiently large, the splitting $K_v \hookrightarrow \mathrm{Mp}(\mathbb{V}_v)$ induces an embedding

$$\mathrm{Mp}(\mathbb{V})_S \times K^S \hookrightarrow \mathrm{Mp}(\mathbb{V})_{S'} \times K^{S'}.$$

Then $\mathrm{Mp}(\mathbb{V})_{\mathbb{A}}$ is defined by

$$\mathrm{Mp}(\mathbb{V})_{\mathbb{A}} := \varinjlim_{S} \left(\mathrm{Mp}(\mathbb{V})_S \times K^S \right).$$

There exists a unique splitting

$$\begin{array}{ccc} & & \mathrm{Mp}(\mathbb{V})_{\mathbb{A}} \\ & \nearrow & \downarrow \\ \mathrm{Sp}(\mathbb{V})(F) & \longrightarrow & \mathrm{Sp}(\mathbb{V})(\mathbb{A}) \end{array}$$

given as follows. Fix a complete polarization $\mathbb{V} = \mathbb{X} \oplus \mathbb{Y}$ over F. Recall that the metaplectic group $\mathrm{Mp}(\mathbb{V}_v) = \mathrm{Sp}(\mathbb{V}_v) \times \mathbb{C}^1$ is determined by the 2-cocycle $z_{\mathbb{Y}_v}$. Moreover, for almost all v, there exists a function $s_{\mathbb{Y}_v} : K_v \to \mathbb{C}^1$ such that

$$z_{\mathbb{Y}_v}(k_1, k_2) = s_{\mathbb{Y}_v}(k_1 k_2) s_{\mathbb{Y}_v}(k_1)^{-1} s_{\mathbb{Y}_v}(k_2)^{-1}$$

for $k_1, k_2 \in K_v$.

LEMMA 3.2.4. *Let $\gamma \in \mathrm{Sp}(\mathbb{V})(F)$. Then we have*

$$s_{\mathbb{Y}_v}(\gamma) = 1$$

for almost all v.

PROOF. By the Bruhat decomposition, we may write $\gamma = p_1 w p_2$ with some

$$p_i = \begin{pmatrix} a_i & b_i \\ & {}^t a_i^{-1} \end{pmatrix}, \quad w = \begin{pmatrix} & & \mathbf{1}_k & \\ & \mathbf{1}_{n-k} & & \\ -\mathbf{1}_k & & & \\ & & & \mathbf{1}_{n-k} \end{pmatrix},$$

where $a_i \in \mathrm{GL}_n(F)$ and $b_i \in \mathrm{M}_n(F)$. By Theorem 4.1 of [**63**], we have

$$z_{\mathbb{Y}_v}(p_1, g) = z_{\mathbb{Y}_v}(g, p_2) = 1$$

for all v and $g \in \mathrm{Sp}(\mathbb{V}_v)$, so that

$$(p_1 w p_2, 1) = (p_1, 1) \cdot (w, 1) \cdot (p_2, 1) \quad \text{in } \mathrm{Mp}(\mathbb{V}_v).$$

On the other hand, for almost all v, we have $p_i \in K_v$ and

$$\omega_{\psi_v}(p_i,1)\varphi_v^0 = \varphi_v^0, \quad \omega_{\psi_v}(w,1)\varphi_v^0 = \varphi_v^0,$$

where ψ_v is a non-trivial character of F_v of order zero and φ_v^0 is the characteristic function of $\mathfrak{o}_v^n$. Thus we obtain

$$\omega_{\psi_v}(\gamma,1)\varphi_v^0 = \omega_{\psi_v}(p_1,1)\omega_{\psi_v}(w,1)\omega_{\psi_v}(p_2,1)\varphi_v^0 = \varphi_v^0$$

for almost all v. □

For $\gamma \in \mathrm{Sp}(\mathbb{V})(F)$, let $(\gamma,1)$ be the element in $\prod_v \mathrm{Mp}(\mathbb{V}_v)$ such that $(\gamma,1)_v = (\gamma,1)$ for all v. By Lemma 3.2.4, we have $(\gamma,1)_v \in K_v$ for almost all v. Hence, if S is a sufficiently large finite set of places of F, then $(\gamma,1)$ maps to an element $i(\gamma)$ in $\mathrm{Mp}(\mathbb{V})_S \times K^S$.

LEMMA 3.2.5. *The map*

$$i : \mathrm{Sp}(\mathbb{V})(F) \longrightarrow \mathrm{Mp}(\mathbb{V})_{\mathbb{A}}$$

is a homomorphism.

PROOF. Let $\gamma_1, \gamma_2 \in \mathrm{Sp}(\mathbb{V})(F)$. For each v, we have

$$(\gamma_1,1)_v \cdot (\gamma_2,1)_v = (\gamma_1\gamma_2, z_{\mathbb{Y}_v}(\gamma_1,\gamma_2)) \quad \text{in } \mathrm{Mp}(\mathbb{V}_v).$$

Choose a finite set S of places of F such that

$$\gamma_1,\gamma_2 \in K_v, \quad s_{\mathbb{Y}_v}(\gamma_1) = s_{\mathbb{Y}_v}(\gamma_2) = s_{\mathbb{Y}_v}(\gamma_1\gamma_2) = 1$$

for $v \notin S$. Then we have

$$z_{\mathbb{Y}_v}(\gamma_1,\gamma_2) = 1$$

for $v \notin S$. Moreover, by the product formula for the Weil index, we have

$$\prod_{v\in S} z_{\mathbb{Y}_v}(\gamma_1,\gamma_2) = 1.$$

Hence the image of $(\gamma_1,1)\cdot(\gamma_2,1)$ in $\mathrm{Mp}(\mathbb{V})_S \times K^S$ is equal to $i(\gamma_1\gamma_2)$. □

Fix a non-trivial additive character ψ of $\mathbb{A}/F$. We have the global Weil representation ω_ψ of $\mathrm{Mp}(\mathbb{V})_{\mathbb{A}}$ on the Schwartz space $\mathcal{S}(\mathbb{X}(\mathbb{A}))$. For each $\varphi \in \mathcal{S}(\mathbb{X}(\mathbb{A}))$, the associated theta function on $\mathrm{Mp}(\mathbb{V})_{\mathbb{A}}$ is defined by

$$\Theta_\varphi(g) := \sum_{x\in\mathbb{X}} \omega_\psi(g)\varphi(x).$$

Then Θ_φ is a left $\mathrm{Sp}(\mathbb{V})(F)$-invariant slowly increasing smooth function on $\mathrm{Mp}(\mathbb{V})_{\mathbb{A}}$.

3.3. Reductive dual pairs

In this section, we consider the reductive dual pair $(\mathrm{GU}(V), \mathrm{GU}(W))$, where V is a skew-Hermitian right B-space of dimension two and W is a Hermitian left B-space of dimension one.

3.3.1. Reductive dual pairs; examples. Recall that in §2.2, we have constructed the 2-dimensional skew-Hermitian right B-space $V = B_1 \otimes_E B_2$ with the skew-Hermitian form $\langle \cdot, \cdot \rangle : V \times V \to B$. Let $W = B$ be the 1-dimensional Hermitian left B-space with the Hermitian form $\langle \cdot, \cdot \rangle : W \times W \to B$ defined by

$$\langle x, y \rangle = xy^*.$$

These forms satisfy that

$$\langle v\boldsymbol{\alpha}, v'\boldsymbol{\beta} \rangle = \boldsymbol{\alpha}^* \langle v, v' \rangle \boldsymbol{\beta} \qquad \langle v', v \rangle = -\langle v, v' \rangle^*$$
$$\langle \boldsymbol{\alpha} w, \boldsymbol{\beta} w' \rangle = \boldsymbol{\alpha} \langle w, w' \rangle \boldsymbol{\beta}^* \qquad \langle w', w \rangle = \langle w, w' \rangle^*$$

for $v, v' \in V$, $w, w' \in W$ and $\boldsymbol{\alpha}, \boldsymbol{\beta} \in B$. We let $\mathrm{GL}(V)$ act on V on the left and let $\mathrm{GL}(W)$ act on W on the right. Let $\mathrm{GU}(V)$ and $\mathrm{GU}(W)$ be the quaternionic unitary similitude groups of V and W with the similitude characters $\nu : \mathrm{GU}(V) \to F^\times$ and $\nu : \mathrm{GU}(W) \to F^\times$ respectively:

$$\mathrm{GU}(V) := \{ g \in \mathrm{GL}(V) \mid \langle gv, gv' \rangle = \nu(g) \cdot \langle v, v' \rangle \text{ for all } v, v' \in V \},$$
$$\mathrm{GU}(W) := \{ g \in \mathrm{GL}(W) \mid \langle wg, w'g \rangle = \nu(g) \cdot \langle w, w' \rangle \text{ for all } w, w' \in W \}.$$

Let $\mathrm{U}(V) := \ker \nu$ and $\mathrm{U}(W) := \ker \nu$ be the unitary groups of V and W respectively.

Put

$$\mathbb{V} := V \otimes_B W.$$

Then $\mathbb{V}$ is an F-space equipped with a symplectic form

$$\langle\!\langle \cdot, \cdot \rangle\!\rangle := \frac{1}{2} \operatorname{tr}_{B/F} \left(\langle \cdot, \cdot \rangle \otimes \langle \cdot, \cdot \rangle^* \right).$$

If we identify $\mathrm{Res}_{B/F}(V)$ with $\mathbb{V}$ via the map $v \mapsto v \otimes 1$, then the associated symplectic form on $\mathbb{V}$ is given by

$$\langle\!\langle \cdot, \cdot \rangle\!\rangle = \frac{1}{2} \operatorname{tr}_{B/F} \langle \cdot, \cdot \rangle = \frac{1}{2} \operatorname{tr}_{E/F} (\cdot, \cdot),$$

where $(\cdot, \cdot) = \mathrm{pr} \circ \langle \cdot, \cdot \rangle$ is the associated E-skew-Hermitian form. We let $\mathrm{GL}(V) \times \mathrm{GL}(W)$ act on $\mathbb{V}$ on the right:

$$(v \otimes w) \cdot (g, h) := g^{-1} v \otimes wh.$$

This gives a natural homomorphism

$$\mathrm{G}(\mathrm{U}(V) \times \mathrm{U}(W)) \longrightarrow \mathrm{Sp}(\mathbb{V}),$$

where

$$\mathrm{G}(\mathrm{U}(V) \times \mathrm{U}(W)) := \{ (g, h) \in \mathrm{GU}(V) \times \mathrm{GU}(W) \mid \nu(g) = \nu(h) \}.$$

3.3.2. Splittings. Recall that

$$\mathbb{V} = \mathbf{e}_1 E + \mathbf{e}_2 E + \mathbf{e}_3 E + \mathbf{e}_4 E.$$

Let $\mathbb{V} = \mathbb{X} + \mathbb{Y}$ be the complete polarization given by

$$\mathbb{X} = F\mathbf{e}_1 + F\mathbf{e}_2 + F\mathbf{e}_3 + F\mathbf{e}_4, \quad \mathbb{Y} = F\mathbf{e}_1^* + F\mathbf{e}_2^* + F\mathbf{e}_3^* + F\mathbf{e}_4^*.$$

We first suppose that F is a local field. In Appendix C, we define a function

$$s : \mathrm{G}(\mathrm{U}(V) \times \mathrm{U}(W))^0 \longrightarrow \mathbb{C}^1$$

such that

$$z_{\mathbb{Y}}(g_1, g_2) = s(g_1 g_2) s(g_1)^{-1} s(g_2)^{-1},$$

so that the map

$$\begin{aligned} \iota : \mathrm{G}(\mathrm{U}(V) \times \mathrm{U}(W))^0 &\longrightarrow \mathrm{Mp}(\mathbb{V})_{\mathbb{Y}} \\ g &\longmapsto (g, s(g)) \end{aligned}$$

is a homomorphism. Thus we have a commutative diagram

$$\begin{array}{ccc} & & \mathrm{Mp}(\mathbb{V}) \\ & \overset{\iota}{\nearrow} & \downarrow \\ \mathrm{G}(\mathrm{U}(V) \times \mathrm{U}(W))^0 & \longrightarrow & \mathrm{Sp}(\mathbb{V}) \end{array} .$$

If $\mathbb{V} = \mathbb{X}' + \mathbb{Y}'$ is another complete polarization, we choose $g_0 \in \mathrm{Sp}(\mathbb{V})$ such that $\mathbb{Y}' = \mathbb{Y}g_0$ and define a function

$$s' : \mathrm{G}(\mathrm{U}(V) \times \mathrm{U}(W))^0 \longrightarrow \mathbb{C}^1$$

by

$$\begin{aligned} s'(g) &= s(g) \cdot z_{\mathbb{Y}}(g_0, g g_0^{-1}) \cdot z_{\mathbb{Y}}(g, g_0^{-1}) \\ &= s(g) \cdot z_{\mathbb{Y}}(g_0 g g_0^{-1}, g_0)^{-1} \cdot z_{\mathbb{Y}}(g_0, g). \end{aligned}$$

By Lemma 3.2.2, we have

$$z_{\mathbb{Y}'}(g_1, g_2) = s'(g_1 g_2) s'(g_1)^{-1} s'(g_2)^{-1},$$

so that the map

$$\begin{aligned} \mathrm{G}(\mathrm{U}(V) \times \mathrm{U}(W))^0 &\longrightarrow \mathrm{Mp}(\mathbb{V})_{\mathbb{Y}'} \\ g &\longmapsto (g, s'(g)) \end{aligned}$$

is a homomorphism.

We next suppose that F is a number field. For each place v of F, we have defined a function

$$s_v : \mathrm{G}(\mathrm{U}(V_v) \times \mathrm{U}(W_v))^0 \longrightarrow \mathbb{C}^1$$

with associated homomorphism

$$\iota_v : \mathrm{G}(\mathrm{U}(V_v) \times \mathrm{U}(W_v))^0 \longrightarrow \mathrm{Mp}(\mathbb{V}_v).$$

LEMMA 3.3.1. *The homomorphisms ι_v induce a homomorphism*

$$\iota : \mathrm{G}(\mathrm{U}(V) \times \mathrm{U}(W))^0(\mathbb{A}) \longrightarrow \mathrm{Mp}(\mathbb{V})_{\mathbb{A}}.$$

Moreover, the diagram

$$\begin{array}{ccc} \mathrm{G}(\mathrm{U}(V) \times \mathrm{U}(W))^0(F) & \hookrightarrow & \mathrm{G}(\mathrm{U}(V) \times \mathrm{U}(W))^0(\mathbb{A}) \\ \downarrow & & \downarrow \iota \\ \mathrm{Sp}(\mathbb{V})(F) & \xrightarrow{\quad i \quad} & \mathrm{Mp}(\mathbb{V})_{\mathbb{A}} \end{array}$$

is commutative.

PROOF. Recall that, for almost all v, K_v is the maximal compact subgroup of $\mathrm{Sp}(\mathbb{V}_v)$ and $s_{\mathbb{Y}_v} : K_v \to \mathbb{C}^1$ is the function which defines the splitting $K_v \hookrightarrow \mathrm{Mp}(\mathbb{V}_v)$. Put

$$\mathbf{K}_v := \mathrm{G}(\mathrm{U}(V_v) \times \mathrm{U}(W_v))^0 \cap K_v.$$

Then $\mathbf{K}_v$ is a maximal compact subgroup of $\mathrm{G}(\mathrm{U}(V_v) \times \mathrm{U}(W_v))^0$ for almost all v. By Lemma C.4.3, we have

$$s_v|_{\mathbf{K}_v} = s_{\mathbb{Y}_v}|_{\mathbf{K}_v}$$

for almost all v. Hence, for $g = (g_v) \in \mathrm{G}(\mathrm{U}(V)\times\mathrm{U}(W))^0(\mathbb{A})$, the element $(\iota_v(g_v)) \in \prod_v \mathrm{Mp}(\mathbb{V}_v)$ maps to an element $\iota(g)$ in $\mathrm{Mp}(\mathbb{V})_S \times K^S$ if S is sufficiently large. This proves the first assertion.

Let $\gamma \in \mathrm{G}(\mathrm{U}(V) \times \mathrm{U}(W))^0(F)$. By Proposition C.4.4, we have

$$\prod_v s_v(\gamma) = 1.$$

Hence, if S is sufficiently large, the image of $(\iota_v(\gamma))$ in $\mathrm{Mp}(\mathbb{V})_S \times K^S$ is equal to that of $(\gamma, 1)$. This proves the second assertion. □

3.3.3. Weil representation for the above reductive dual pair. If F is local, we get the Weil representation $\omega_\psi \circ \iota$ of $\mathrm{G}(\mathrm{U}(V) \times \mathrm{U}(W))^0$ on $\mathcal{S}(\mathbb{X})$, where ω_ψ is the Weil representation of $\mathrm{Mp}(\mathbb{V})_\mathbb{Y}$ and $\iota : \mathrm{G}(\mathrm{U}(V) \times \mathrm{U}(W))^0 \to \mathrm{Mp}(\mathbb{V})_\mathbb{Y}$ is the above homomorphism. Similarly, if F is global, we get the global Weil representation $\omega_\psi \circ \iota$ of $\mathrm{G}(\mathrm{U}(V) \times \mathrm{U}(W))^0(\mathbb{A})$ on $\mathcal{S}(\mathbb{X}(\mathbb{A}))$. If there is no confusion, we suppress ι from the notation.

CHAPTER 4

The Rallis Inner Product formula and the Jacquet–Langlands Correspondence

In this chapter, we generalize the result of Shimizu [**68**] and realize the Jacquet–Langlands correspondence from $B^\times$ to $B_1^\times \times B_2^\times$ as theta lifts, where B, B_1, B_2 are quaternion algebras over a number field F such that $B \neq \mathrm{M}_2(F)$ and $B = B_1 \cdot B_2$ in the Brauer group. More precisely, we consider the quaternionic unitary similitude groups

$$\mathrm{GU}(V)^0 \cong (B_1^\times \times B_2^\times)/F^\times, \quad \mathrm{GU}(W) \cong B^\times,$$

where V is the skew-hermitian B-space given in §2.2 and W is the hermitian B-space given in §3.3.1. Then we have a Weil representation of $\mathrm{G}(\mathrm{U}(V) \times \mathrm{U}(W))^0(\mathbb{A})$ on the Schwartz space $\mathcal{S}(\mathbb{X}(\mathbb{A}))$, where $\mathbb{X}$ is the maximal isotropic subspace of the symplectic F-space $\mathbb{V} = V \otimes_B W$ given in §2.2. For any irreducible unitary automorphic representation π_B of $B^\times(\mathbb{A})$, we can define a space $\Theta(\pi_B)$ of automorphic forms on $B_1^\times(\mathbb{A}) \times B_2^\times(\mathbb{A})$ with an equivariant surjective map

$$\begin{aligned} \pi_B \otimes \mathcal{S}(\mathbb{X}(\mathbb{A})) &\longrightarrow \Theta(\pi_B), \\ f \otimes \varphi &\longmapsto \theta_\varphi(f) \end{aligned}$$

where $\theta_\varphi(f)$ is the theta lift of f, i.e. the integral of f against the theta function associated to φ. The main result of this section (Proposition 4.2.3) says that if the Jacquet–Langlands transfers π_{B_1}, π_{B_2} of π_B to $B_1^\times(\mathbb{A})$, $B_2^\times(\mathbb{A})$ exist, then

$$\Theta(\pi_B) = \pi_{B_1} \boxtimes \pi_{B_2}.$$

In fact, this easily follows from the local theta correspondence for unramified representations and the strong multiplicity one theorem, once we know that $\Theta(\pi_B)$ is non-zero.

To prove the non-vanishing of $\Theta(\pi_B)$, we will compute the inner product $\langle \theta_\varphi(f), \theta_\varphi(f) \rangle$ for $f \in \pi_B$ and $\varphi \in \mathcal{S}(\mathbb{X}(\mathbb{A}))$ by using a see-saw diagram

$$\begin{array}{ccc} \mathrm{GU}(W^\square) & & \mathrm{G}(\mathrm{U}(V) \times \mathrm{U}(V)) \\ | & \times & | \\ \mathrm{G}(\mathrm{U}(W) \times \mathrm{U}(W)) & & \mathrm{GU}(V) \end{array},$$

where $W^\square = W \oplus W$ is the B-space equipped with a hermitian form

$$\langle (x, x'), (y, y') \rangle = \langle x, y \rangle - \langle x', y' \rangle.$$

Indeed, by the associated see-saw identity, we have

$$\begin{aligned} \langle \theta_\varphi(f), \theta_\varphi(f) \rangle &= \langle \theta_{\varphi \otimes \bar{\varphi}}(f \otimes \bar{f})|_{\mathrm{GU}(V)}, \mathbf{1} \rangle \\ &= \langle \theta_{\varphi \otimes \bar{\varphi}}(\mathbf{1})|_{\mathrm{G}(\mathrm{U}(W) \times \mathrm{U}(W))}, \bar{f} \otimes f \rangle, \end{aligned}$$

where

- $\varphi \otimes \bar{\varphi} \in \mathcal{S}(\mathbb{X}^\square)$;
- $\mathbb{X}^\square = \mathbb{X} \oplus \mathbb{X}$ is a maximal isotropic subspace of $\mathbb{V}^\square = \mathbb{V} \oplus \mathbb{V}$;
- $\theta_{\varphi\otimes\bar{\varphi}}(f \otimes \bar{f})$ is the theta lift of $f \otimes \bar{f}$ to $\mathrm{G}(\mathrm{U}(V) \times \mathrm{U}(V))(\mathbb{A})$;
- $\theta_{\varphi\otimes\bar{\varphi}}(\mathbf{1})$ is the theta lift of the constant function $\mathbf{1}$ to $\mathrm{GU}(W^\square)(\mathbb{A})$.

On the other hand, $W^\square$ always has a complete polarization $W^\square = W^\triangledown \oplus W^\vartriangle$, where

$$W^\triangledown = \{(x, -x) \mid x \in W\}, \quad W^\vartriangle = \{(x, x) \mid x \in W\}.$$

Since $\mathbb{V}^\triangledown = V \otimes W^\triangledown$ is also a maximal isotropic subspace of $\mathbb{V}^\square = V \otimes_B W^\square$, we have a partial Fourier transform

$$\mathcal{S}(\mathbb{X}^\square(\mathbb{A})) \longrightarrow \mathcal{S}(\mathbb{V}^\triangledown(\mathbb{A})),$$

which relates the two models of the Weil representation. By the Poisson summation formula, the two associated theta distributions on the Schwartz spaces are the same, and in particular, we have

$$\theta_{\varphi\otimes\bar{\varphi}}(\mathbf{1}) = \theta_{\hat{\varphi}}(\mathbf{1}),$$

where $\hat{\varphi} \in S(\mathbb{V}^\triangledown(\mathbb{A}))$ is the image of $\varphi \otimes \bar{\varphi}$ under the partial Fourier transform. However, the Siegel–Weil formula says that $\theta_{\hat{\varphi}}(\mathbf{1})$ is equal to the Siegel Eisenstein series on $\mathrm{GU}(W^\square)(\mathbb{A})$ associated to $\hat{\varphi}$. Combining this with the doubling method, we obtain the Rallis inner product formula (Proposition 4.4.2)

$$\langle \theta_\varphi(f), \theta_\varphi(f) \rangle = 2 \cdot \frac{L^S(1, \pi, \mathrm{ad})}{\zeta^S(2)^2} \cdot \prod_{v \in S} Z_v,$$

where

- S is a sufficiently large finite set of places of F;
- π is the Jacquet–Langlands transfer of π_B to $\mathrm{GL}_2(\mathbb{A})$;
- $L^S(s, \pi, \mathrm{ad})$ is the partial adjoint L-function of π;
- $\zeta^S(s)$ is the partial Dedekind zeta function of F;
- Z_v is the local zeta integral.

Thus, we may reduce the non-vanishing of $\Theta(\pi_B)$ to that of Z_v, which will be proved in Lemma 4.3.3.

4.1. Setup

Let F be a number field and B a quaternion algebra over F. As in Appendix D, we consider the following spaces:

- $V = B_1 \otimes_E B_2$ is the 2-dimensional right skew-hermitian B-space.
- $W = B$ is the 1-dimensional left hermitian B-space.
- $W^\square = W \oplus W$ is the 2-dimensional left hermitian B-space.
- $\mathbb{V} = V \otimes_B W$ is the 8-dimensional symplectic F-space.
- $\mathbb{V}^\square = V \otimes_B W^\square = \mathbb{V} \oplus \mathbb{V}$ is the 16-dimensional F-space.
- $W^\square = W^\triangledown \oplus W^\vartriangle$ is the complete polarization over B.
- $\mathbb{V} = \mathbb{X} \oplus \mathbb{Y}$ is the complete polarization over F.
- $\mathbb{V}_v = \mathbb{X}'_v \oplus \mathbb{Y}'_v$ is the complete polarization over F_v.
- $\mathbb{V}^\square = \mathbb{V}^\triangledown \oplus \mathbb{V}^\vartriangle$ is the complete polarization over F.
- $\mathbb{V}^\square = \mathbb{X}^\square \oplus \mathbb{Y}^\square$ is the complete polarization over F.
- $\mathbb{V}_v^\square = \mathbb{X}_v'^\square \oplus \mathbb{Y}_v'^\square$ is the complete polarization over F_v.

We have a natural map

$$\iota : \mathrm{G}(\mathrm{U}(W) \times \mathrm{U}(W)) \longrightarrow \mathrm{GU}(W^{\square})$$

and a see-saw diagram

$$\begin{array}{ccc} \mathrm{GU}(W^{\square}) & & \mathrm{G}(\mathrm{U}(V) \times \mathrm{U}(V)) \\ | & \times & | \\ \mathrm{G}(\mathrm{U}(W) \times \mathrm{U}(W)) & & \mathrm{GU}(V) \end{array}.$$

4.1.1. Partial Fourier transform. Fix a non-trivial character $\psi = \otimes_v \psi_v$ of $\mathbb{A}/F$. Recall that $\mathbf{e}_1, \dots, \mathbf{e}_4$ is a basis of $\mathbb{X}$ over F. For each place v of F, this basis and the self-dual measure on F_v with respect to ψ_v define a Haar measure dx_v on $\mathbb{X}_v$. Then the product measure $dx = \prod_v dx_v$ is the Tamagawa measure on $\mathbb{X}(\mathbb{A})$. We define a hermitian inner product $\langle \cdot, \cdot \rangle$ on $\mathcal{S}(\mathbb{X}(\mathbb{A}))$ by

$$\langle \varphi_1, \varphi_2 \rangle := \int_{\mathbb{X}(\mathbb{A})} \varphi_1(x) \overline{\varphi_2(x)} \, dx.$$

Recall that $\mathbb{V}^{\square} = \mathbb{V}^{\nabla} \oplus \mathbb{V}^{\triangle} = \mathbb{X}^{\square} \oplus \mathbb{Y}^{\square}$. We define a partial Fourier transform

$$\begin{aligned} \mathcal{S}(\mathbb{X}^{\square}(\mathbb{A})) &\longrightarrow \mathcal{S}(\mathbb{V}^{\nabla}(\mathbb{A})) \\ \varphi &\longmapsto \hat{\varphi} \end{aligned}$$

by

$$\hat{\varphi}(u) = \int_{(\mathbb{V}^{\triangle} \cap \mathbb{Y}^{\square} \backslash \mathbb{V}^{\triangle})(\mathbb{A})} \varphi(x) \psi\left(\frac{1}{2}\left(\langle\!\langle x, y \rangle\!\rangle - \langle\!\langle u, v \rangle\!\rangle\right)\right) dv,$$

where we write $u + v = x + y$ with $u \in \mathbb{V}^{\nabla}(\mathbb{A})$, $v \in \mathbb{V}^{\triangle}(\mathbb{A})$, $x \in \mathbb{X}^{\square}(\mathbb{A})$, $y \in \mathbb{Y}^{\square}(\mathbb{A})$, and dv is the Tamagawa measure.

Lemma 4.1.1 ([**48**, p. 182, (13)]). *If* $\varphi = \varphi_1 \otimes \bar{\varphi}_2 \in \mathcal{S}(\mathbb{X}^{\square}(\mathbb{A}))$ *with* $\varphi_i \in \mathcal{S}(\mathbb{X}(\mathbb{A}))$, *then we have*

$$\hat{\varphi}(0) = \langle \varphi_1, \varphi_2 \rangle.$$

Proof. We include a proof for convenience. Since $\mathbb{V}^{\triangle} \cap \mathbb{Y}^{\square} = \mathbb{Y}^{\triangle}$, we have

$$\hat{\varphi}(u) = \int_{\mathbb{X}^{\triangle}(\mathbb{A})} \varphi(x) \psi\left(\frac{1}{2}\left(\langle\!\langle x, y \rangle\!\rangle - \langle\!\langle u, v \rangle\!\rangle\right)\right) dv.$$

We write

$$v = (v_0, v_0), \quad u = (u_0, -u_0), \quad u_0 = x_0 + y_0$$

with $v_0, x_0 \in \mathbb{X}(\mathbb{A})$ and $y_0 \in \mathbb{Y}(\mathbb{A})$, so that

$$x = (v_0 + x_0, v_0 - x_0), \quad y = (y_0, -y_0).$$

We have

$$\begin{aligned} \langle\!\langle x, y \rangle\!\rangle &= \langle\!\langle v_0 + x_0, y_0 \rangle\!\rangle - \langle\!\langle v_0 - x_0, -y_0 \rangle\!\rangle = 2\langle\!\langle v_0, y_0 \rangle\!\rangle, \\ \langle\!\langle u, v \rangle\!\rangle &= 2\langle\!\langle u_0, v_0 \rangle\!\rangle = 2\langle\!\langle y_0, v_0 \rangle\!\rangle, \end{aligned}$$

and hence

$$\hat{\varphi}(u) = \int_{\mathbb{X}(\mathbb{A})} \varphi(v_0 + x_0, v_0 - x_0) \psi(2\langle\!\langle v_0, y_0 \rangle\!\rangle) \, dv_0,$$

where dv_0 is the Tamagawa measure on $\mathbb{X}(\mathbb{A})$. In particular, we have

$$\hat{\varphi}(0) = \int_{\mathbb{X}(\mathbb{A})} \varphi(v_0, v_0)\, dv_0. \qquad \square$$

For each place v of F, we define a hermitian inner product $\langle\cdot,\cdot\rangle$ on $\mathcal{S}(\mathbb{X}_v)$ with respect to the Haar measure dx_v on $\mathbb{X}_v$ given above. Fix a Haar measure on $\mathbb{X}'_v$ and define a hermitian inner product $\langle\cdot,\cdot\rangle$ on $\mathcal{S}(\mathbb{X}'_v)$ similarly. For $\varphi' \in \mathcal{S}(\mathbb{X}'_v)$, we define its partial Fourier transform $\varphi \in \mathcal{S}(\mathbb{X}_v)$ by

$$\varphi(x) = \int_{\mathbb{Y}_v \cap \mathbb{Y}'_v \backslash \mathbb{Y}_v} \varphi'(x')\psi_v\left(\frac{1}{2}\left(\langle\langle x', y'\rangle\rangle - \langle\langle x, y\rangle\rangle\right)\right) dy,$$

where we write $x + y = x' + y'$ with $x \in \mathbb{X}_v$, $y \in \mathbb{Y}_v$, $x' \in \mathbb{X}'_v$, $y' \in \mathbb{Y}'_v$, and we normalize a Haar measure dy so that

$$\langle \varphi_1, \varphi_2\rangle = \langle \varphi'_1, \varphi'_2\rangle$$

holds for $\varphi'_1, \varphi'_2 \in \mathcal{S}(\mathbb{X}'_v)$ and their partial Fourier transforms $\varphi_1, \varphi_2 \in \mathcal{S}(\mathbb{X}_v)$.

4.1.2. Weil representations. Fix a place v of F and suppress the subscript v from the notation. In Appendices C, D, we have defined the maps

- $\hat{s} : \mathrm{G}(\mathrm{U}(V) \times \mathrm{U}(W^\square)) \to \mathbb{C}^1$ such that $z_{\mathbb{Y}^\triangle} = \partial \hat{s}$,
- $s : \mathrm{GU}(V)^0 \times \mathrm{GU}(W) \to \mathbb{C}^1$ such that $z_{\mathbb{Y}} = \partial s$,
- $s' : \mathrm{GU}(V)^0 \times \mathrm{GU}(W) \to \mathbb{C}^1$ such that $z_{\mathbb{Y}'} = \partial s'$.

Let ω_ψ and $\omega_\psi^\square$ be the Weil representations of $\mathrm{Mp}(\mathbb{V})$ and $\mathrm{Mp}(\mathbb{V}^\square)$ with respect to ψ, respectively. Composing these with $\hat{s}$, s, s', we obtain:

- a representation $\omega_\psi^\square$ of $\mathrm{G}(\mathrm{U}(V) \times \mathrm{U}(W^\square))$ on $\mathcal{S}(\mathbb{V}^\triangledown)$,
- a representation ω_ψ of $\mathrm{G}(\mathrm{U}(V)^0 \times \mathrm{U}(W))$ on $\mathcal{S}(\mathbb{X})$,
- a representation ω_ψ of $\mathrm{G}(\mathrm{U}(V)^0 \times \mathrm{U}(W))$ on $\mathcal{S}(\mathbb{X}')$.

By §D.4, the partial Fourier transform

$$\mathcal{S}(\mathbb{V}^\triangledown) \cong \mathcal{S}(\mathbb{X}^\square) = \mathcal{S}(\mathbb{X}) \otimes \mathcal{S}(\mathbb{X})$$

induces an isomorphism

$$\omega_\psi^\square \circ (\mathrm{id} \otimes \iota) \cong \omega_\psi \otimes \bar{\omega}_\psi$$

as representations of $\mathrm{G}(\mathrm{U}(V)^0 \times \mathrm{U}(W) \times \mathrm{U}(W))$. By definition, the partial Fourier transform $\mathcal{S}(\mathbb{X}') \cong \mathcal{S}(\mathbb{X})$ is $\mathrm{G}(\mathrm{U}(V)^0 \times \mathrm{U}(W))$-equivariant.

4.2. The Jacquet–Langlands–Shimizu correspondence

Let F be a number field and B a quaternion algebra over F. We assume that B is division. Set

$$G = \mathrm{GU}(W), \qquad H = \mathrm{GU}(V), \qquad H^0 = \mathrm{GU}(V)^0,$$
$$G_1 = \mathrm{U}(W), \qquad H_1 = \mathrm{U}(V), \qquad H_1^0 = \mathrm{U}(V)^0.$$

Recall that $G \cong B^\times$ and

$$1 \longrightarrow F^\times \stackrel{i}{\longrightarrow} B_1^\times \times B_2^\times \longrightarrow H^0 \longrightarrow 1,$$

where B_1 and B_2 are quaternion algebras over F such that $B_1 \cdot B_2 = B$ in the Brauer group and $i(z) = (z, z^{-1})$.

Put $(\mathbb{A}^\times)^+ = \nu(G(\mathbb{A})) \cap \nu(H^0(\mathbb{A}))$,

$$G(\mathbb{A})^+ = \{g \in G(\mathbb{A}) \,|\, \nu(g) \in (\mathbb{A}^\times)^+\},$$
$$H^0(\mathbb{A})^+ = \{h \in H^0(\mathbb{A}) \,|\, \nu(h) \in (\mathbb{A}^\times)^+\}.$$

For each place v of F, we define $(F_v^\times)^+$, G_v^+, $(H_v^0)^+$ similarly. We have $(F_v^\times)^+ = F_v^\times$ if v is either finite or complex. If v is real, then we have

$$(F_v^\times)^+ = \begin{cases} \mathbb{R}^\times & \text{if } B_v,\ B_{1,v},\ B_{2,v} \text{ are split,} \\ \mathbb{R}_+^\times & \text{otherwise.} \end{cases}$$

We have $(\mathbb{A}^\times)^+ = \prod_v' (F_v^\times)^+$, $G(\mathbb{A})^+ = \prod_v' G_v^+$, and $H^0(\mathbb{A})^+ = \prod_v' (H_v^0)^+$.

Let π be an irreducible unitary cuspidal automorphic representation of $\mathrm{GL}_2(\mathbb{A})$. We assume that its Jacquet–Langlands transfers π_B, π_{B_1}, π_{B_2} to $B^\times(\mathbb{A})$, $B_1^\times(\mathbb{A})$, $B_2^\times(\mathbb{A})$ exist. We regard π_B and $\pi_{B_1} \boxtimes \pi_{B_2}$ as irreducible unitary automorphic representations of $G(\mathbb{A})$ and $H^0(\mathbb{A})$ respectively.

We define a theta distribution $\Theta : \mathcal{S}(\mathbb{X}(\mathbb{A})) \to \mathbb{C}$ by

$$\Theta(\varphi) = \sum_{x \in \mathbb{X}(F)} \varphi(x)$$

for $\varphi \in \mathcal{S}(\mathbb{X}(\mathbb{A}))$. Let $\varphi \in \mathcal{S}(\mathbb{X}(\mathbb{A}))$ and $f \in \pi_B$. For $h \in H^0(\mathbb{A})^+$, choose $g \in G(\mathbb{A})^+$ such that $\nu(g) = \nu(h)$ and put

$$\theta_\varphi(f)(h) := \int_{G_1(F)\backslash G_1(\mathbb{A})} \Theta(\omega_\psi(g_1 g h)\varphi) f(g_1 g)\, dg_1. \tag{4.1}$$

Here $dg_1 = \prod_v dg_{1,v}$ is the Tamagawa measure on $G_1(\mathbb{A})$ and we may assume that the volume of a hyperspecial maximal compact subgroup of $G_{1,v}$ with respect to $dg_{1,v}$ is 1 for almost all v. Note that $\mathrm{vol}(G_1(F)\backslash G_1(\mathbb{A})) = 1$. Using Eichler's norm theorem, one can see that $\theta_\varphi(f)(\gamma h) = \theta_\varphi(f)(h)$ for $\gamma \in H^0(F) \cap H^0(\mathbb{A})^+$ and $h \in H^0(\mathbb{A})^+$. Since $H^0(\mathbb{A}) = H^0(F)H^0(\mathbb{A})^+$, $\theta_\varphi(f)$ defines an automorphic form on $H^0(\mathbb{A})$. Let $\Theta(\pi_B)$ be the automorphic representation of $H^0(\mathbb{A})$ generated by $\theta_\varphi(f)$ for all $\varphi \in \mathcal{S}(\mathbb{X}(\mathbb{A}))$ and $f \in \pi_B$.

LEMMA 4.2.1. *The automorphic representation $\Theta(\pi_B)$ is cuspidal.*

PROOF. If both B_1 and B_2 are division, then H^0 is anisotropic modulo center and the assertion is obvious. Hence we may assume that either B_1 or B_2 is split. Then there exists a complete polarization $V = \tilde{X} \oplus \tilde{Y}$ over B. As in §C.3, we regard V as a left B-space. Choosing a basis $\tilde{\mathbf{v}}, \tilde{\mathbf{v}}^*$ of V such that $\tilde{X} = B\tilde{\mathbf{v}}$, $\tilde{Y} = B\tilde{\mathbf{v}}^*$, $\langle \tilde{\mathbf{v}}, \tilde{\mathbf{v}}^* \rangle = 1$, we may write

$$H = \left\{ h \in \mathrm{GL}_2(B) \,\middle|\, h \begin{pmatrix} & 1 \\ -1 & \end{pmatrix} {}^t h^* = \nu(h) \cdot \begin{pmatrix} & 1 \\ -1 & \end{pmatrix} \right\}.$$

Put

$$\mathbf{n}(b) := \begin{pmatrix} 1 & b \\ & 1 \end{pmatrix} \in H$$

for $b \in F$. It remains to show that

$$\int_{F\backslash\mathbb{A}} \theta_\varphi(f)(\mathbf{n}(b))\, db = 0$$

for all $\varphi \in \mathcal{S}(\mathbb{X}(\mathbb{A}))$ and $f \in \pi_B$.

Let $\mathbb{V} = \tilde{\mathbb{X}} \oplus \tilde{\mathbb{Y}}$ be another complete polarization over F given by $\tilde{\mathbb{X}} = W \otimes_B \tilde{X}$ and $\tilde{\mathbb{Y}} = W \otimes_B \tilde{Y}$, where we regard W as a right B-space. As in [**40**, §5], we define a Weil representation $\tilde{\omega}_\psi$ of $G_1(\mathbb{A}) \times H_1(\mathbb{A})$ with respect to ψ on $\mathcal{S}(\tilde{\mathbb{X}}(\mathbb{A})) \cong \mathcal{S}(W(\mathbb{A}))$. Note that

$$\tilde{\omega}_\psi(g_1)\tilde{\varphi}(x) = \tilde{\varphi}(g_1^{-1}x),$$
$$\tilde{\omega}_\psi(\mathbf{n}(b))\tilde{\varphi}(x) = \psi(\tfrac{1}{2}\langle x, x\rangle b)\tilde{\varphi}(x)$$

for $\tilde{\varphi} \in \mathcal{S}(W(\mathbb{A}))$, $x \in W(\mathbb{A})$, $g_1 \in G_1(\mathbb{A})$, and $b \in \mathbb{A}$. Let $\tilde{\varphi} \in \mathcal{S}(W(\mathbb{A}))$ be the partial Fourier transform of $\varphi \in \mathcal{S}(\mathbb{X}(\mathbb{A}))$. Then we have

$$\Theta(\omega_\psi(\mathbf{g})\varphi) = \chi(\mathbf{g}) \sum_{x \in W(F)} \tilde{\omega}_\psi(\mathbf{g})\tilde{\varphi}(x)$$

for $\mathbf{g} \in G_1(\mathbb{A}) \times H_1^0(\mathbb{A})$ with some character χ of $G_1(\mathbb{A}) \times H_1^0(\mathbb{A})$ trivial on $G_1(F) \times H_1^0(F)$. One can see that $\chi(g_1) = \chi(\mathbf{n}(b)) = 1$ for $g_1 \in G_1(\mathbb{A})$ and $b \in \mathbb{A}$. Since W is anisotropic, we have

$$\begin{aligned}
&\int_{F\backslash\mathbb{A}} \theta_\varphi(f)(\mathbf{n}(b))\, db \\
&= \int_{F\backslash\mathbb{A}} \int_{G_1(F)\backslash G_1(\mathbb{A})} \sum_{x \in W(F)} \psi(\tfrac{1}{2}\langle x, x\rangle b)\tilde{\omega}_\psi(g_1)\tilde{\varphi}(x) f(g_1)\, dg_1\, db \\
&= \int_{G_1(F)\backslash G_1(\mathbb{A})} \tilde{\omega}_\psi(g_1)\tilde{\varphi}(0) f(g_1)\, dg_1 \\
&= \tilde{\varphi}(0) \int_{G_1(F)\backslash G_1(\mathbb{A})} f(g_1)\, dg_1.
\end{aligned}$$

Since π is cuspidal, the restriction of π_B to $G_1(\mathbb{A})$ is orthogonal to the trivial representation of $G_1(\mathbb{A})$, so that this integral vanishes. This completes the proof. □

LEMMA 4.2.2. *The automorphic representation $\Theta(\pi_B)$ is non-zero.*

The proof of this lemma will be given in §4.4 below.

PROPOSITION 4.2.3. *We have*

$$\Theta(\pi_B) = \pi_{B_1} \boxtimes \pi_{B_2}.$$

PROOF. Since $\Theta(\pi_B)$ is cuspidal and non-zero, the assertion follows from the local theta correspondence for unramified representations and the strong multiplicity one theorem. □

4.3. The doubling method

4.3.1. Degenerate principal series representations.

Set

$$G^\square = \mathrm{GU}(W^\square), \quad G_1^\square = \mathrm{U}(W^\square).$$

Choosing a basis $\mathbf{w}, \mathbf{w}^*$ of $W^\square$ such that $W^\triangledown = B\mathbf{w}$, $W^\triangle = B\mathbf{w}^*$, $\langle \mathbf{w}, \mathbf{w}^*\rangle = 1$, we may write

$$G^\square = \left\{ g \in \mathrm{GL}_2(B) \,\middle|\, g \begin{pmatrix} & 1 \\ 1 & \end{pmatrix} {}^t g^* = \nu(g) \cdot \begin{pmatrix} & 1 \\ 1 & \end{pmatrix} \right\}.$$

Let P and P_1 be the Siegel parabolic subgroups of $G^\square$ and $G_1^\square$ given by

$$P = \left\{ \begin{pmatrix} a & * \\ & \nu \cdot (a^*)^{-1} \end{pmatrix} \in G^\square \;\middle|\; a \in B^\times,\ \nu \in F^\times \right\}$$

and $P_1 = P \cap G_1^\square$ respectively. Let δ_P and δ_{P_1} denote the modulus characters of $P(\mathbb{A})$ and $P_1(\mathbb{A})$ respectively. We have

$$\delta_P\left(\begin{pmatrix} a & * \\ & \nu \cdot (a^*)^{-1} \end{pmatrix}\right) = |\nu(a)|^3 \cdot |\nu|^{-3}$$

and $\delta_{P_1} = \delta_P|_{P_1(\mathbb{A})}$. Put

$$d(\nu) := \begin{pmatrix} 1 & \\ & \nu \end{pmatrix} \in P$$

for $\nu \in F^\times$. We fix a maximal compact subgroup K of $G^\square(\mathbb{A})$ such that $G^\square(\mathbb{A}) = P(\mathbb{A})K$ and $G_1^\square(\mathbb{A}) = P_1(\mathbb{A})K_1$, where $K_1 = K \cap G_1^\square(\mathbb{A})$.

For $s \in \mathbb{C}$, we consider a degenerate principal series representation $\mathcal{I}(s) := \operatorname{Ind}_P^{G^\square}(\delta_P^{s/3})$ of $G^\square(\mathbb{A})$ consisting of smooth functions $\mathcal{F}$ on $G^\square(\mathbb{A})$ which satisfy

$$\mathcal{F}\left(\begin{pmatrix} a & * \\ & \nu \cdot (a^*)^{-1} \end{pmatrix} g\right) = |\nu(a)|^{s+\frac{3}{2}} \cdot |\nu|^{-s-\frac{3}{2}} \cdot \mathcal{F}(g).$$

We define a degenerate principal series representation $\mathcal{I}_1(s) := \operatorname{Ind}_{P_1}^{G_1^\square}(\delta_{P_1}^{s/3})$ of $G_1^\square(\mathbb{A})$ similarly. Then the restriction $\mathcal{I}(s) \to \mathcal{I}_1(s)$ to $G_1^\square(\mathbb{A})$ as functions is a $G_1^\square(\mathbb{A})$-equivariant isomorphism. For each place v of F, we define degenerate principal series representations $\mathcal{I}_v(s)$ and $\mathcal{I}_{1,v}(s)$ of $G_v^\square$ and $G_{1,v}^\square$ similarly.

For $\varphi \in \mathcal{S}(\mathbb{V}^\triangledown(\mathbb{A}))$, we define $\mathcal{F}_\varphi \in \mathcal{I}(\frac{1}{2})$ by

$$\mathcal{F}_\varphi(g) = |\nu(g)|^{-2} \cdot (\omega_\psi^\square(d(\nu(g)^{-1})g)\varphi)(0).$$

One can see that the map $\varphi \mapsto \mathcal{F}_\varphi$ is $\mathrm{G}(\mathrm{U}(V) \times \mathrm{U}(W^\square))(\mathbb{A})$-equivariant, where $\mathrm{GU}(V)(\mathbb{A})$ acts trivially on $\mathcal{I}(\frac{1}{2})$.

4.3.2. Eisenstein series. For a holomorphic section $\mathcal{F}_s$ of $\mathcal{I}(s)$, we define an Eisenstein series $E(\mathcal{F}_s)$ on $G^\square(\mathbb{A})$ by (the meromorphic continuation of)

$$E(g, \mathcal{F}_s) = \sum_{\gamma \in P(F) \backslash G^\square(F)} \mathcal{F}_s(\gamma g).$$

For a holomorphic section $\mathcal{F}_{1,s}$ of $\mathcal{I}_1(s)$, we define an Eisenstein series $E(\mathcal{F}_{1,s})$ on $G_1^\square(\mathbb{A})$ similarly. If $\mathcal{F}_s$ is a holomorphic section of $\mathcal{I}(s)$, then $\mathcal{F}_s|_{G_1^\square(\mathbb{A})}$ is a holomorphic section of $\mathcal{I}_1(s)$ and $E(\mathcal{F}_s)|_{G_1^\square(\mathbb{A})} = E(\mathcal{F}_s|_{G_1^\square(\mathbb{A})})$. By [**79**, Theorem 3.1], $E(\mathcal{F}_s)$ is holomorphic at $s = \frac{1}{2}$. In particular, the map

$$E : \mathcal{I}(\tfrac{1}{2}) \longrightarrow \mathcal{A}(G^\square)$$

given by $E(\mathcal{F}) := E(\mathcal{F}_s)|_{s=\frac{1}{2}}$ is $G^\square(\mathbb{A})$-equivariant, where $\mathcal{A}(G^\square)$ is the space of automorphic forms on $G^\square(\mathbb{A})$ and $\mathcal{F}_s$ is the holomorphic section of $\mathcal{I}(s)$ such that $\mathcal{F}_{\frac{1}{2}} = \mathcal{F}$ and $\mathcal{F}_s|_K$ is independent of s.

4.3.3. Doubling zeta integrals. Let $\langle \cdot, \cdot \rangle$ be the invariant hermitian inner product on π_B given by

$$\langle f_1, f_2 \rangle := \int_{Z_G(\mathbb{A})G(F)\backslash G(\mathbb{A})} f_1(g)\overline{f_2(g)}\, dg$$

for $f_1, f_2 \in \pi_B$. Here Z_G is the center of G and dg is the Tamagawa measure on $Z_G(\mathbb{A})\backslash G(\mathbb{A})$. Note that $\mathrm{vol}(Z_G(\mathbb{A})G(F)\backslash G(\mathbb{A})) = 2$. Fix an isomorphism $\pi_B \cong \otimes_v \pi_{B,v}$. For each place v of F, we choose an invariant hermitian inner product $\langle \cdot, \cdot \rangle$ on $\pi_{B,v}$ so that $\langle f_1, f_2 \rangle = \prod_v \langle f_{1,v}, f_{2,v} \rangle$ and $\langle f_{1,v}, f_{2,v} \rangle = 1$ for almost all v for $f_1 = \otimes_v f_{1,v}, f_2 = \otimes_v f_{2,v} \in \pi_B$. Set

$$\mathbf{G} = \mathrm{G}(\mathrm{U}(W) \times \mathrm{U}(W)) = \{(g_1, g_2) \in G \times G \mid \nu(g_1) = \nu(g_2)\}.$$

Then the doubling zeta integral of Piatetski-Shapiro and Rallis [**58**] is given by

$$Z(\mathcal{F}_s, f_1, f_2) = \int_{Z(\mathbb{A})\mathbf{G}(F)\backslash \mathbf{G}(\mathbb{A})} E(\iota(g_1, g_2), \mathcal{F}_s) f_1(g_1)\overline{f_2(g_2)}\, d\mathbf{g}$$

for a holomorphic section $\mathcal{F}_s$ of $\mathcal{I}(s)$ and $f_1, f_2 \in \pi_B$. Here Z is the center of $G^\square$ and $d\mathbf{g}$ is the Tamagawa measure on $Z(\mathbb{A})\backslash \mathbf{G}(\mathbb{A})$. Note that $\mathrm{vol}(Z(\mathbb{A})\mathbf{G}(F)\backslash \mathbf{G}(\mathbb{A})) = 2$. For each place v of F, put

$$Z(\mathcal{F}_{s,v}, f_{1,v}, f_{2,v}) = \int_{G_{1,v}} \mathcal{F}_{s,v}(\iota(g_{1,v}, 1)) \langle \pi_{B,v}(g_{1,v}) f_{1,v}, f_{2,v} \rangle \, dg_{1,v}$$

for a holomorphic section $\mathcal{F}_{s,v}$ of $\mathcal{I}_v(s)$ and $f_{1,v}, f_{2,v} \in \pi_{B,v}$. Note that, for fixed $f_{1,v}$ and $f_{2,v}$, this integral depends only on the holomorphic section $\mathcal{F}_{s,v}|_{G^\square_{1,v}}$ of $\mathcal{I}_{1,v}(s)$.

LEMMA 4.3.1. *We have*

$$Z(\mathcal{F}_s, f_1, f_2) = \frac{L^S(s+\frac{1}{2}, \pi, \mathrm{ad})}{\zeta^S(s+\frac{3}{2})\zeta^S(2s+1)} \cdot \prod_{v \in S} Z(\mathcal{F}_{s,v}, f_{1,v}, f_{2,v})$$

for a holomorphic section $\mathcal{F}_s = \otimes_v \mathcal{F}_{s,v}$ of $\mathcal{I}(s)$ and $f_1 = \otimes_v f_{1,v}, f_2 = \otimes_v f_{2,v} \in \pi_B$. Here S is a sufficiently large finite set of places of F.

PROOF. The assertion follows from the doubling method [**58**]. Indeed, as in [**58**], [**22**, §6.2], we unfold the Eisenstein series $E(\iota(g_1, g_2), \mathcal{F}_s)$ and see that only the open $\mathbf{G}$-orbit $P\backslash P\mathbf{G}$ in $P\backslash G^\square$ contributes to the integral $Z(\mathcal{F}_s, f_1, f_2)$. Hence we have

$$Z(\mathcal{F}_s, f_1, f_2) = \int_{Z(\mathbb{A})G^\triangle(F)\backslash \mathbf{G}(\mathbb{A})} \mathcal{F}_s(\iota(g_1, g_2)) f_1(g_1)\overline{f_2(g_2)}\, d\mathbf{g},$$

where $G^\triangle = \{(g, g) \mid g \in G\}$. We have $\mathcal{F}_s(\iota(g_1, g_2)) = \mathcal{F}_s(\iota(g_2^{-1} g_1, 1))$ for $(g_1, g_2) \in \mathbf{G}$. Writing $g = g_2$ and $g' = g_2^{-1} g_1$, we have

$$\begin{aligned}
Z(\mathcal{F}_s, f_1, f_2) &= \int_{G_1(\mathbb{A})} \int_{Z_G(\mathbb{A})G(F)\backslash G(\mathbb{A})} \mathcal{F}_s(\iota(g', 1)) f_1(gg')\overline{f_2(g)}\, dg\, dg' \\
&= \int_{G_1(\mathbb{A})} \mathcal{F}_s(\iota(g', 1)) \langle \pi_B(g') f_1, f_2 \rangle \, dg' \\
&= \prod_v Z(\mathcal{F}_{s,v}, f_{1,v}, f_{2,v}).
\end{aligned}$$

By [**58**], we have

$$Z(\mathcal{F}_{s,v}, f_{1,v}, f_{2,v}) = \frac{L(s+\frac{1}{2}, \pi_v, \mathrm{ad})}{\zeta_v(s+\frac{3}{2})\zeta_v(2s+1)}$$

for almost all v. This completes the proof. □

4.3.4. Local zeta integrals.

LEMMA 4.3.2. *The integral $Z(\mathcal{F}_v, f_{1,v}, f_{2,v})$ is absolutely convergent for $\mathcal{F}_v \in \mathcal{I}_v(\frac{1}{2})$ and $f_{1,v}, f_{2,v} \in \pi_{B,v}$.*

PROOF. If B_v is split, then the lemma is proved in [**16**, Lemma 6.5]. If B_v is division, then $G_{1,v}$ is compact and the assertion is obvious. □

LEMMA 4.3.3. *There exist $\varphi_v \in \mathcal{S}(\mathbb{V}_v^\triangledown)$ and $f_{1,v}, f_{2,v} \in \pi_{B,v}$ such that*

$$Z(\mathcal{F}_{\varphi_v}, f_{1,v}, f_{2,v}) \neq 0.$$

PROOF. If B_v is split, then the lemma is proved in [**16**, Lemma 6.6]. Assume that B_v is division. As in [**43**, Theorem 3.2.2], [**45**, Proposition 7.2.1], one can see that there exist $\mathcal{F}_v \in \mathcal{I}_v(\frac{1}{2})$ and $f_{1,v}, f_{2,v} \in \pi_{B,v}$ such that $Z(\mathcal{F}_v, f_{1,v}, f_{2,v}) \neq 0$. On the other hand, by [**78**, Theorems 1.2, 9.2], the map

$$\begin{aligned} \mathcal{S}(\mathbb{V}_v^\triangledown) &\longrightarrow \mathcal{I}_{1,v}(\tfrac{1}{2}) \\ \varphi_v &\longmapsto \mathcal{F}_{\varphi_v}|_{G_{1,v}^\square} \end{aligned}$$

is surjective . This yields the lemma. □

If φ_v is the partial Fourier transform of $\varphi_{1,v} \otimes \bar{\varphi}_{2,v} \in \mathcal{S}(\mathbb{X}_v^\square)$ with $\varphi_{i,v} \in \mathcal{S}(\mathbb{X}_v)$, then we have

$$Z(\mathcal{F}_{\varphi_v}, f_{1,v}, f_{2,v}) = \int_{G_{1,v}} \langle \omega_\psi(g_{1,v})\varphi_{1,v}, \varphi_{2,v}\rangle \langle \pi_{B,v}(g_{1,v}) f_{1,v}, f_{2,v}\rangle \, dg_{1,v}. \tag{4.2}$$

This will be used later to explicate the Rallis inner product formula.

4.4. The Rallis inner product formula

4.4.1. Theta integrals. Recall that $\mathrm{G}(\mathrm{U}(V) \times \mathrm{U}(W^\square))(\mathbb{A})$ acts on $\mathcal{S}(\mathbb{V}^\triangledown(\mathbb{A}))$ via the Weil representation $\omega_\psi^\square$. We define a $G_1^\square(\mathbb{A})$-equivariant and $H_1(\mathbb{A})$-invariant map

$$I : \mathcal{S}(\mathbb{V}^\triangledown(\mathbb{A})) \longrightarrow \mathcal{A}(G_1^\square)$$

as follows. Here $\mathcal{A}(G_1^\square)$ is the space of automorphic forms on $G_1^\square(\mathbb{A})$.

Let $\Theta : \mathcal{S}(\mathbb{V}^\triangledown(\mathbb{A})) \to \mathbb{C}$ be the theta distribution given by

$$\Theta(\varphi) = \sum_{x \in \mathbb{V}^\triangledown(F)} \varphi(x)$$

for $\varphi \in \mathcal{S}(\mathbb{V}^\triangledown(\mathbb{A}))$. Let dh_1 be the Haar measure on $H_1(\mathbb{A})$ such that

$$\mathrm{vol}(H_1(F)\backslash H_1(\mathbb{A})) = 1.$$

First we assume that either B_1 or B_2 is split. Then the integral

$$\int_{H_1(F)\backslash H_1(\mathbb{A})} \Theta(\omega_\psi^\square(g_1 h_1)\varphi)\, dh_1 \tag{4.3}$$

may not be convergent. Following Yamana [**79**, §2], we choose a place $v \in \Sigma_B$ and a 'regularizing element' z_0 in the Bernstein center of $H_{1,v}$ or the universal enveloping algebra of the complexified Lie algebra of $H_{1,v}$. Then the integral

$$I(g_1, \varphi) := \int_{H_1(F)\backslash H_1(\mathbb{A})} \Theta(\omega_\psi^\square(g_1 h_1)(z_0 \cdot \varphi))\, dh_1$$

is absolutely convergent for all $g_1 \in G_1^\square(\mathbb{A})$ and $\varphi \in \mathcal{S}(\mathbb{V}^\triangledown(\mathbb{A}))$, and defines an automorphic form on $G_1^\square(\mathbb{A})$. Note that $I(g_1, \varphi) = (4.3)$ if the right-hand side is absolutely convergent for all g_1. In particular, $I(g_1, \varphi)$ does not depend on choice of v and z_0. Next we assume that both B_1 and B_2 are division. Then $H_1(F)\backslash H_1(\mathbb{A})$ is compact. For $\varphi \in \mathcal{S}(\mathbb{V}^\triangledown(\mathbb{A}))$, we define an automorphic form $I(\varphi)$ on $G_1^\square(\mathbb{A})$ by

$$I(g_1, \varphi) := \int_{H_1(F)\backslash H_1(\mathbb{A})} \Theta(\omega_\psi^\square(g_1 h_1)(z_0 \cdot \varphi))\, dh_1,$$

where we write z_0 for the identity operator for uniformity.

Similarly, we define a $G_1^\square(\mathbb{A})$-equivariant and $H_1^0(\mathbb{A})$-invariant map

$$I^0 : \mathcal{S}(\mathbb{V}^\triangledown(\mathbb{A})) \longrightarrow \mathcal{A}(G_1^\square)$$

by

$$I^0(g_1, \varphi) := \int_{H_1^0(F)\backslash H_1^0(\mathbb{A})} \Theta(\omega_\psi^\square(g_1 h_1^0)(z_0 \cdot \varphi))\, dh_1^0,$$

where dh_1^0 is the Tamagawa measure on $H_1^0(\mathbb{A})$. Note that $\mathrm{vol}(H_1^0(F)\backslash H_1^0(\mathbb{A})) = 2$.

LEMMA 4.4.1. *We have*

$$I^0 = 2 \cdot I.$$

PROOF. The lemma follows from [**41**, Proposition 4.2] with slight modifications. We include a proof for convenience. For each place $v \notin \Sigma_B$, we consider the space $\mathrm{Hom}_{H_{1,v}^0}(\mathcal{S}(\mathbb{V}_v^\triangledown), \mathbb{C})$ with the natural action of $H_{1,v}^0\backslash H_{1,v}$. Let $V_v^\dagger$ and $(W_v^\square)^\dagger$ be the 4-dimensional quadratic F_v-space and the 4-dimensional symplectic F_v-space associated to V_v and $W_v^\square$ respectively. Since $\dim V_v^\dagger > \frac{1}{2} \dim (W_v^\square)^\dagger$, we have $\mathrm{Hom}_{H_{1,v}}(\mathcal{S}(\mathbb{V}_v^\triangledown), \mathrm{sgn}_v) = \{0\}$ by [**62**, p. 399], where sgn_v is the non-trivial character of $H_{1,v}^0\backslash H_{1,v}$. Hence $H_{1,v}$ acts trivially on $\mathrm{Hom}_{H_{1,v}^0}(\mathcal{S}(\mathbb{V}_v^\triangledown), \mathbb{C})$. On the other hand, we have $H_{1,v}^0 = H_{1,v}$ for all $v \in \Sigma_B$. Hence $H_1(\mathbb{A})$ acts trivially on $\mathrm{Hom}_{H_1^0(\mathbb{A})}(\mathcal{S}(\mathbb{V}^\triangledown(\mathbb{A})), \mathbb{C})$, so that

$$\begin{aligned} I(g_1, \varphi) &= \int_{H_1^0(\mathbb{A}) H_1(F)\backslash H_1(\mathbb{A})} I^0(g_1, \omega_\psi^\square(\dot{h}_1)\varphi)\, d\dot{h}_1 \\ &= \int_{H_1^0(\mathbb{A}) H_1(F)\backslash H_1(\mathbb{A})} I^0(g_1, \varphi)\, d\dot{h}_1 \\ &= \frac{1}{2} \cdot I^0(g_1, \varphi), \end{aligned}$$

where $d\dot{h}_1$ is the Haar measure on $H_1^0(\mathbb{A})\backslash H_1(\mathbb{A})$ such that

$$\mathrm{vol}(H_1^0(\mathbb{A}) H_1(F)\backslash H_1(\mathbb{A})) = \tfrac{1}{2}.$$

□

4.4.2. The Siegel–Weil formula. The Siegel–Weil formula [**79**, Theorem 3.4] due to Yamana says that $I(\varphi) = E(\mathcal{F}_\varphi)|_{G_1^\square(\mathbb{A})}$ for $\varphi \in \mathcal{S}(\mathbb{V}^\triangledown(\mathbb{A}))$. Hence, by Lemma 4.4.1, we have

$$I^0(\varphi) = 2 \cdot E(\mathcal{F}_\varphi)|_{G_1^\square(\mathbb{A})} \tag{4.4}$$

for $\varphi \in \mathcal{S}(\mathbb{V}^\triangledown(\mathbb{A}))$.

4.4.3. The Rallis inner product formula. Recall that π_B is the Jacquet–Langlands transfer to $B^\times(\mathbb{A})$ of an irreducible unitary cuspidal automorphic representation π of $\mathrm{GL}_2(\mathbb{A})$. Recall also that we have fixed an isomorphism $\pi_B \cong \otimes_v \pi_{B,v}$ and an invariant hermitian inner product $\langle\cdot,\cdot\rangle$ on $\pi_{B,v}$ as in §4.3.3. This inner product is used in the definition of the integral $Z(\mathcal{F}_{\varphi_v}, f_{1,v}, f_{2,v})$ (see (4.2)). Let Z_{H^0} be the center of H^0 and dh^0 the Tamagawa measure on $Z_H^0(\mathbb{A})\backslash H^0(\mathbb{A})$. Note that $\mathrm{vol}(Z_{H^0}(\mathbb{A})H^0(F)\backslash H^0(\mathbb{A})) = 4$.

PROPOSITION 4.4.2. *Let $\varphi = \otimes_v \varphi_v \in \mathcal{S}(\mathbb{V}^\triangledown(\mathbb{A}))$ be the partial Fourier transform of $\varphi_1 \otimes \bar{\varphi}_2 \in \mathcal{S}(\mathbb{X}^\square(\mathbb{A}))$ with $\varphi_i = \otimes_v \varphi_{i,v} \in \mathcal{S}(\mathbb{X}(\mathbb{A}))$. Let $f_1 = \otimes_v f_{1,v}, f_2 = \otimes_v f_{2,v} \in \pi_B$. Then we have*

$$\int_{Z_{H^0}(\mathbb{A})H^0(F)\backslash H^0(\mathbb{A})} \theta_{\varphi_1}(f_1)(h^0) \cdot \overline{\theta_{\varphi_2}(f_2)(h^0)}\, dh^0 = 2 \cdot \frac{L^S(1,\pi,\mathrm{ad})}{\zeta^S(2)^2} \cdot \prod_{v \in S} Z(\mathcal{F}_{\varphi_v}, f_{1,v}, f_{2,v}).$$

Here S is a sufficiently large finite set of places of F.

PROOF. Put $(F^\times)^+ = F^\times \cap (\mathbb{A}^\times)^+$,

$$G(F)^+ = G(F) \cap G(\mathbb{A})^+,$$
$$H^0(F)^+ = H^0(F) \cap H^0(\mathbb{A})^+.$$

Set $\mathcal{C} = (\mathbb{A}^\times)^2(F^\times)^+\backslash(\mathbb{A}^\times)^+$. Then the similitude characters induce isomorphisms

$$Z_G(\mathbb{A})G_1(\mathbb{A})G(F)^+\backslash G(\mathbb{A})^+ \cong \mathcal{C},$$
$$Z_{H^0}(\mathbb{A})H_1^0(\mathbb{A})H^0(F)^+\backslash H^0(\mathbb{A})^+ \cong \mathcal{C}.$$

Fix cross sections $c \mapsto g_c$ and $c \mapsto h_c$ of $G(\mathbb{A})^+ \to \mathcal{C}$ and $H^0(\mathbb{A}) \to \mathcal{C}$ respectively. Since

$$\mathbf{G}(\mathbb{A}) = Z(\mathbb{A}) \cdot \mathbf{G}(F) \cdot (G_1 \times G_1)(\mathbb{A}) \cdot \{(g_c, g_c) \mid c \in \mathcal{C}\},$$

we have

$$Z(\mathcal{F}_{\varphi,s}, f_1, f_2) = 2 \int_{\mathcal{C}} \int_{G_1(F)\backslash G_1(\mathbb{A})} \int_{G_1(F)\backslash G_1(\mathbb{A})} E(\iota(g_1 g_c, g_2 g_c), \mathcal{F}_{\varphi,s}) f_1(g_1 g_c) \overline{f_2(g_2 g_c)}\, dg_1\, dg_2\, dc,$$

where dg_1, dg_2 are the Tamagawa measures on $G_1(\mathbb{A})$ and dc is the Haar measure on $\mathcal{C}$ such that $\mathrm{vol}(\mathcal{C}) = 1$. For each $c \in \mathcal{C}$, put $\varphi_c = \omega_\psi^\square(\iota(g_c, g_c), h_c)\varphi$. Since

$E(g\iota(g_c, g_c), \mathcal{F}_\varphi) = E(g, \mathcal{F}_{\varphi_c})$, we have

$$\begin{aligned}
&Z(\mathcal{F}_\varphi, f_1, f_2)\\
&= 2\int_{\mathcal{C}}\int_{G_1(F)\backslash G_1(\mathbb{A})}\int_{G_1(F)\backslash G_1(\mathbb{A})} E(\iota(g_1, g_2), \mathcal{F}_{\varphi_c}) f_1(g_1 g_c)\overline{f_2(g_2 g_c)}\, dg_1\, dg_2\, dc\\
&= \int_{\mathcal{C}}\int_{G_1(F)\backslash G_1(\mathbb{A})}\int_{G_1(F)\backslash G_1(\mathbb{A})} I^0(\iota(g_1, g_2), \varphi_c) f_1(g_1 g_c)\overline{f_2(g_2 g_c)}\, dg_1\, dg_2\, dc\\
&= \int_{\mathcal{C}}\int_{G_1(F)\backslash G_1(\mathbb{A})}\int_{G_1(F)\backslash G_1(\mathbb{A})}\int_{H_1^0(F)\backslash H_1^0(\mathbb{A})} \Theta(\omega_\psi^\square(\iota(g_1, g_2)h_1^0)(z_0\cdot\varphi_c))\\
&\qquad\qquad\qquad \times f_1(g_1 g_c)\overline{f_2(g_2 g_c)}\, dh_1^0\, dg_1\, dg_2\, dc\\
&= \int_{\mathcal{C}}\int_{H_1^0(F)\backslash H_1^0(\mathbb{A})}\int_{G_1(F)\backslash G_1(\mathbb{A})}\int_{G_1(F)\backslash G_1(\mathbb{A})} \Theta(\omega_\psi^\square(\iota(g_1, g_2)h_1^0)\varphi_c) * z_0\\
&\qquad\qquad\qquad \times f_1(g_1 g_c)\overline{f_2(g_2 g_c)}\, dg_1\, dg_2\, dh_1^0\, dc
\end{aligned}$$

by the Siegel–Weil formula (4.4). On the other hand, we have

$$\Theta(\omega_\psi^\square(\iota(g_1, g_2)h_1^0)\varphi_c) = \Theta(\omega_\psi(g_1 g_c h_1^0 h_c)\varphi_1)\cdot\overline{\Theta(\omega_\psi(g_2 g_c h_1^0 h_c)\varphi_2)}.$$

Hence we have

$$\begin{aligned}
&\int_{G_1(F)\backslash G_1(\mathbb{A})}\int_{G_1(F)\backslash G_1(\mathbb{A})} \Theta(\omega_\psi^\square(\iota(g_1, g_2)h_1^0)\varphi_c) * z_0\cdot f_1(g_1 g_c)\overline{f_2(g_2 g_c)}\, dg_1\, dg_2\\
&\qquad\qquad = \left(\theta_{\varphi_1}(f_1)(h_1^0 h_c)\cdot\overline{\theta_{\varphi_2}(f_2)(h_1^0 h_c)}\right) * z_0.
\end{aligned}$$

By Lemma 4.2.1, the function $h_1^0 \mapsto \theta_{\varphi_1}(f_1)(h_1^0 h_c)\cdot\overline{\theta_{\varphi_2}(f_2)(h_1^0 h_c)}$ is integrable over $H_1^0(F)\backslash H_1^0(\mathbb{A})$, so that

$$\begin{aligned}
&\int_{H_1^0(F)\backslash H_1^0(\mathbb{A})} \left(\theta_{\varphi_1}(f_1)(h_1^0 h_c)\cdot\overline{\theta_{\varphi_2}(f_2)(h_1^0 h_c)}\right) * z_0\, dh_1^0\\
&\qquad\qquad = \int_{H_1^0(F)\backslash H_1^0(\mathbb{A})} \theta_{\varphi_1}(f_1)(h_1^0 h_c)\cdot\overline{\theta_{\varphi_2}(f_2)(h_1^0 h_c)}\, dh_1^0
\end{aligned}$$

and hence

$$Z(\mathcal{F}_\varphi, f_1, f_2) = \int_{\mathcal{C}}\int_{H_1^0(F)\backslash H_1^0(\mathbb{A})} \theta_{\varphi_1}(f_1)(h_1^0 h_c)\cdot\overline{\theta_{\varphi_2}(f_2)(h_1^0 h_c)}\, dh_1^0\, dc.$$

Since $H^0(\mathbb{A}) = Z_{H^0}(\mathbb{A})\cdot H^0(F)\cdot H_1^0(\mathbb{A})\cdot\{h_c \mid c\in\mathcal{C}\}$, this integral is equal to

$$\frac{1}{2}\int_{Z_{H^0}(\mathbb{A})H^0(F)\backslash H^0(\mathbb{A})} \theta_{\varphi_1}(f_1)(h^0)\cdot\overline{\theta_{\varphi_2}(f_2)(h^0)}\, dh^0.$$

Now the assertion follows from this and Lemma 4.3.1. □

Now Lemma 4.2.2 follows from Proposition 4.4.2 and Lemma 4.3.3.

CHAPTER 5

Schwartz Functions

Recall that, in Chapter 4, we have defined the theta lift $\theta_\varphi(f_B)$ for an automorphic form $f_B \in \pi_B$ and a Schwartz function $\varphi \in \mathcal{S}(\mathbb{X}(\mathbb{A}))$. For our applications, we will study arithmetic properties of $\theta_\varphi(f_B)$ for some particular choice of f_B and φ. The purpose of this chapter is to choose φ. Note that f_B will be chosen in §6 below.

More precisely, we will pick a decomposable Schwartz function $\varphi = \otimes_v \varphi_v$, where $\varphi_v \in \mathcal{S}(\mathbb{X}_v)$ depends not only on the representation $\pi_{B,v}$ but also on some auxiliary data as described in §5.5 below. We require that φ_v satisfies some equivariance properties, but it is not easy to give a direct definition since the Weil representation on $\mathcal{S}(\mathbb{X}_v)$ is complicated. On the other hand, if we take a different maximal isotropic subspace $\mathbb{X}'_v$ of the symplectic F_v-space $\mathbb{V}_v$ as in §5.2 below, then the Weil representation on $\mathcal{S}(\mathbb{X}'_v)$ has a simple description (see §5.3). In particular, it is easy to find a Schwartz function $\varphi'_v \in \mathcal{S}(\mathbb{X}'_v)$ with the required properties (see §5.6). Since we have a partial Fourier transform (which is explicated in §5.4 below)

$$\mathcal{S}(\mathbb{X}'_v) \longrightarrow \mathcal{S}(\mathbb{X}_v),$$

which relates the two models of the Weil representation, we can define φ_v as the image of φ'_v. We also compute φ_v explicitly in §5.7 for the sequel [**31**] of this paper.

5.1. Notation

Let F be a number field. Let $\mathfrak{o}$ be the integer ring of F and $\mathfrak{d}$ the different of F over $\mathbb{Q}$. Let D be the discriminant of F. For each finite place v of F, let $\mathfrak{o}_v$ be the integer ring of F_v, $\mathfrak{p}_v = \varpi_v \mathfrak{o}_v$ the maximal ideal of $\mathfrak{o}_v$, ϖ_v a uniformizer of $\mathfrak{o}_v$, and q_v the cardinality of the residue field $\mathfrak{o}_v/\mathfrak{p}_v$. Let d_v be the non-negative integer such that $\mathfrak{d} \otimes_{\mathfrak{o}} \mathfrak{o}_v = \varpi_v^{d_v} \mathfrak{o}_v$. Then we have $|D| = \prod_{v \in \Sigma_{\mathrm{fin}}} q_v^{d_v}$.

Let $\psi_0 = \otimes_v \psi_{0,v}$ be the non-trivial character of $\mathbb{A}_\mathbb{Q}/\mathbb{Q}$ given by

- $\psi_{0,\infty}(x) = e^{2\pi\sqrt{-1}x}$ for $x \in \mathbb{R}$,
- $\psi_{0,p}(x) = e^{-2\pi\sqrt{-1}x}$ for $x \in \mathbb{Q}_p$.

Let $\psi = \otimes_v \psi_v$ be the non-trivial character of $\mathbb{A}/F$ defined by $\psi = \psi_0 \circ \mathrm{tr}_{F/\mathbb{Q}}$. We call ψ the standard additive character of $\mathbb{A}/F$. If v is a real place of F, then $\psi_v(x) = e^{2\pi\sqrt{-1}x}$ for $x \in F_v$. If v is a complex place of F, then $\psi_v(x) = e^{2\pi\sqrt{-1}(x+\bar{x})}$ for $x \in F_v$, where $\bar{x}$ is the complex conjugate of x. If v is a finite place of F, then ψ_v is trivial on $\varpi_v^{-d_v}\mathfrak{o}_v$ but non-trivial on $\varpi_v^{-d_v-1}\mathfrak{o}_v$. For each place v of F, we define a Fourier transform

$$\begin{aligned} \mathcal{S}(F_v) &\longrightarrow \mathcal{S}(F_v) \\ \phi &\longmapsto \hat{\phi} \end{aligned}$$

by

$$\hat{\phi}(x) = \int_{F_v} \phi(y)\psi_v(xy)\,dy,$$

where dy is the self-dual Haar measure on F_v with respect to ψ_v.

Let $V = B_1 \otimes_E B_2$ be the 2-dimensional right skew-hermitian B-space given in §2.2 and $W = B$ the 1-dimensional left hermitian B-space given in §3.3.1. Recall that

$$\begin{aligned} E &= F + F\mathbf{i}, & B &= E + E\mathbf{j}, & B_1 &= E + E\mathbf{j}_1, & B_2 &= E + E\mathbf{j}_2, \\ u &= \mathbf{i}^2, & J &= \mathbf{j}^2, & J_1 &= \mathbf{j}_1^2, & J_2 &= \mathbf{j}_2^2, \end{aligned}$$

where $J = J_1 J_2$. Let $\mathbb{V} = V \otimes_B W$ be the 8-dimensional symplectic F-space. We identify $\mathbb{V}$ with $\mathrm{Res}_{B/F}(V)$ via the map $v \mapsto v \otimes 1$. In §2.2, we chose a complete polarization $\mathbb{V} = \mathbb{X} \oplus \mathbb{Y}$ over F. Let $\mathbf{e}_1, \dots, \mathbf{e}_4$ and $\mathbf{e}_1^*, \dots, \mathbf{e}_4^*$ be the bases of $\mathbb{X}$ and $\mathbb{Y}$, respectively, given by (2.3), (2.15).

5.2. Complete polarizations

In Appendix C, we also choose a complete polarization $\mathbb{V}_v = \mathbb{X}'_v \oplus \mathbb{Y}'_v$ over F_v for each place v of F. Note that in picking the polarization, we use the assumption that for any place v of F, at least one of u, J, J_1, J_2 is a square in F_v. In this section, we recall the choice of this polarization. Later, we will pick a Schwartz function on $\mathbb{X}'_v$ and then transfer it to a Schwartz function on $\mathbb{X}_v$ by a partial Fourier transform. From now on, we fix a place v of F and suppress the subscript v from the notation.

5.2.1. The case $u \in (F^\times)^2$. Choose $t \in F^\times$ such that $u = t^2$. We define an isomorphism $\mathfrak{i} : B \to \mathrm{M}_2(F)$ of quaternion F-algebras by

$$(5.1)\quad \mathfrak{i}(1) = \begin{pmatrix} 1 & \\ & 1 \end{pmatrix}, \quad \mathfrak{i}(\mathbf{i}) = \begin{pmatrix} t & \\ & -t \end{pmatrix}, \quad \mathfrak{i}(\mathbf{j}) = \begin{pmatrix} & 1 \\ J & \end{pmatrix}, \quad \mathfrak{i}(\mathbf{ij}) = \begin{pmatrix} & t \\ -tJ & \end{pmatrix}.$$

Put

$$e = \frac{1}{2} + \frac{1}{2t}\mathbf{i}, \quad e' = \frac{1}{2}\mathbf{j} + \frac{1}{2t}\mathbf{ij}, \quad e'' = \frac{1}{2J}\mathbf{j} - \frac{1}{2tJ}\mathbf{ij}, \quad e^* = \frac{1}{2} - \frac{1}{2t}\mathbf{i},$$

so that

$$\mathfrak{i}(e) = \begin{pmatrix} 1 & 0 \\ 0 & 0 \end{pmatrix}, \quad \mathfrak{i}(e') = \begin{pmatrix} 0 & 1 \\ 0 & 0 \end{pmatrix}, \quad \mathfrak{i}(e'') = \begin{pmatrix} 0 & 0 \\ 1 & 0 \end{pmatrix}, \quad \mathfrak{i}(e^*) = \begin{pmatrix} 0 & 0 \\ 0 & 1 \end{pmatrix}.$$

Let $W^\dagger := eW$ be the 2-dimensional F-space associated to W equipped with a non-degenerate symplectic form $\langle \cdot, \cdot \rangle^\dagger$ defined by

$$(5.2)\qquad \langle x, y \rangle^* = \langle x, y \rangle^\dagger \cdot e'$$

for $x, y \in W^\dagger$. Then the restriction to $W^\dagger$ induces a natural isomorphism $\mathrm{GU}(W) \cong \mathrm{GSp}(W^\dagger)$. We have

$$\langle e, e \rangle^\dagger = \langle e', e' \rangle^\dagger = 0, \quad \langle e, e' \rangle^\dagger = 1,$$

and

$$\begin{bmatrix} e \cdot \boldsymbol{\alpha} \\ e' \cdot \boldsymbol{\alpha} \end{bmatrix} = \mathfrak{i}(\boldsymbol{\alpha}) \cdot \begin{bmatrix} e \\ e' \end{bmatrix}$$

for $\boldsymbol{\alpha} \in B$. We take a complete polarization $W^\dagger = X \oplus Y$ given by

$$X = Fe, \quad Y = Fe'.$$

Similarly, let $V^\dagger := Ve$ be the 4-dimensional F-space associated to V equipped with a non-degenerate symmetric bilinear form $\langle\cdot,\cdot\rangle^\dagger$ defined by

$$\frac{1}{2}\cdot\langle x,y\rangle = \langle x,y\rangle^\dagger\cdot e'' \tag{5.3}$$

for $x,y\in V^\dagger$. Then the restriction to $V^\dagger$ induces a natural isomorphism $\mathrm{GU}(V)\cong \mathrm{GO}(V^\dagger)$. We take a complete polarization $\mathbb{V}=\mathbb{X}'\oplus\mathbb{Y}'$ given by

$$\mathbb{X}' = V^\dagger\otimes X,\quad \mathbb{Y}' = V^\dagger\otimes Y.$$

We identify $\mathbb{X}'$ with $V^\dagger$ via the map $v\mapsto v\otimes e$. Put

$$\begin{aligned}
&\mathbf{v}_1 = 2\mathbf{e}_1 e = \mathbf{e}_1 + t\mathbf{e}_1^*, && \mathbf{v}_1^* = -\frac{1}{t}\mathbf{e}_1e^* = -\frac{1}{2t}\mathbf{e}_1 + \frac{1}{2}\mathbf{e}_1^*,\\
&\mathbf{v}_2 = 2\mathbf{e}_2 e = \mathbf{e}_2 - tJ_1\mathbf{e}_2^*, && \mathbf{v}_2^* = \frac{1}{tJ_1}\mathbf{e}_2e^* = \frac{1}{2tJ_1}\mathbf{e}_2 + \frac{1}{2}\mathbf{e}_2^*,\\
&\mathbf{v}_3 = -\frac{1}{tJ_1}\mathbf{e}_2e'' = -\frac{1}{2tJ}\mathbf{e}_3 + \frac{1}{2J_1}\mathbf{e}_3^*, && \mathbf{v}_3^* = -2\mathbf{e}_2e' = -J_1\mathbf{e}_3 - tJ\mathbf{e}_3^*,\\
&\mathbf{v}_4 = -\frac{1}{t}\mathbf{e}_1e'' = -\frac{1}{2tJ}\mathbf{e}_4 - \frac{1}{2}\mathbf{e}_4^*, && \mathbf{v}_4^* = 2\mathbf{e}_1e' = \mathbf{e}_4 - tJ\mathbf{e}_4^*.
\end{aligned} \tag{5.4}$$

Then $\mathbf{v}_1,\dots,\mathbf{v}_4$ and $\mathbf{v}_1^*,\dots,\mathbf{v}_4^*$ are bases of $\mathbb{X}'$ and $\mathbb{Y}'$, respectively, such that $\langle\langle \mathbf{v}_i,\mathbf{v}_j^*\rangle\rangle = \delta_{ij}$.

We may identify the quadratic space $V^\dagger$ with the space $\mathrm{M}_2(F)$ equipped with a non-degenerate symmetric bilinear form

$$\mathrm{tr}(xy^*) = x_1y_4 - x_2y_3 - x_3y_2 + x_4y_1 \tag{5.5}$$

for $x = \left(\begin{smallmatrix} x_1 & x_2\\ x_3 & x_4\end{smallmatrix}\right)$, $y = \left(\begin{smallmatrix} y_1 & y_2\\ y_3 & y_4\end{smallmatrix}\right)$. Indeed, the basis $\mathbf{v}_1,\dots,\mathbf{v}_4$ of $V^\dagger$ gives rise to an isomorphism $V^\dagger\cong \mathrm{M}_2(F)$ of quadratic spaces by

$$\mathbf{v}_1\mapsto\begin{pmatrix}1&0\\0&0\end{pmatrix},\quad \mathbf{v}_2\mapsto\begin{pmatrix}0&1\\0&0\end{pmatrix},\quad \mathbf{v}_3\mapsto\begin{pmatrix}0&0\\1&0\end{pmatrix},\quad \mathbf{v}_4\mapsto\begin{pmatrix}0&0\\0&1\end{pmatrix}.$$

Under this identification, we have

$$\boldsymbol{\alpha}_1\mathbf{v} = \mathbf{v}\cdot\mathfrak{i}_1(\boldsymbol{\alpha}_1)^*,\quad \boldsymbol{\alpha}_2\mathbf{v} = \mathfrak{i}_2(\boldsymbol{\alpha}_2)\cdot\mathbf{v}$$

for $\boldsymbol{\alpha}_i\in B_i$ and $\mathbf{v}\in V^\dagger\cong\mathrm{M}_2(F)$, where $\mathfrak{i}_1 : B_1\to\mathrm{M}_2(F)$ and $\mathfrak{i}_2 : B_2\to\mathrm{M}_2(F)$ are isomorphisms of quaternion F-algebras given by

$$\begin{aligned}
\mathfrak{i}_1(a+b\mathbf{i}+c\mathbf{j}_1+d\mathbf{i}\mathbf{j}_1) &= \begin{pmatrix} a-bt & -(c-dt)\\ -J_1(c+dt) & a+bt\end{pmatrix},\\
\mathfrak{i}_2(a+b\mathbf{i}+c\mathbf{j}_2+d\mathbf{i}\mathbf{j}_2) &= \begin{pmatrix} a+bt & -\frac{1}{2tJ_1}(c+dt)\\ -2tJ(c-dt) & a-bt\end{pmatrix}.
\end{aligned} \tag{5.6}$$

5.2.2. The case $J\in(F^\times)^2$. Choose $t\in F^\times$ such that $J=t^2$. We define an isomorphism $\mathfrak{i} : B\to\mathrm{M}_2(F)$ of quaternion F-algebras by

$$\mathfrak{i}(1) = \begin{pmatrix}1&\\&1\end{pmatrix},\quad \mathfrak{i}(\mathbf{i}) = \begin{pmatrix}&1\\u&\end{pmatrix},\quad \mathfrak{i}(\mathbf{j}) = \begin{pmatrix}t&\\&-t\end{pmatrix},\quad \mathfrak{i}(\mathbf{ij}) = \begin{pmatrix}&-t\\tu&\end{pmatrix}. \tag{5.7}$$

Put

$$e = \frac{1}{2}+\frac{1}{2t}\mathbf{j},\quad e' = \frac{1}{2}\mathbf{i}-\frac{1}{2t}\mathbf{ij},\quad e'' = \frac{1}{2u}\mathbf{i}+\frac{1}{2tu}\mathbf{ij},\quad e^* = \frac{1}{2}-\frac{1}{2t}\mathbf{j},$$

so that

$$\mathfrak{i}(e) = \begin{pmatrix} 1 & 0 \\ 0 & 0 \end{pmatrix}, \quad \mathfrak{i}(e') = \begin{pmatrix} 0 & 1 \\ 0 & 0 \end{pmatrix}, \quad \mathfrak{i}(e'') = \begin{pmatrix} 0 & 0 \\ 1 & 0 \end{pmatrix}, \quad \mathfrak{i}(e^*) = \begin{pmatrix} 0 & 0 \\ 0 & 1 \end{pmatrix}.$$

Let $W^\dagger := eW$ be the 2-dimensional F-space associated to W equipped with a non-degenerate symplectic form $\langle \cdot, \cdot \rangle^\dagger$ defined by (5.2). We have

$$\langle e, e \rangle^\dagger = \langle e', e' \rangle^\dagger = 0, \quad \langle e, e' \rangle^\dagger = 1,$$

and

$$\begin{bmatrix} e \cdot \boldsymbol{\alpha} \\ e' \cdot \boldsymbol{\alpha} \end{bmatrix} = \mathfrak{i}(\boldsymbol{\alpha}) \cdot \begin{bmatrix} e \\ e' \end{bmatrix}$$

for $\boldsymbol{\alpha} \in B$. We take a complete polarization $W^\dagger = X \oplus Y$ given by

$$X = Fe, \quad Y = Fe'.$$

Similarly, let $V^\dagger := Ve$ be the 4-dimensional F-space associated to V equipped with a non-degenerate symmetric bilinear form $\langle \cdot, \cdot \rangle^\dagger$ defined by (5.3). We take a complete polarization $\mathbb{V} = \mathbb{X}' \oplus \mathbb{Y}'$ given by

$$\mathbb{X}' = V^\dagger \otimes X, \quad \mathbb{Y}' = V^\dagger \otimes Y.$$

We identify $\mathbb{X}'$ with $V^\dagger$ via the map $v \mapsto v \otimes e$. Put

$$\begin{aligned}
\tilde{\mathbf{v}}_1 &= \mathbf{e}_1 e = \frac{1}{2}\mathbf{e}_1 + \frac{1}{2t}\mathbf{e}_4, & \tilde{\mathbf{v}}_1^* &= \frac{2}{u}\mathbf{e}_1 e' = \mathbf{e}_1^* + t\mathbf{e}_4^*, \\
\tilde{\mathbf{v}}_2 &= \mathbf{e}_1 e'' = \frac{1}{2}\mathbf{e}_1^* - \frac{t}{2}\mathbf{e}_4^*, & \tilde{\mathbf{v}}_2^* &= -2\mathbf{e}_1 e^* = -\mathbf{e}_1 + \frac{1}{t}\mathbf{e}_4, \\
\tilde{\mathbf{v}}_3 &= \mathbf{e}_2 e = \frac{1}{2}\mathbf{e}_2 + \frac{J_1}{2t}\mathbf{e}_3, & \tilde{\mathbf{v}}_3^* &= -\frac{2}{uJ_1}\mathbf{e}_2 e' = \mathbf{e}_2^* + \frac{t}{J_1}\mathbf{e}_3^*, \\
\tilde{\mathbf{v}}_4 &= \mathbf{e}_2 e'' = -\frac{J_1}{2}\mathbf{e}_2^* + \frac{t}{2}\mathbf{e}_3^*, & \tilde{\mathbf{v}}_4^* &= \frac{2}{J_1}\mathbf{e}_2 e^* = \frac{1}{J_1}\mathbf{e}_2 - \frac{1}{t}\mathbf{e}_3.
\end{aligned}$$

Then $\tilde{\mathbf{v}}_1, \dots, \tilde{\mathbf{v}}_4$ and $\tilde{\mathbf{v}}_1^*, \dots, \tilde{\mathbf{v}}_4^*$ are bases of $\mathbb{X}'$ and $\mathbb{Y}'$, respectively, such that $\langle\langle \tilde{\mathbf{v}}_i, \tilde{\mathbf{v}}_j^* \rangle\rangle = \delta_{ij}$.

We need to use two coordinate systems given as follows:

5.2.2.1. *The case (i).* We fix $s \in F^\times$ and define bases $\mathbf{v}_1, \dots, \mathbf{v}_4$ and $\mathbf{v}_1^*, \dots, \mathbf{v}_4^*$ of $\mathbb{X}'$ and $\mathbb{Y}'$, respectively, such that $\langle\langle \mathbf{v}_i, \mathbf{v}_j^* \rangle\rangle = \delta_{ij}$ by

$$\begin{aligned}
&\mathbf{v}_1 = \tilde{\mathbf{v}}_1, \quad \mathbf{v}_2 = \tilde{\mathbf{v}}_2, \quad \mathbf{v}_3 = \frac{1}{s}\tilde{\mathbf{v}}_3, \quad \mathbf{v}_4 = \frac{1}{s}\tilde{\mathbf{v}}_4, \\
&\mathbf{v}_1^* = \tilde{\mathbf{v}}_1^*, \quad \mathbf{v}_2^* = \tilde{\mathbf{v}}_2^*, \quad \mathbf{v}_3^* = s\tilde{\mathbf{v}}_3^*, \quad \mathbf{v}_4^* = s\tilde{\mathbf{v}}_4^*.
\end{aligned} \tag{5.8}$$

We may identify the quadratic space $V^\dagger$ with the space B_1 equipped with a non-degenerate symmetric bilinear form

$$-\frac{1}{4}\operatorname{tr}_{B_1/F}(xy^*).$$

Indeed, since

$$\langle \tilde{\mathbf{v}}_1, \tilde{\mathbf{v}}_1 \rangle^\dagger = \frac{u}{2}, \quad \langle \tilde{\mathbf{v}}_2, \tilde{\mathbf{v}}_2 \rangle^\dagger = -\frac{1}{2}, \quad \langle \tilde{\mathbf{v}}_3, \tilde{\mathbf{v}}_3 \rangle^\dagger = -\frac{uJ_1}{2}, \quad \langle \tilde{\mathbf{v}}_4, \tilde{\mathbf{v}}_4 \rangle^\dagger = \frac{J_1}{2},$$

and $\langle \tilde{\mathbf{v}}_i, \tilde{\mathbf{v}}_j \rangle^\dagger = 0$ if $i \neq j$, the basis $\tilde{\mathbf{v}}_1, \dots, \tilde{\mathbf{v}}_4$ of $V^\dagger$ gives rise to an isomorphism $V^\dagger \cong B_1$ of quadratic spaces by

$$\tilde{\mathbf{v}}_1 \mapsto \mathbf{i}, \quad \tilde{\mathbf{v}}_2 \mapsto 1, \quad \tilde{\mathbf{v}}_3 \mapsto \mathbf{j}_1\mathbf{i}, \quad \tilde{\mathbf{v}}_4 \mapsto \mathbf{j}_1.$$

Under this identification, we have

$$\boldsymbol{\alpha}_1 \mathbf{v} = \boldsymbol{\alpha}_1 \cdot \mathbf{v}, \quad \boldsymbol{\alpha}_2 \mathbf{v} = \mathbf{v} \cdot \mathfrak{i}_2(\boldsymbol{\alpha}_2)^*$$

for $\boldsymbol{\alpha}_i \in B_i$ and $\mathbf{v} \in V^\dagger \cong B_1$, where $\mathfrak{i}_2 : B_2 \to B_1$ is an isomorphism of quaternion F-algebras given by

$$\mathfrak{i}_2(\alpha + \beta \mathbf{j}_2) = \alpha^\rho + \frac{t\beta^\rho}{J_1}\mathbf{j}_1 \tag{5.9}$$

for $\alpha, \beta \in E$.

5.2.2.2. *The case (ii).* Assume that $J_1 \in (F^\times)^2$. We choose $t_1 \in F^\times$ such that $J_1 = t_1^2$ and define bases $\mathbf{v}_1, \dots, \mathbf{v}_4$ and $\mathbf{v}_1^*, \dots, \mathbf{v}_4^*$ of $\mathbb{X}'$ and $\mathbb{Y}'$, respectively, such that $\langle\!\langle \mathbf{v}_i, \mathbf{v}_j^* \rangle\!\rangle = \delta_{ij}$ by

$$\begin{aligned}
\mathbf{v}_1 &= \tilde{\mathbf{v}}_1 + \frac{1}{t_1}\tilde{\mathbf{v}}_3 = \frac{1}{2}\mathbf{e}_1 + \frac{1}{2t_1}\mathbf{e}_2 + \frac{t_1}{2t}\mathbf{e}_3 + \frac{1}{2t}\mathbf{e}_4, \\
\mathbf{v}_2 &= \tilde{\mathbf{v}}_2 + \frac{1}{t_1}\tilde{\mathbf{v}}_4 = \frac{1}{2}\mathbf{e}_1^* - \frac{t_1}{2}\mathbf{e}_2^* + \frac{t}{2t_1}\mathbf{e}_3^* - \frac{t}{2}\mathbf{e}_4^*, \\
\mathbf{v}_3 &= \tilde{\mathbf{v}}_2 - \frac{1}{t_1}\tilde{\mathbf{v}}_4 = \frac{1}{2}\mathbf{e}_1^* + \frac{t_1}{2}\mathbf{e}_2^* - \frac{t}{2t_1}\mathbf{e}_3^* - \frac{t}{2}\mathbf{e}_4^*, \\
\mathbf{v}_4 &= \frac{1}{u}\tilde{\mathbf{v}}_1 - \frac{1}{t_1 u}\tilde{\mathbf{v}}_3 = \frac{1}{2u}\mathbf{e}_1 - \frac{1}{2t_1 u}\mathbf{e}_2 - \frac{t_1}{2tu}\mathbf{e}_3 + \frac{1}{2tu}\mathbf{e}_4, \\
\mathbf{v}_1^* &= \frac{1}{2}\tilde{\mathbf{v}}_1^* + \frac{t_1}{2}\tilde{\mathbf{v}}_3^* = \frac{1}{2}\mathbf{e}_1^* + \frac{t_1}{2}\mathbf{e}_2^* + \frac{t}{2t_1}\mathbf{e}_3^* + \frac{t}{2}\mathbf{e}_4^*, \\
\mathbf{v}_2^* &= \frac{1}{2}\tilde{\mathbf{v}}_2^* + \frac{t_1}{2}\tilde{\mathbf{v}}_4^* = -\frac{1}{2}\mathbf{e}_1 + \frac{1}{2t_1}\mathbf{e}_2 - \frac{t_1}{2t}\mathbf{e}_3 + \frac{1}{2t}\mathbf{e}_4, \\
\mathbf{v}_3^* &= \frac{1}{2}\tilde{\mathbf{v}}_2^* - \frac{t_1}{2}\tilde{\mathbf{v}}_4^* = -\frac{1}{2}\mathbf{e}_1 - \frac{1}{2t_1}\mathbf{e}_2 + \frac{t_1}{2t}\mathbf{e}_3 + \frac{1}{2t}\mathbf{e}_4, \\
\mathbf{v}_4^* &= \frac{u}{2}\tilde{\mathbf{v}}_1^* - \frac{t_1 u}{2}\tilde{\mathbf{v}}_3^* = \frac{u}{2}\mathbf{e}_1^* - \frac{t_1 u}{2}\mathbf{e}_2^* - \frac{tu}{2t_1}\mathbf{e}_3^* + \frac{tu}{2}\mathbf{e}_4^*.
\end{aligned} \tag{5.10}$$

We may identify the quadratic space $V^\dagger$ with the space $\mathrm{M}_2(F)$ equipped with the non-degenerate symmetric bilinear form (5.5). Indeed, the basis $\mathbf{v}_1, \dots, \mathbf{v}_4$ of $V^\dagger$ gives rise to an isomorphism $V^\dagger \cong \mathrm{M}_2(F)$ of quadratic spaces by

$$\mathbf{v}_1 \mapsto \begin{pmatrix} 1 & 0 \\ 0 & 0 \end{pmatrix}, \quad \mathbf{v}_2 \mapsto \begin{pmatrix} 0 & 1 \\ 0 & 0 \end{pmatrix}, \quad \mathbf{v}_3 \mapsto \begin{pmatrix} 0 & 0 \\ 1 & 0 \end{pmatrix}, \quad \mathbf{v}_4 \mapsto \begin{pmatrix} 0 & 0 \\ 0 & 1 \end{pmatrix}.$$

Under this identification, we have

$$\boldsymbol{\alpha}_1 \mathbf{v} = \mathfrak{i}_1(\boldsymbol{\alpha}_1) \cdot \mathbf{v}, \quad \boldsymbol{\alpha}_2 \mathbf{v} = \mathbf{v} \cdot \mathfrak{i}_2(\boldsymbol{\alpha}_2)^*$$

for $\boldsymbol{\alpha}_i \in B_i$ and $\mathbf{v} \in V^\dagger \cong \mathrm{M}_2(F)$, where $\mathfrak{i}_1 : B_1 \to \mathrm{M}_2(F)$ and $\mathfrak{i}_2 : B_2 \to \mathrm{M}_2(F)$ are isomorphisms of quaternion F-algebras given by

$$\begin{aligned}
\mathfrak{i}_1(a + b\mathbf{i} + c\mathbf{j}_1 + d\mathbf{i}\mathbf{j}_1) &= \begin{pmatrix} a + ct_1 & b - dt_1 \\ u(b + dt_1) & a - ct_1 \end{pmatrix}, \\
\mathfrak{i}_2(a + b\mathbf{i} + c\mathbf{j}_2 + d\mathbf{i}\mathbf{j}_2) &= \begin{pmatrix} a - c\frac{t}{t_1} & -u(b + d\frac{t}{t_1}) \\ -(b - d\frac{t}{t_1}) & a + c\frac{t}{t_1} \end{pmatrix}.
\end{aligned} \tag{5.11}$$

5.2.3. The case $J_1 \in (F^\times)^2$ or $J_2 \in (F^\times)^2$. We only consider the case $J_1 \in (F^\times)^2$; we switch the roles of B_1 and B_2 in the other case. Choose $t \in F^\times$ such that $J_1 = t^2$. We define isomorphisms $\mathfrak{i}_1 : B_1 \to \mathrm{M}_2(F)$ and $\mathfrak{i}_2 : B_2 \to B$ of quaternion F-algebras by

(5.12)
$$\mathfrak{i}_1(1) = \begin{pmatrix} 1 & \\ & 1 \end{pmatrix}, \quad \mathfrak{i}_1(\mathbf{i}) = \begin{pmatrix} & 2 \\ \frac{u}{2} & \end{pmatrix}, \quad \mathfrak{i}_1(\mathbf{j}_1) = \begin{pmatrix} t & \\ & -t \end{pmatrix}, \quad \mathfrak{i}_1(\mathbf{i}\mathbf{j}_1) = \begin{pmatrix} & -2t \\ \frac{tu}{2} & \end{pmatrix},$$

and

$$\mathfrak{i}_2(\alpha + \beta \mathbf{j}_2) = \alpha + \frac{\beta}{t}\mathbf{j} \tag{5.13}$$

for $\alpha, \beta \in E$. Put

$$\mathbf{v} := \frac{1}{2}\mathbf{e}_1 + \frac{1}{2t}\mathbf{e}_2, \quad \mathbf{v}^* := \mathbf{e}_1^* + t\mathbf{e}_2^* = \frac{1}{u}\mathbf{e}_1\mathbf{i} - \frac{1}{tu}\mathbf{e}_2\mathbf{i}.$$

Then $\mathbf{v}, \mathbf{v}^*$ is a basis of V over B such that

$$\langle \mathbf{v}, \mathbf{v} \rangle = \langle \mathbf{v}^*, \mathbf{v}^* \rangle = 0, \quad \langle \mathbf{v}, \mathbf{v}^* \rangle = 1.$$

Moreover, we have

$$\begin{bmatrix} \boldsymbol{\alpha}_i \cdot \mathbf{v} & \boldsymbol{\alpha}_i \cdot \mathbf{v}^* \end{bmatrix} = \begin{bmatrix} \mathbf{v} & \mathbf{v}^* \end{bmatrix} \cdot \mathfrak{i}_i(\boldsymbol{\alpha}_i)$$

for $\boldsymbol{\alpha}_i \in B_i$. Here we identify $\mathfrak{i}_2(\boldsymbol{\alpha}_2)$ with the scalar matrix $\mathfrak{i}_2(\boldsymbol{\alpha}_2) \cdot \mathbf{1}_2$ in $\mathrm{M}_2(B)$. Let $V' := V$, regarded as a left B-space via $\boldsymbol{\alpha} \cdot x' := (x \cdot \boldsymbol{\alpha}^*)'$, where for an element $x \in V$, we write x' for the corresponding element in V'. We have a natural skew-hermitian form $\langle \cdot, \cdot \rangle'$ on V' defined by $\langle x', y' \rangle' = \langle x, y \rangle$. Let $\mathrm{GL}(V')$ act on V' on the right. We may identify $\mathrm{GU}(V)$ with $\mathrm{GU}(V')$ via the isomorphism

$$\begin{aligned} \mathrm{GL}(V) &\longrightarrow \mathrm{GL}(V'). \\ g &\longmapsto [x' \mapsto (g^{-1} \cdot x)'] \end{aligned}$$

Under this identification, we have

$$\begin{bmatrix} \mathbf{v}' \cdot \boldsymbol{\alpha}_i \\ (\mathbf{v}^*)' \cdot \boldsymbol{\alpha}_i \end{bmatrix} = {}^t(\mathfrak{i}_i(\boldsymbol{\alpha}_i)^{-1})^* \cdot \begin{bmatrix} \mathbf{v}' \\ (\mathbf{v}^*)' \end{bmatrix}$$

for $\boldsymbol{\alpha}_i \in B_i$. We take a complete polarization $V' = X' \oplus Y'$ given by

$$X' = B \cdot \mathbf{v}', \quad Y' = B \cdot (\mathbf{v}^*)'.$$

Similarly, let $W' := W$, regarded as a right B-space via $x' \cdot \boldsymbol{\alpha} := (\boldsymbol{\alpha}^* \cdot x)'$. We have a natural hermitian form $\langle \cdot, \cdot \rangle'$ on W' defined by $\langle x', y' \rangle' = \langle x, y \rangle$. Let $\mathrm{GL}(W')$ act on W' on the left. We may identify $\mathrm{GU}(W)$ with $\mathrm{GU}(W')$ via the isomorphism

$$\begin{aligned} \mathrm{GL}(W) &\longrightarrow \mathrm{GL}(W'). \\ g &\longmapsto [x' \mapsto (x \cdot g^{-1})'] \end{aligned}$$

We now consider an F-space $\mathbb{V}' := W' \otimes_B V'$ equipped with a non-degenerate symplectic form

$$\langle\!\langle \cdot, \cdot \rangle\!\rangle' := \frac{1}{2} \mathrm{tr}_{B/F}(\langle \cdot, \cdot \rangle' \otimes \langle \cdot, \cdot \rangle'^*).$$

Let $\mathrm{GL}(\mathbb{V}')$ act on $\mathbb{V}'$ on the right. We identify $\mathbb{V}$ with $\mathbb{V}'$ via the map $\mathbf{x} = x \otimes y \mapsto \mathbf{x}' = y' \otimes x'$. Then by Lemma C.3.1, we may identify $\mathrm{GSp}(\mathbb{V})$ with $\mathrm{GSp}(\mathbb{V}')$ via the isomorphism

$$\begin{aligned} \mathrm{GL}(\mathbb{V}) &\longrightarrow \mathrm{GL}(\mathbb{V}'), \\ \mathbf{g} &\longmapsto [\mathbf{x}' \mapsto (\mathbf{x}\cdot\mathbf{g})'] \end{aligned}$$

which induces a commutative diagram

$$\begin{array}{ccc} \mathrm{GU}(V)\times\mathrm{GU}(W) & \longrightarrow & \mathrm{GSp}(\mathbb{V}) \\ \downarrow & & \downarrow \\ \mathrm{GU}(W')\times\mathrm{GU}(V') & \longrightarrow & \mathrm{GSp}(\mathbb{V}') \end{array}.$$

We take a complete polarization

$$\mathbb{V}' = (W'\otimes_B X') \oplus (W'\otimes_B Y').$$

Under the identification $\mathbb{V} = \mathbb{V}'$, this gives a complete polarization $\mathbb{V} = \mathbb{X}' \oplus \mathbb{Y}'$, where

$$\mathbb{X}' = (\mathbf{v}\cdot B)\otimes_B W, \quad \mathbb{Y}' = (\mathbf{v}^*\cdot B)\otimes_B W.$$

We identify $\mathbb{X}'$ with W via the map $w \mapsto \mathbf{v}\otimes w$. We fix $s \in F^\times$ and put

$$\begin{aligned} \mathbf{v}_1 &= \mathbf{v} = \frac{1}{2}\mathbf{e}_1 + \frac{1}{2t}\mathbf{e}_2, & \mathbf{v}_1^* &= \mathbf{v}^* = \mathbf{e}_1^* + t\mathbf{e}_2^*, \\ \mathbf{v}_2 &= \frac{1}{u}\mathbf{v}\mathbf{i} = \frac{1}{2}\mathbf{e}_1^* - \frac{t}{2}\mathbf{e}_2^*, & \mathbf{v}_2^* &= -\mathbf{v}^*\mathbf{i} = -\mathbf{e}_1 + \frac{1}{t}\mathbf{e}_2, \\ \mathbf{v}_3 &= \frac{1}{s}\mathbf{v}\mathbf{j} = \frac{1}{2s}\mathbf{e}_4 + \frac{t}{2s}\mathbf{e}_3, & \mathbf{v}_3^* &= -\frac{s}{J}\mathbf{v}^*\mathbf{j} = s\mathbf{e}_4^* + \frac{s}{t}\mathbf{e}_3^*, \\ \mathbf{v}_4 &= \frac{1}{su}\mathbf{v}\mathbf{i}\mathbf{j} = -\frac{J}{2s}\mathbf{e}_4^* + \frac{J}{2st}\mathbf{e}_3^*, & \mathbf{v}_4^* &= \frac{s}{J}\mathbf{v}^*\mathbf{i}\mathbf{j} = \frac{s}{J}\mathbf{e}_4 - \frac{st}{J}\mathbf{e}_3. \end{aligned} \tag{5.14}$$

Then $\mathbf{v}_1,\dots,\mathbf{v}_4$ and $\mathbf{v}_1^*,\dots,\mathbf{v}_4^*$ are bases of $\mathbb{X}'$ and $\mathbb{Y}'$, respectively, such that $\langle\langle \mathbf{v}_i, \mathbf{v}_j^*\rangle\rangle = \delta_{ij}$.

5.3. Weil representations

Recall that we have the Weil representation ω_ψ of $\mathrm{G}(\mathrm{U}(V)^0\times\mathrm{U}(W))$ on $\mathcal{S}(\mathbb{X})$ obtained from the map $s : \mathrm{GU}(V)^0\times\mathrm{GU}(W)\to\mathbb{C}^1$ such that $z_{\mathbb{Y}} = \partial s$ given in Appendix C. This Weil representation is unitary with respect to the hermitian inner product $\langle\cdot,\cdot\rangle$ on $\mathcal{S}(\mathbb{X})$ given by

$$\langle\varphi_1,\varphi_2\rangle = \int_{\mathbb{X}} \varphi_1(x)\overline{\varphi_2(x)}\,dx,$$

where $dx = dx_1\cdots dx_4$ for $x = x_1\mathbf{e}_1+\dots+x_4\mathbf{e}_4$ with the self-dual Haar measure dx_i on F with respect to ψ. The map s is defined in terms of another map s' : $\mathrm{GU}(V)^0\times\mathrm{GU}(W)\to\mathbb{C}^1$ such that $z_{\mathbb{Y}'} = \partial s'$ given in Appendix C, based on [**40**]. Thus we obtain the Weil representation ω_ψ of $\mathrm{G}(\mathrm{U}(V)^0\times\mathrm{U}(W))$ on $\mathcal{S}(\mathbb{X}')$ from s', as in [**40**, §5], [**27**, §5]. This Weil representation is unitary with respect to the hermitian inner product $\langle\cdot,\cdot\rangle$ on $\mathcal{S}(\mathbb{X}')$ given in terms of certain Haar measure on $\mathbb{X}'$. In this section, we define this Haar measure on $\mathbb{X}'$ and give explicit formulas for the Weil representation on $\mathcal{S}(\mathbb{X}')$.

5.3.1. The case $u \in (F^\times)^2$. Recall that we identified $\mathbb{X}'$ with $V^\dagger$. We take the self-dual Haar measure on $V^\dagger$ with respect to the pairing $(x, y) \mapsto \psi(\langle x, y\rangle^\dagger)$. More explicitly, this measure is given by

$$dx = dx_1 \cdots dx_4$$

for $x = x_1\mathbf{v}_1 + \cdots + x_4\mathbf{v}_4 \in \mathbb{X}'$, where $\mathbf{v}_1, \ldots, \mathbf{v}_4$ is the basis of $\mathbb{X}'$ given by (5.4) and dx_i is the self-dual Haar measure on F with respect to ψ.

We identity $\mathrm{GU}(W) \cong B^\times$ with $\mathrm{GL}_2(F)$ via the isomorphism $\mathfrak{i}$ given by (5.1). Then $\mathrm{U}(W) \cong \mathrm{SL}_2(F)$ acts on $\mathcal{S}(\mathbb{X}')$ by

$$\omega_\psi\begin{pmatrix} a & \\ & a^{-1}\end{pmatrix}\varphi(x) = |a|^2\varphi(ax), \qquad a \in F^\times,$$

$$\omega_\psi\begin{pmatrix} 1 & b \\ & 1\end{pmatrix}\varphi(x) = \psi\left(\frac{1}{2}b\langle x, x\rangle^\dagger\right)\varphi(x), \qquad b \in F,$$

$$\omega_\psi\begin{pmatrix} & -1 \\ 1 & \end{pmatrix}\varphi(x) = \int_{\mathbb{X}'}\varphi(y)\psi(-\langle x, y\rangle^\dagger)\,dy.$$

This action extends to an action of $\mathrm{G}(\mathrm{U}(V)^0 \times \mathrm{U}(W))$ by

$$\omega_\psi(g, h) = \omega_\psi(g \cdot d(\nu)^{-1}) \circ L(h) = L(h) \circ \omega_\psi(d(\nu)^{-1} \cdot g)$$

for $g \in \mathrm{GU}(W) \cong \mathrm{GL}_2(F)$ and $h \in \mathrm{GU}(V)^0 \cong \mathrm{GO}(V^\dagger)^0$ such that $\nu(g) = \nu(h) =: \nu$, where $d(\nu) = \left(\begin{smallmatrix} 1 & \\ & \nu\end{smallmatrix}\right)$ and

$$L(h)\varphi(x) = |\nu|^{-1}\varphi(h^{-1}x).$$

5.3.2. The case $J \in (F^\times)^2$. Recall that we identified $\mathbb{X}'$ with $V^\dagger$. We take the self-dual Haar measure on $V^\dagger$ with respect to the pairing $(x, y) \mapsto \psi(\langle x, y\rangle^\dagger)$. More explicitly, according the coordinate system, this measure is given as follows:

(i)
$$dx = \left|\frac{uJ_1}{4s^2}\right| dx_1 \cdots dx_4$$
for $x = x_1\mathbf{v}_1 + \cdots + x_4\mathbf{v}_4 \in \mathbb{X}'$, where $\mathbf{v}_1, \ldots, \mathbf{v}_4$ is the basis of $\mathbb{X}'$ given by (5.8) and dx_i is the self-dual Haar measure on F with respect to ψ.

(ii)
$$dx = dx_1 \cdots dx_4$$
for $x = x_1\mathbf{v}_1 + \cdots + x_4\mathbf{v}_4 \in \mathbb{X}'$, where $\mathbf{v}_1, \ldots, \mathbf{v}_4$ is the basis of $\mathbb{X}'$ given by (5.10) and dx_i is the self-dual Haar measure on F with respect to ψ.

We identity $\mathrm{GU}(W) \cong B^\times$ with $\mathrm{GL}_2(F)$ via the isomorphism $\mathfrak{i}$ given by (5.7). Then $\mathrm{U}(W) \cong \mathrm{SL}_2(F)$ acts on $\mathcal{S}(\mathbb{X}')$ by

$$\omega_\psi\begin{pmatrix} a & \\ & a^{-1}\end{pmatrix}\varphi(x) = |a|^2\varphi(ax), \qquad a \in F^\times,$$

$$\omega_\psi\begin{pmatrix} 1 & b \\ & 1\end{pmatrix}\varphi(x) = \psi\left(\frac{1}{2}b\langle x, x\rangle^\dagger\right)\varphi(x), \qquad b \in F,$$

$$\omega_\psi\begin{pmatrix} & -1 \\ 1 & \end{pmatrix}\varphi(x) = \gamma_{B_1}\int_{\mathbb{X}'}\varphi(y)\psi(-\langle x, y\rangle^\dagger)\,dy,$$

where

$$\gamma_{B_1} = \begin{cases} 1 & \text{if } B_1 \text{ is split}, \\ -1 & \text{if } B_1 \text{ is ramified}. \end{cases}$$

This action extends to an action of $\mathrm{G}(\mathrm{U}(V)^0 \times \mathrm{U}(W))$ by

$$\omega_\psi(g,h) = \omega_\psi(g \cdot d(\nu)^{-1}) \circ L(h) = L(h) \circ \omega_\psi(d(\nu)^{-1} \cdot g)$$

for $g \in \mathrm{GU}(W) \cong \mathrm{GL}_2(F)$ and $h \in \mathrm{GU}(V)^0 \cong \mathrm{GO}(V^\dagger)^0$ such that $\nu(g) = \nu(h) =: \nu$, where $d(\nu) = \left(\begin{smallmatrix} 1 & \\ & \nu \end{smallmatrix}\right)$ and

$$L(h)\varphi(x) = |\nu|^{-1}\varphi(h^{-1}x).$$

5.3.3. The case $J_1 \in (F^\times)^2$ or $J_2 \in (F^\times)^2$. We only consider the case $J_1 \in (F^\times)^2$; we switch the roles of B_1 and B_2 in the other case. Recall that we identified $\mathbb{X}'$ with W. We take the self-dual Haar measure on W with respect to the pairing $(x,y) \mapsto \psi(\frac{1}{2}\operatorname{tr}_{B/F}\langle x,y\rangle)$. More explicitly, this measure is given by

$$dx = \left|\frac{J}{s^2u}\right| dx_1 \cdots dx_4$$

for $x = x_1\mathbf{v}_1 + \cdots + x_4\mathbf{v}_4 \in \mathbb{X}'$, where $\mathbf{v}_1, \ldots, \mathbf{v}_4$ is the basis of $\mathbb{X}'$ given by (5.14) and dx_i is the self-dual Haar measure on F with respect to ψ.

We identity $\mathrm{GU}(V)^0 \cong (B_1^\times \times B_2^\times)/F^\times$ with the group

$$\left\{ g \in \mathrm{GL}_2(B) \;\middle|\; {}^t g^* \begin{pmatrix} & 1 \\ -1 & \end{pmatrix} g = \nu(g) \begin{pmatrix} & 1 \\ -1 & \end{pmatrix} \right\}$$

via the map $(\boldsymbol{\alpha}_1, \boldsymbol{\alpha}_2) \mapsto \mathfrak{i}_1(\boldsymbol{\alpha}_1)\mathfrak{i}_2(\boldsymbol{\alpha}_2)$, where $\mathfrak{i}_1$ and $\mathfrak{i}_2$ are the isomorphisms given by (5.12), (5.13). Then $\mathrm{U}(V)^0$ acts on $\mathcal{S}(\mathbb{X}')$ via the identification $\mathrm{U}(V)^0 \cong \mathrm{U}(V')^0$ followed by the Weil representation of $\mathrm{U}(V')^0$ on $\mathcal{S}(W' \otimes_B X')$ given in [**40**, §5]. Hence $\mathrm{U}(V)^0$ acts on $\mathcal{S}(\mathbb{X}')$ by

$$\begin{aligned}
\omega_\psi \begin{pmatrix} a & \\ & (a^{-1})^* \end{pmatrix} \varphi(x) &= |\nu(a)|^{-1}\varphi(a^{-1}x), && a \in B^\times, \\
\omega_\psi \begin{pmatrix} 1 & \\ b & 1 \end{pmatrix} \varphi(x) &= \psi\left(-\frac{1}{2}b\langle x,x\rangle\right)\varphi(x), && b \in F, \\
\omega_\psi \begin{pmatrix} & -1 \\ 1 & \end{pmatrix} \varphi(x) &= \gamma_B \int_{\mathbb{X}'} \varphi(y)\psi\left(-\frac{1}{2}\operatorname{tr}_{B/F}\langle x,y\rangle\right) dy,
\end{aligned}$$

where

$$\gamma_B = \begin{cases} 1 & \text{if } B \text{ is split}, \\ -1 & \text{if } B \text{ is ramified}. \end{cases}$$

This action extends to an action of $\mathrm{G}(\mathrm{U}(V)^0 \times \mathrm{U}(W))$ by

$$\omega_\psi(g,h) = \omega_\psi(h \cdot d(\nu)^{-1}) \circ R(g) = R(g) \circ \omega_\psi(d(\nu)^{-1} \cdot h)$$

for $g \in \mathrm{GU}(W)$ and $h \in \mathrm{GU}(V)^0$ such that $\nu(g) = \nu(h) =: \nu$, where $d(\nu) = \left(\begin{smallmatrix} 1 & \\ & \nu \end{smallmatrix}\right)$ and

$$R(g)\varphi(x) = |\nu|\varphi(xg).$$

5.4. Partial Fourier transforms

Recall that the partial Fourier transform $\varphi \in \mathcal{S}(\mathbb{X})$ of $\varphi' \in \mathcal{S}(\mathbb{X}')$ is given by

$$\varphi(x) = \int_{\mathbb{Y}/\mathbb{Y}\cap\mathbb{Y}'} \varphi'(x')\psi\left(\frac{1}{2}\left(\langle\!\langle x',y'\rangle\!\rangle - \langle\!\langle x,y\rangle\!\rangle\right)\right) d\mu_{\mathbb{Y}/\mathbb{Y}\cap\mathbb{Y}'}(y),$$

where for $x \in \mathbb{X}$ and $y \in \mathbb{Y}$, we write $x + y = x' + y'$ with $x' = x'(x, y) \in \mathbb{X}'$ and $y' = y'(x, y) \in \mathbb{Y}'$, and we take the Haar measure $\mu_{\mathbb{Y}/\mathbb{Y}\cap\mathbb{Y}'}$ on $\mathbb{Y}/\mathbb{Y}\cap\mathbb{Y}'$ so that the map

$$\begin{aligned} \mathcal{S}(\mathbb{X}') &\longrightarrow \mathcal{S}(\mathbb{X}) \\ \varphi' &\longmapsto \varphi \end{aligned}$$

respects the hermitian inner products (given in terms of the Haar measures on $\mathbb{X}$ and $\mathbb{X}'$ given in §5.3). By construction, this partial Fourier transform is a unitary equivalence between the Weil representations of $\mathrm{G}(\mathrm{U}(V)^0 \times \mathrm{U}(W))$ on $\mathcal{S}(\mathbb{X})$ and $\mathcal{S}(\mathbb{X}')$. In this section, we explicate the Haar measure $\mu_{\mathbb{Y}/\mathbb{Y}\cap\mathbb{Y}'}$ and the partial Fourier transform $\mathcal{S}(\mathbb{X}') \to \mathcal{S}(\mathbb{X})$.

We write

$$\begin{aligned} x &= x_1\mathbf{e}_1 + \cdots + x_4\mathbf{e}_4 \in \mathbb{X}, & y &= y_1\mathbf{e}_1^* + \cdots + y_4\mathbf{e}_4^* \in \mathbb{Y}, \\ x' &= x_1'\mathbf{v}_1 + \cdots + x_4'\mathbf{v}_4 \in \mathbb{X}', & y' &= y_1'\mathbf{v}_1^* + \cdots + y_4'\mathbf{v}_4^* \in \mathbb{Y}', \end{aligned}$$

where $\mathbf{v}_1, \ldots, \mathbf{v}_4$ and $\mathbf{v}_1^*, \ldots, \mathbf{v}_4^*$ are the bases of $\mathbb{X}'$ and $\mathbb{Y}'$, respectively, given in §5.2. Let dx_i, dy_j, dx_i', dy_j' be the self-dual Haar measures on F with respect to ψ.

5.4.1. The case $u \in (F^\times)^2$. Recall that $\mathbf{v}_i, \mathbf{v}_j^*$ are given by (5.4). Note that $\mathbb{Y} \cap \mathbb{Y}' = \{0\}$. We define a Haar measure $\mu_{\mathbb{Y}/\mathbb{Y}\cap\mathbb{Y}'}$ on $\mathbb{Y}$ by

$$d\mu_{\mathbb{Y}/\mathbb{Y}\cap\mathbb{Y}'}(y) = |4u|^{-\frac{1}{2}}\, dy_1 \cdots dy_4$$

for $y = y_1\mathbf{e}_1^* + \cdots + y_4\mathbf{e}_4^*$. We will see below that the partial Fourier transform with respect to this Haar measure is an isometry.

If $x + y = x' + y'$, then we have

$$\begin{aligned} x_1' &= \frac{1}{2t}(y_1 + tx_1), & y_1' &= y_1 - tx_1, \\ x_2' &= -\frac{1}{2tJ_1}(y_2 - tJ_1x_2), & y_2' &= y_2 + tJ_1x_2, \\ x_3' &= J_1(y_3 - tJ_2x_3), & y_3' &= -\frac{1}{2tJ}(y_3 + tJ_2x_3), \\ x_4' &= -(y_4 + tJx_4), & y_4' &= -\frac{1}{2tJ}(y_4 - tJx_4). \end{aligned}$$

Namely, putting

$$a_1 = t, \quad a_2 = -tJ_1, \quad a_3 = -tJ_2, \quad a_4 = tJ, \quad b_1 = b_2 = 1, \quad b_3 = b_4 = -2tJ,$$

we have

$$x_i' = \frac{b_i}{2a_i}(y_i + a_ix_i), \quad y_i' = \frac{1}{b_i}(y_i - a_ix_i),$$

so that

$$\begin{aligned} x_i'y_i' - x_iy_i &= x_i'\left(\frac{2a_i}{b_i^2}x_i' - \frac{2a_i}{b_i}x_i\right) - x_i\left(\frac{2a_i}{b_i}x_i' - a_ix_i\right) \\ &= \frac{2a_i}{b_i^2}(x_i')^2 - \frac{4a_i}{b_i}x_ix_i' + a_ix_i^2. \end{aligned}$$

Hence, if $\varphi'(x') = \prod_{i=1}^4 \varphi_i'(x_i')$ with $\varphi_i' \in \mathcal{S}(F)$, then we have

$$\varphi(x) = |4u|^{-\frac{1}{2}} \prod_{i=1}^4 \varphi_i(x_i),$$

where

$$\begin{aligned}\varphi_i(x_i) &= \int_F \varphi_i'(x_i')\psi\left(\frac{1}{2}(x_i'y_i' - x_iy_i)\right) dy_i \\ &= \left|\frac{2a_i}{b_i}\right| \psi\left(\frac{a_i}{2}x_i^2\right) \int_F \varphi_i'(x_i')\psi\left(\frac{a_i}{b_i^2}(x_i')^2 - \frac{2a_i}{b_i}x_ix_i'\right) dx_i'.\end{aligned}$$

Since

$$\prod_{i=1}^{4} \frac{2a_i}{b_i} = 4u,$$

the partial Fourier transform with respect to $\mu_{\mathbb{Y}/\mathbb{Y}\cap\mathbb{Y}'}$ is an isometry.

5.4.2. The case $J \in (F^\times)^2$.

5.4.2.1. *The case (i).* Recall that $\mathbf{v}_i, \mathbf{v}_j^*$ are given by (5.8). Note that $\mathbb{Y}\cap\mathbb{Y}' = F\mathbf{v}_1^* + F\mathbf{v}_3^*$. Let $\mu_{\mathbb{Y}}$ and $\mu_{\mathbb{Y}\cap\mathbb{Y}'}$ be the Haar measures on $\mathbb{Y}$ and $\mathbb{Y}\cap\mathbb{Y}'$, respectively, defined by

$$d\mu_{\mathbb{Y}}(y) = dy_1 \cdots dy_4, \quad d\mu_{\mathbb{Y}\cap\mathbb{Y}'}(y') = dy_1'\, dy_3'$$

for $y = y_1\mathbf{e}_1^* + \cdots + y_4\mathbf{e}_4^*$ and $y' = y_1'\mathbf{v}_1^* + y_3'\mathbf{v}_3^*$. We define a Haar measure $\mu_{\mathbb{Y}/\mathbb{Y}\cap\mathbb{Y}'}$ on $\mathbb{Y}/\mathbb{Y}\cap\mathbb{Y}'$ by

$$\mu_{\mathbb{Y}/\mathbb{Y}\cap\mathbb{Y}'} = \left|\frac{uJ_1}{4s^2J}\right|^{\frac{1}{2}} \frac{\mu_{\mathbb{Y}}}{\mu_{\mathbb{Y}\cap\mathbb{Y}'}}.$$

We will see below that the partial Fourier transform with respect to this Haar measure is an isometry.

If $x + y = x' + y'$, then we have

$$\begin{aligned}
x_1' &= x_1 + tx_4, & y_1' &= \frac{1}{2}\left(y_1 + \frac{1}{t}y_4\right), \\
x_2' &= y_1 - \frac{1}{t}y_4, & y_2' &= -\frac{1}{2}(x_1 - tx_4), \\
x_3' &= s\left(x_2 + \frac{t}{J_1}x_3\right), & y_3' &= \frac{1}{2s}\left(y_2 + \frac{J_1}{t}y_3\right), \\
x_4' &= -\frac{s}{J_1}\left(y_2 - \frac{J_1}{t}y_3\right), & y_4' &= \frac{J_1}{2s}\left(x_2 - \frac{t}{J_1}x_3\right),
\end{aligned}$$

so that

$$x_1'y_1' - x_2'y_2' = x_1y_1 + x_4y_4, \quad x_3'y_3' - x_4'y_4' = x_2y_2 + x_3y_3.$$

Also, we have

$$\begin{aligned} dx_1'\, dx_3'\, dy_2'\, dy_4' &= |J|\, dx_1 \cdots dx_4, \\ dx_2'\, dx_4'\, dy_1'\, dy_3' &= |J|^{-1}\, dy_1 \cdots dy_4. \end{aligned}$$

Hence, if $\varphi'(x') = \prod_{i=1}^4 \varphi_i'(x_i')$ with $\varphi_i' \in \mathcal{S}(F)$, then we have

$$\begin{aligned}\varphi(x) &= \left|\frac{uJJ_1}{4s^2}\right|^{\frac{1}{2}} \varphi_1'(x_1')\varphi_3'(x_3') \int_F\int_F \varphi_2'(x_2')\varphi_4'(x_4')\psi(x_2'y_2' + x_4'y_4')\, dx_2'\, dx_4' \\ &= \left|\frac{uJJ_1}{4s^2}\right|^{\frac{1}{2}} \varphi_1'(x_1')\hat{\varphi}_2'(y_2')\varphi_3'(x_3')\hat{\varphi}_4'(y_4').\end{aligned}$$

In particular, the partial Fourier transform with respect to $\mu_{\mathbb{Y}/\mathbb{Y}\cap\mathbb{Y}'}$ is an isometry.

5.4.2.2. *The case (ii).* Recall that $\mathbf{v}_i, \mathbf{v}_j^*$ are given by (5.10). Note that $\mathbb{Y}\cap\mathbb{Y}' = F\mathbf{v}_1^* + F\mathbf{v}_4^*$. Let $\mu_{\mathbb{Y}}$ and $\mu_{\mathbb{Y}\cap\mathbb{Y}'}$ be the Haar measures on $\mathbb{Y}$ and $\mathbb{Y}\cap\mathbb{Y}'$, respectively, defined by

$$d\mu_{\mathbb{Y}}(y) = dy_1 \cdots dy_4, \quad d\mu_{\mathbb{Y}\cap\mathbb{Y}'}(y') = dy_1'\, dy_4'$$

for $y = y_1\mathbf{e}_1^* + \cdots + y_4\mathbf{e}_4^*$ and $y' = y_1'\mathbf{v}_1^* + y_4'\mathbf{v}_4^*$. We define a Haar measure $\mu_{\mathbb{Y}/\mathbb{Y}\cap\mathbb{Y}'}$ on $\mathbb{Y}/\mathbb{Y}\cap\mathbb{Y}'$ by

$$\mu_{\mathbb{Y}/\mathbb{Y}\cap\mathbb{Y}'} = |uJ|^{-\frac{1}{2}} \frac{\mu_{\mathbb{Y}}}{\mu_{\mathbb{Y}\cap\mathbb{Y}'}}.$$

We will see below that the partial Fourier transform with respect to this Haar measure is an isometry.

If $x + y = x' + y'$, then we have

$$\begin{aligned}
x_1' &= \frac{1}{2}\left(x_1 + t_1x_2 + \frac{t}{t_1}x_3 + tx_4\right), & y_1' &= \frac{1}{2}\left(y_1 + \frac{1}{t_1}y_2 + \frac{t_1}{t}y_3 + \frac{1}{t}y_4\right),\\
x_2' &= \frac{1}{2}\left(y_1 - \frac{1}{t_1}y_2 + \frac{t_1}{t}y_3 - \frac{1}{t}y_4\right), & y_2' &= -\frac{1}{2}\left(x_1 - t_1x_2 + \frac{t}{t_1}x_3 - tx_4\right),\\
x_3' &= \frac{1}{2}\left(y_1 + \frac{1}{t_1}y_2 - \frac{t_1}{t}y_3 - \frac{1}{t}y_4\right), & y_3' &= -\frac{1}{2}\left(x_1 + t_1x_2 - \frac{t}{t_1}x_3 - tx_4\right),\\
x_4' &= \frac{u}{2}\left(x_1 - t_1x_2 - \frac{t}{t_1}x_3 + tx_4\right), & y_4' &= \frac{1}{2u}\left(y_1 - \frac{1}{t_1}y_2 - \frac{t_1}{t}y_3 + \frac{1}{t}y_4\right),
\end{aligned}$$

so that

$$x_1'y_1' - x_2'y_2' - x_3'y_3' + x_4'y_4' = x_1y_1 + x_2y_2 + x_3y_3 + x_4y_4.$$

Also, we have

$$\begin{aligned}
dx_1'\, dx_4'\, dy_2'\, dy_3' &= |uJ|\, dx_1 \cdots dx_4,\\
dx_2'\, dx_3'\, dy_1'\, dy_4' &= |uJ|^{-1}\, dy_1 \cdots dy_4.
\end{aligned}$$

Hence, if $\varphi'(x') = \prod_{i=1}^4 \varphi_i'(x_i')$ with $\varphi_i' \in \mathcal{S}(F)$, then we have

$$\begin{aligned}
\varphi(x) &= |uJ|^{\frac{1}{2}} \varphi_1'(x_1')\varphi_4'(x_4') \int_F \int_F \varphi_2'(x_2')\varphi_3'(x_3')\psi(x_2'y_2' + x_3'y_3')\, dx_2'\, dx_3'\\
&= |uJ|^{\frac{1}{2}} \varphi_1'(x_1')\hat{\varphi}_2'(y_2')\hat{\varphi}_3'(y_3')\varphi_4'(x_4').
\end{aligned}$$

In particular, the partial Fourier transform with respect to $\mu_{\mathbb{Y}/\mathbb{Y}\cap\mathbb{Y}'}$ is an isometry.

5.4.3. The case $J_1 \in (F^\times)^2$ or $J_2 \in (F^\times)^2$. We only consider the case $J_1 \in (F^\times)^2$; we switch the roles of B_1 and B_2 in the other case. Recall that $\mathbf{v}_i, \mathbf{v}_j^*$ are given by (5.14). Note that $\mathbb{Y}\cap\mathbb{Y}' = F\mathbf{v}_1^* + F\mathbf{v}_3^*$. Let $\mu_{\mathbb{Y}}$ and $\mu_{\mathbb{Y}\cap\mathbb{Y}'}$ be the Haar measures on $\mathbb{Y}$ and $\mathbb{Y}\cap\mathbb{Y}'$, respectively, defined by

$$d\mu_{\mathbb{Y}}(y) = dy_1 \cdots dy_4, \quad d\mu_{\mathbb{Y}\cap\mathbb{Y}'}(y') = dy_1'\, dy_3'$$

for $y = y_1\mathbf{e}_1^* + \cdots + y_4\mathbf{e}_4^*$ and $y' = y_1'\mathbf{v}_1^* + y_3'\mathbf{v}_3^*$. We define a Haar measure $\mu_{\mathbb{Y}/\mathbb{Y}\cap\mathbb{Y}'}$ on $\mathbb{Y}/\mathbb{Y}\cap\mathbb{Y}'$ by

$$\mu_{\mathbb{Y}/\mathbb{Y}\cap\mathbb{Y}'} = |s^2u|^{-\frac{1}{2}} \frac{\mu_{\mathbb{Y}}}{\mu_{\mathbb{Y}\cap\mathbb{Y}'}}.$$

We will see below that the partial Fourier transform with respect to this Haar measure is an isometry.

If $x+y = x'+y'$, then we have

$$x_1' = x_1 + tx_2, \qquad y_1' = \frac{1}{2}\left(y_1 + \frac{1}{t}y_2\right),$$
$$x_2' = y_1 - \frac{1}{t}y_2, \qquad y_2' = -\frac{1}{2}(x_1 - tx_2),$$
$$x_3' = s\left(x_4 + \frac{1}{t}x_3\right), \qquad y_3' = \frac{1}{2s}(y_4 + ty_3),$$
$$x_4' = -\frac{s}{J}(y_4 - ty_3), \qquad y_4' = \frac{J}{2s}\left(x_4 - \frac{1}{t}x_3\right),$$

so that

$$x_1'y_1' - x_2'y_2' = x_1y_1 + x_2y_2, \quad x_3'y_3' - x_4'y_4' = x_3y_3 + x_4y_4.$$

Also, we have

$$dx_1'\,dx_3'\,dy_2'\,dy_4' = |J|\,dx_1\cdots dx_4,$$
$$dx_2'\,dx_4'\,dy_1'\,dy_3' = |J|^{-1}\,dy_1\cdots dy_4.$$

Hence, if $\varphi'(x') = \prod_{i=1}^4 \varphi_i'(x_i')$ with $\varphi_i' \in \mathcal{S}(F)$, then we have

$$\begin{aligned}\varphi(x) &= \left|\frac{J^2}{s^2u}\right|^{\frac{1}{2}} \varphi_1'(x_1')\varphi_3'(x_3')\int_F\int_F \varphi_2'(x_2')\varphi_4'(x_4')\psi(x_2'y_2' + x_4'y_4')\,dx_2'\,dx_4' \\ &= \left|\frac{J^2}{s^2u}\right|^{\frac{1}{2}} \varphi_1'(x_1')\hat{\varphi}_2'(y_2')\varphi_3'(x_3')\hat{\varphi}_4'(y_4').\end{aligned}$$

In particular, the partial Fourier transform with respect to $\mu_{\mathbb{Y}/\mathbb{Y}\cap\mathbb{Y}'}$ is an isometry.

5.5. Automorphic representations

Suppose that F is a totally real number field. Let $\pi_B \cong \otimes_v \pi_{B,v}$ be an irreducible unitary cuspidal automorphic representation of $B^\times(\mathbb{A})$ satisfying the following conditions:

- For $v \in \Sigma_{\mathrm{fin}} \smallsetminus \Sigma_{B,\mathrm{fin}}$,

(ur) $\pi_{B,v} = \mathrm{Ind}(\chi_v \otimes \mu_v)$ is a principal series representation, where χ_v and μ_v are unitary unramified; or

(rps) $\pi_{B,v} = \mathrm{Ind}(\chi_v \otimes \mu_v)$ is a principal series representation, where χ_v is unitary unramified and μ_v is unitary ramified; or

(st) $\pi_{B,v} = \mathrm{St} \otimes \chi_v$ is a twist of the Steinberg representation, where χ_v is unitary unramified.

- For $v \in \Sigma_{B,\mathrm{fin}}$,

(1d) $\pi_{B,v} = \chi_v \circ \nu_v$ is a 1-dimensional representation, where χ_v is unitary unramified.

- For $v \in \Sigma_\infty \smallsetminus \Sigma_{B,\infty}$,

(ds) $\pi_{B,v} = \mathrm{DS}_{k_v}$ is the irreducible unitary (limit of) discrete series representation of weight k_v.

- For $v \in \Sigma_{B,\infty}$,

(fd) $\pi_{B,v} = \mathrm{Sym}^{k_v}$ is the irreducible unitary (k_v+1)-dimensional representation.

We assume that $\pi_{B,v}$ is unramified for all finite places v of F such that F_v is ramified or of residual characteristic 2. By Proposition 7.1.4, we may assume that the following conditions (which are relevant to the choice of the polarization $\mathbb{V}_v = \mathbb{X}_v' \oplus \mathbb{Y}_v'$) are satisfied:

- If $v \notin \Sigma_B$, then $J \in (F_v^\times)^2$ except in the case (ur).
- If $v \in \Sigma_B$, then either $J_1 \in (F_v^\times)^2$ or $J_2 \in (F_v^\times)^2$.

In fact, Proposition 7.1.4 (applied as at the beginning of §7.2; see also Remark 7.1.3) enables us to impose more precise ramification conditions as described in the following table:

	π	B	B_1, B_2	E	u	F	J	J_1, J_2
ur	ur.p.s.	spl	spl, spl	spl	sq unit	ur	int	ints
				spl	sq unit	ram	sq unit	sq units
				inert	nsq unit	ur	$*$	$*$
				ram	*unif*	ur	sq unit	sq units
rps	r.p.s.	spl	spl, spl	spl	sq unit	ur	sq unit	sq units
st	St	spl	spl, spl	spl	sq unit	ur	sq unit	sq units
			ram, ram	inert	nsq unit	ur	*sq of unif*	$**$
1d	St	ram	spl, ram	inert	nsq unit	ur	*unif*	sq unit, *unif*
			ram, spl	inert	nsq unit	ur	*unif*	*unif*, sq unit
ds	d.s.	spl	spl, spl	$\mathbb{C}$	$-$	$\mathbb{R}$	$+$	$+, +$
			ram, ram	$\mathbb{C}$	$-$	$\mathbb{R}$	$+$	$-, -$
fd	d.s.	ram	spl, ram	$\mathbb{C}$	$-$	$\mathbb{R}$	$-$	$+, -$
			ram, spl	$\mathbb{C}$	$-$	$\mathbb{R}$	$-$	$-, +$

· All places above 2 fall into the case (ur) with E being split.
· In the case (ur) with E being inert, we need to consider separately the case $J \in (F_v^\times)^2$ and the case $J_1 \in (F_v^\times)^2$ or $J_2 \in (F_v^\times)^2$.
· $*$ indicates integers with even valuation.
· $**$ indicates uniformizers such that the ratio J_1/J_2 is a square unit.
· int : integer, sq unit : square unit, nsq unit : non-square unit
· *unif* : uniformizer, *sq of unif* : square of a uniformizer
· $+$: positive, $-$: negative

Here $\pi \cong \otimes_v \pi_v$ is the Jacquet–Langlands transfer of π_B to $\mathrm{GL}_2(\mathbb{A})$. These conditions will be very useful in the computation of the partial Fourier transform. From now on, we fix a place v of F and suppress the subscript v from the notation.

5.6. Schwartz functions on $\mathbb{X}'$

In this section, we pick a Schwartz function $\varphi' \in \mathcal{S}(\mathbb{X}')$ such that $\langle \varphi', \varphi' \rangle = 1$, together with maximal compact subgroups $\mathcal{K}, \mathcal{K}_1, \mathcal{K}_2$ of $B^\times, B_1^\times, B_2^\times$, respectively. Also, we study equivariance properties of φ' under the action of $\mathcal{K}$ and $\mathcal{K}_1 \times \mathcal{K}_2$, regarded as subgroups of $\mathrm{GU}(W) \cong B^\times$ and $\mathrm{GU}(V)^0 \cong (B_1^\times \times B_2^\times)/F^\times$, respectively.

We need to introduce some notation. For any set A, let $\mathbb{I}_A$ denote the characteristic function of A. If F is non-archimedean, then for any positive integer n, we define a subalgebra R_n of $\mathrm{M}_2(F)$ by

$$R_n = \left\{ \begin{pmatrix} a & b \\ c & d \end{pmatrix} \in \mathrm{M}_2(\mathfrak{o}) \,\middle|\, c \in \varpi^n \mathfrak{o} \right\}.$$

Note that R_1 is an Iwahori subalgebra of $\mathrm{M}_2(F)$. If $F = \mathbb{R}$, then we choose an isomorphism $E \cong \mathbb{C}$ such that

$$\frac{\mathbf{i}}{\sqrt{-1}} > 0,$$

i.e., $\mathbf{i} = |u|^{\frac{1}{2}}\sqrt{-1}$. Put

$$\frac{\partial}{\partial z} = \frac{1}{2}\left(\frac{\partial}{\partial x} + \frac{1}{\mathbf{i}}\frac{\partial}{\partial y}\right), \quad \frac{\partial}{\partial z^\rho} = \frac{1}{2}\left(\frac{\partial}{\partial x} - \frac{1}{\mathbf{i}}\frac{\partial}{\partial y}\right)$$

for $z = x + y\mathbf{i}$. For any integer k, we define a character χ_k of $\mathbb{C}^\times$ by

$$\chi_k(\alpha) = \left(\frac{\alpha}{\sqrt{\alpha\alpha^\rho}}\right)^k.$$

Put

$$\mathtt{H} = \begin{pmatrix} 1 & 0 \\ 0 & -1 \end{pmatrix}, \quad \mathtt{X} = \begin{pmatrix} 0 & 1 \\ 0 & 0 \end{pmatrix}, \quad \mathtt{Y} = \begin{pmatrix} 0 & 0 \\ 1 & 0 \end{pmatrix}.$$

5.6.1. The case (ur).

5.6.1.1. *The case when E is split and F is unramified.* In this case, we have:

- F is non-archimedean,
- ψ is of order zero,
- $u = t^2$ for some $t \in \mathfrak{o}^\times$,
- $J, J_1, J_2 \in \mathfrak{o}$.

We define maximal orders $\mathfrak{o}_B, \mathfrak{o}_{B_1}, \mathfrak{o}_{B_2}$ in B, B_1, B_2, respectively, by

$$\mathfrak{o}_B = \mathfrak{i}^{-1}(\mathrm{M}_2(\mathfrak{o})), \quad \mathfrak{o}_{B_1} = \mathfrak{i}_1^{-1}(\mathrm{M}_2(\mathfrak{o})), \quad \mathfrak{o}_{B_2} = \mathfrak{i}_2^{-1}(\mathrm{M}_2(\mathfrak{o})),$$

where $\mathfrak{i}, \mathfrak{i}_1, \mathfrak{i}_2$ are the isomorphisms given by (5.1), (5.6). Put

$$\mathcal{K} = \mathfrak{o}_B^\times, \quad \mathcal{K}_1 = \mathfrak{o}_{B_1}^\times, \quad \mathcal{K}_2 = \mathfrak{o}_{B_2}^\times.$$

We take the complete polarization $\mathbb{V} = \mathbb{X}' \oplus \mathbb{Y}'$ and identify $\mathbb{X}'$ with $V^\dagger \cong \mathrm{M}_2(F)$ as in §5.2.1. We define $\varphi' \in \mathcal{S}(\mathbb{X}')$ by $\varphi' = \mathbb{I}_{\mathrm{M}_2(\mathfrak{o})}$, i.e.,

$$\varphi'(x) = \mathbb{I}_\mathfrak{o}(x_1)\mathbb{I}_\mathfrak{o}(x_2)\mathbb{I}_\mathfrak{o}(x_3)\mathbb{I}_\mathfrak{o}(x_4)$$

for $x = x_1\mathbf{v}_1 + \cdots + x_4\mathbf{v}_4$, where $\mathbf{v}_1, \ldots, \mathbf{v}_4$ is the basis of $\mathbb{X}'$ given by (5.4). Then we have

$$\omega_\psi(k, (k_1, k_2))\varphi' = \varphi'$$

for $k \in \mathcal{K}$, $k_1 \in \mathcal{K}_1$, $k_2 \in \mathcal{K}_2$ such that $\nu(k) = \nu(k_1)\nu(k_2)$.

5.6.1.2. *The case when E is split and F is ramified.* In this case, we have:

- F is non-archimedean,
- $u = t^2$ for some $t \in \mathfrak{o}^\times$,
- $J, J_1, J_2 \in (\mathfrak{o}^\times)^2$.

Let d be the non-negative integer such that ψ is trivial on $\varpi^{-d}\mathfrak{o}$ but non-trivial on $\varpi^{-d-1}\mathfrak{o}$. We define maximal orders $\mathfrak{o}_B, \mathfrak{o}_{B_1}, \mathfrak{o}_{B_2}$ in B, B_1, B_2, respectively, by

$$\mathfrak{o}_B = \mathfrak{i}^{-1}\left(\left(\begin{smallmatrix} 1 & \\ & \varpi^d \end{smallmatrix}\right) \mathrm{M}_2(\mathfrak{o}) \left(\begin{smallmatrix} 1 & \\ & \varpi^{-d} \end{smallmatrix}\right)\right), \quad \mathfrak{o}_{B_1} = \mathfrak{i}_1^{-1}(\mathrm{M}_2(\mathfrak{o})), \quad \mathfrak{o}_{B_2} = \mathfrak{i}_2^{-1}(\mathrm{M}_2(\mathfrak{o})),$$

where $\mathfrak{i}, \mathfrak{i}_1, \mathfrak{i}_2$ are the isomorphisms given by (5.1), (5.6). Put

$$\mathcal{K} = \mathfrak{o}_B^\times, \quad \mathcal{K}_1 = \mathfrak{o}_{B_1}^\times, \quad \mathcal{K}_2 = \mathfrak{o}_{B_2}^\times.$$

We take the complete polarization $\mathbb{V} = \mathbb{X}' \oplus \mathbb{Y}'$ and identify $\mathbb{X}'$ with $V^\dagger \cong \mathrm{M}_2(F)$ as in §5.2.1. We define $\varphi' \in \mathcal{S}(\mathbb{X}')$ by $\varphi' = q^d \cdot \mathbb{I}_{\mathrm{M}_2(\mathfrak{o})}$, i.e.,

$$\varphi'(x) = q^d \cdot \mathbb{I}_\mathfrak{o}(x_1)\mathbb{I}_\mathfrak{o}(x_2)\mathbb{I}_\mathfrak{o}(x_3)\mathbb{I}_\mathfrak{o}(x_4)$$

for $x = x_1\mathbf{v}_1 + \cdots + x_4\mathbf{v}_4$, where $\mathbf{v}_1, \ldots, \mathbf{v}_4$ is the basis of $\mathbb{X}'$ given by (5.4). Then we have

$$\omega_\psi(k, (k_1, k_2))\varphi' = \varphi'$$

for $k \in \mathcal{K}$, $k_1 \in \mathcal{K}_1$, $k_2 \in \mathcal{K}_2$ such that $\nu(k) = \nu(k_1)\nu(k_2)$.

5.6.1.3. *The case when E is inert and $J \in (F^\times)^2$.* In this case, we have:

- F is non-archimedean,
- ψ is of order zero,
- $2 \in \mathfrak{o}^\times$,
- $u \in \mathfrak{o}^\times \smallsetminus (\mathfrak{o}^\times)^2$,
- $J = t^2$ for some $t \in \mathfrak{o}$,
- $J_1 \in s^2\mathfrak{o}^\times$ for some $s \in \mathfrak{o}$,
- $J_2 \in \mathfrak{o}$.

We define maximal orders $\mathfrak{o}_B, \mathfrak{o}_{B_1}, \mathfrak{o}_{B_2}$ in B, B_1, B_2, respectively, by

$$\mathfrak{o}_B = \mathfrak{i}^{-1}(\mathrm{M}_2(\mathfrak{o})), \quad \mathfrak{o}_{B_1} = \mathfrak{o} + \mathfrak{o}\mathbf{i} + \mathfrak{o}\frac{\mathbf{j}_1}{s} + \mathfrak{o}\frac{\mathbf{i}\mathbf{j}_1}{s}, \quad \mathfrak{o}_{B_2} = \mathfrak{i}_2^{-1}(\mathfrak{o}_{B_1}),$$

where $\mathfrak{i}, \mathfrak{i}_2$ are the isomorphisms given by (5.7), (5.9). Put

$$\mathcal{K} = \mathfrak{o}_B^\times, \quad \mathcal{K}_1 = \mathfrak{o}_{B_1}^\times, \quad \mathcal{K}_2 = \mathfrak{o}_{B_2}^\times.$$

We take the complete polarization $\mathbb{V} = \mathbb{X}' \oplus \mathbb{Y}'$ as in §5.2.2 and identify $\mathbb{X}'$ with $V^\dagger \cong B_1$ as in §5.2.2.1. We define $\varphi' \in \mathcal{S}(\mathbb{X}')$ by $\varphi' = \mathbb{I}_{\mathfrak{o}_{B_1}}$, i.e.,

$$\varphi'(x) = \mathbb{I}_{\mathfrak{o}}(x_1)\mathbb{I}_{\mathfrak{o}}(x_2)\mathbb{I}_{\mathfrak{o}}(x_3)\mathbb{I}_{\mathfrak{o}}(x_4)$$

for $x = x_1\mathbf{v}_1 + \cdots + x_4\mathbf{v}_4$, where $\mathbf{v}_1, \ldots, \mathbf{v}_4$ is the basis of $\mathbb{X}'$ given by (5.8). Then we have

$$\omega_\psi(k, (k_1, k_2))\varphi' = \varphi'$$

for $k \in \mathcal{K}$, $k_1 \in \mathcal{K}_1$, $k_2 \in \mathcal{K}_2$ such that $\nu(k) = \nu(k_1)\nu(k_2)$.

5.6.1.4. *The case when E is inert, and $J_1 \in (F^\times)^2$ or $J_2 \in (F^\times)^2$.* We only consider the case $J_1 \in (F^\times)^2$; we switch the roles of B_1 and B_2 in the other case. In this case, we have:

- F is non-archimedean,
- ψ is of order zero,
- $2 \in \mathfrak{o}^\times$,
- $u \in \mathfrak{o}^\times \smallsetminus (\mathfrak{o}^\times)^2$,
- $J_1 = t^2$ for some $t \in \mathfrak{o}$,
- $J \in s^2\mathfrak{o}^\times$ for some $s \in \mathfrak{o}$,
- $J_2 \in \mathfrak{o}$.

We define maximal orders $\mathfrak{o}_B, \mathfrak{o}_{B_1}, \mathfrak{o}_{B_2}$ in B, B_1, B_2, respectively, by

$$\mathfrak{o}_B = \mathfrak{o} + \mathfrak{o}\mathbf{i} + \mathfrak{o}\frac{\mathbf{j}}{s} + \mathfrak{o}\frac{\mathbf{i}\mathbf{j}}{s}, \quad \mathfrak{o}_{B_1} = \mathfrak{i}_1^{-1}(\mathrm{M}_2(\mathfrak{o})), \quad \mathfrak{o}_{B_2} = \mathfrak{i}_2^{-1}(\mathfrak{o}_B),$$

where $\mathfrak{i}_1, \mathfrak{i}_2$ are the isomorphisms given by (5.12), (5.13). Put

$$\mathcal{K} = \mathfrak{o}_B^\times, \quad \mathcal{K}_1 = \mathfrak{o}_{B_1}^\times, \quad \mathcal{K}_2 = \mathfrak{o}_{B_2}^\times.$$

We take the complete polarization $\mathbb{V} = \mathbb{X}' \oplus \mathbb{Y}'$ and identify $\mathbb{X}'$ with $W = B$ as in §5.2.3. We define $\varphi' \in \mathcal{S}(\mathbb{X}')$ by $\varphi' = \mathbb{I}_{\mathfrak{o}_B}$, i.e.,

$$\varphi'(x) = \mathbb{I}_{\mathfrak{o}}(x_1)\mathbb{I}_{\mathfrak{o}}(x_2)\mathbb{I}_{\mathfrak{o}}(x_3)\mathbb{I}_{\mathfrak{o}}(x_4)$$

for $x = x_1\mathbf{v}_1 + \cdots + x_4\mathbf{v}_4$, where $\mathbf{v}_1, \ldots, \mathbf{v}_4$ is the basis of $\mathbb{X}'$ given by (5.14). Then we have

$$\omega_\psi(k, (k_1, k_2))\varphi' = \varphi'$$

for $k \in \mathcal{K}$, $k_1 \in \mathcal{K}_1$, $k_2 \in \mathcal{K}_2$ such that $\nu(k) = \nu(k_1)\nu(k_2)$.

5.6.1.5. *The case when E is ramified.* In this case, we have:

- F is non-archimedean,
- ψ is of order zero,
- $2 \in \mathfrak{o}^\times$,
- $u \in \varpi\mathfrak{o}^\times$,
- $J = t^2$ for some $t \in \mathfrak{o}^\times$,
- $J_1 = t_1^2$ for some $t_1 \in \mathfrak{o}^\times$,
- $J_2 \in (\mathfrak{o}^\times)^2$.

We define maximal orders $\mathfrak{o}_B, \mathfrak{o}_{B_1}, \mathfrak{o}_{B_2}$ in B, B_1, B_2, respectively, by

$$\mathfrak{o}_B = \mathfrak{i}^{-1}(\mathrm{M}_2(\mathfrak{o})), \quad \mathfrak{o}_{B_1} = \mathfrak{i}_1^{-1}(\mathrm{M}_2(\mathfrak{o})), \quad \mathfrak{o}_{B_2} = \mathfrak{i}_2^{-1}(\mathrm{M}_2(\mathfrak{o})),$$

where $\mathfrak{i}, \mathfrak{i}_1, \mathfrak{i}_2$ are the isomorphisms given by (5.7), (5.11). Put

$$\mathcal{K} = \mathfrak{o}_B^\times, \quad \mathcal{K}_1 = \mathfrak{o}_{B_1}^\times, \quad \mathcal{K}_2 = \mathfrak{o}_{B_2}^\times.$$

We take the complete polarization $\mathbb{V} = \mathbb{X}' \oplus \mathbb{Y}'$ as in §5.2.2 and identify $\mathbb{X}'$ with $V^\dagger \cong \mathrm{M}_2(F)$ as in §5.2.2.2. We define $\varphi' \in \mathcal{S}(\mathbb{X}')$ by $\varphi' = \mathbb{I}_{\mathrm{M}_2(\mathfrak{o})}$, i.e.,

$$\varphi'(x) = \mathbb{I}_\mathfrak{o}(x_1)\mathbb{I}_\mathfrak{o}(x_2)\mathbb{I}_\mathfrak{o}(x_3)\mathbb{I}_\mathfrak{o}(x_4)$$

for $x = x_1\mathbf{v}_1 + \cdots + x_4\mathbf{v}_4$, where $\mathbf{v}_1, \ldots, \mathbf{v}_4$ is the basis of $\mathbb{X}'$ given by (5.10). Then we have

$$\omega_\psi(k, (k_1, k_2))\varphi' = \varphi'$$

for $k \in \mathcal{K}$, $k_1 \in \mathcal{K}_1$, $k_2 \in \mathcal{K}_2$ such that $\nu(k) = \nu(k_1)\nu(k_2)$.

5.6.2. The case (rps). In this case, we have:

- F is non-archimedean,
- ψ is of order zero,
- $2 \in \mathfrak{o}^\times$,
- $u = t^2$ for some $t \in \mathfrak{o}^\times$,
- $J, J_1, J_2 \in (\mathfrak{o}^\times)^2$.

We define maximal orders $\mathfrak{o}_B, \mathfrak{o}_{B_1}, \mathfrak{o}_{B_2}$ in B, B_1, B_2 and subalgebras $\mathfrak{o}_{B,n}$, $\mathfrak{o}_{B_1,n}$, $\mathfrak{o}_{B_2,n}$ of B, B_1, B_2, respectively, by

$$\mathfrak{o}_B = \mathfrak{i}^{-1}(\mathrm{M}_2(\mathfrak{o})), \qquad \mathfrak{o}_{B_1} = \mathfrak{i}_1^{-1}(\mathrm{M}_2(\mathfrak{o})), \qquad \mathfrak{o}_{B_2} = \mathfrak{i}_2^{-1}(\mathrm{M}_2(\mathfrak{o})),$$
$$\mathfrak{o}_{B,n} = \mathfrak{i}^{-1}(R_n), \qquad \mathfrak{o}_{B_1,n} = \mathfrak{i}_1^{-1}(R_n), \qquad \mathfrak{o}_{B_2,n} = \mathfrak{i}_2^{-1}(R_n),$$

where $\mathfrak{i}, \mathfrak{i}_1, \mathfrak{i}_2$ are the isomorphisms given by (5.1), (5.6). We define orientations

$$o_B : \mathfrak{o}_{B,n} \longrightarrow \mathfrak{o}/\varpi^n\mathfrak{o},$$
$$o_{B_1} : \mathfrak{o}_{B_1,n} \longrightarrow \mathfrak{o}/\varpi^n\mathfrak{o},$$
$$o_{B_2} : \mathfrak{o}_{B_2,n} \longrightarrow \mathfrak{o}/\varpi^n\mathfrak{o}$$

by

$$o_B\left(\mathfrak{i}^{-1}\left(\begin{smallmatrix} a & b \\ c & d \end{smallmatrix}\right)\right) = d \bmod \varpi^n\mathfrak{o},$$
$$o_{B_1}\left(\mathfrak{i}_1^{-1}\left(\begin{smallmatrix} a & b \\ c & d \end{smallmatrix}\right)\right) = d \bmod \varpi^n\mathfrak{o},$$
$$o_{B_2}\left(\mathfrak{i}_2^{-1}\left(\begin{smallmatrix} a & b \\ c & d \end{smallmatrix}\right)\right) = a \bmod \varpi^n\mathfrak{o}.$$

Put

$$\mathcal{K} = \mathfrak{o}_B^\times, \qquad \mathcal{K}_1 = \mathfrak{o}_{B_1}^\times, \qquad \mathcal{K}_2 = \mathfrak{o}_{B_2}^\times,$$
$$\mathcal{K}_n = \mathfrak{o}_{B,n}^\times, \qquad \mathcal{K}_{1,n} = \mathfrak{o}_{B_1,n}^\times, \qquad \mathcal{K}_{2,n} = \mathfrak{o}_{B_2,n}^\times.$$

We take the complete polarization $\mathbb{V} = \mathbb{X}' \oplus \mathbb{Y}'$ and identify $\mathbb{X}'$ with $V^\dagger \cong \mathrm{M}_2(F)$ as in §5.2.1. For a unitary ramified character μ of $F^\times$ of conductor q^n, i.e., trivial on $1 + \varpi^n\mathfrak{o}$ but non-trivial on $1 + \varpi^{n-1}\mathfrak{o}$ (resp. $\mathfrak{o}^\times$) if $n > 1$ (resp. if $n = 1$), we define $\varphi' = \varphi'_\mu \in \mathcal{S}(\mathbb{X}')$ by

$$\varphi'(x) = q^{\frac{n+1}{2}}(q-1)^{-\frac{1}{2}} \cdot \mathbb{I}_{\mathfrak{o}}(x_1)\mathbb{I}_{\mathfrak{o}}(x_2)\mathbb{I}_{\varpi^n\mathfrak{o}}(x_3)\mathbb{I}_{\mathfrak{o}^\times}(x_4)\mu(x_4)$$

for $x = x_1\mathbf{v}_1 + \cdots + x_4\mathbf{v}_4$, where $\mathbf{v}_1, \dots, \mathbf{v}_4$ is the basis of $\mathbb{X}'$ given by (5.4). Then we have

$$\omega_\psi(k, (k_1, k_2))\varphi' = \boldsymbol{\mu}(k)^{-1}\boldsymbol{\mu}(k_1)\boldsymbol{\mu}(k_2)^{-1}\mu(\nu(k_2))\varphi'$$

for $k \in \mathcal{K}_n$, $k_1 \in \mathcal{K}_{1,n}$, $k_2 \in \mathcal{K}_{2,n}$ such that $\nu(k) = \nu(k_1)\nu(k_2)$, where $\boldsymbol{\mu}$ is the character of $R_n^\times$ (and those of $\mathcal{K}_n, \mathcal{K}_{1,n}, \mathcal{K}_{2,n}$ via $\mathfrak{i}, \mathfrak{i}_1, \mathfrak{i}_2$) defined by

$$\boldsymbol{\mu}(k) := \mu(d)$$

for $k = \left(\begin{smallmatrix} a & b \\ c & d \end{smallmatrix}\right)$.

5.6.3. The case (st).

5.6.3.1. *The case when B_1 and B_2 are split.* In this case, we have:

- F is non-archimedean,
- ψ is of order zero,
- $2 \in \mathfrak{o}^\times$,
- $u = t^2$ for some $t \in \mathfrak{o}^\times$,
- $J, J_1, J_2 \in (\mathfrak{o}^\times)^2$.

We define maximal orders $\mathfrak{o}_B, \mathfrak{o}_{B_1}, \mathfrak{o}_{B_2}$ in B, B_1, B_2 and Iwahori subalgebras $\mathfrak{I}, \mathfrak{I}_1, \mathfrak{I}_2$ of B, B_1, B_2, respectively, by

$$\mathfrak{o}_B = \mathfrak{i}^{-1}(\mathrm{M}_2(\mathfrak{o})), \qquad \mathfrak{o}_{B_1} = \mathfrak{i}_1^{-1}(\mathrm{M}_2(\mathfrak{o})), \qquad \mathfrak{o}_{B_2} = \mathfrak{i}_2^{-1}(\mathrm{M}_2(\mathfrak{o})),$$
$$\mathfrak{I} = \mathfrak{i}^{-1}(R_1), \qquad \mathfrak{I}_1 = \mathfrak{i}_1^{-1}(R_1), \qquad \mathfrak{I}_2 = \mathfrak{i}_2^{-1}(R_1),$$

where $\mathfrak{i}, \mathfrak{i}_1, \mathfrak{i}_2$ are the isomorphisms given by (5.1), (5.6). Put

$$\mathcal{K} = \mathfrak{o}_B^\times, \qquad \mathcal{K}_1 = \mathfrak{o}_{B_1}^\times, \qquad \mathcal{K}_2 = \mathfrak{o}_{B_2}^\times,$$
$$\mathcal{I} = \mathfrak{I}^\times, \qquad \mathcal{I}_1 = \mathfrak{I}_1^\times, \qquad \mathcal{I}_2 = \mathfrak{I}_2^\times.$$

We take the complete polarization $\mathbb{V} = \mathbb{X}' \oplus \mathbb{Y}'$ and identify $\mathbb{X}'$ with $V^\dagger \cong \mathrm{M}_2(F)$ as in §5.2.1. We define $\varphi' \in \mathcal{S}(\mathbb{X}')$ by

$$\varphi'(x) = q^{\frac{1}{2}} \cdot \mathbb{I}_{\mathfrak{o}}(x_1)\mathbb{I}_{\mathfrak{o}}(x_2)\mathbb{I}_{\mathfrak{p}}(x_3)\mathbb{I}_{\mathfrak{o}}(x_4)$$

for $x = x_1\mathbf{v}_1 + \cdots + x_4\mathbf{v}_4$, where $\mathbf{v}_1, \dots, \mathbf{v}_4$ is the basis of $\mathbb{X}'$ given by (5.4). Then we have

$$\omega_\psi(k, (k_1, k_2))\varphi' = \varphi'$$

for $k \in \mathcal{I}$, $k_1 \in \mathcal{I}_1$, $k_2 \in \mathcal{I}_2$ such that $\nu(k) = \nu(k_1)\nu(k_2)$.

5.6.3.2. *The case when B_1 and B_2 are ramified.* In this case, we have:

- F is non-archimedean,
- ψ is of order zero,
- $2 \in \mathfrak{o}^\times$,
- $u \in \mathfrak{o}^\times \smallsetminus (\mathfrak{o}^\times)^2$,
- $J = t^2$ for some $t \in \varpi\mathfrak{o}^\times$,
- $J_1, J_2 \in \varpi\mathfrak{o}^\times$.

We define a maximal order $\mathfrak{o}_B$ in B and an Iwahori subalgebra $\mathfrak{I}$ of B by

$$\mathfrak{o}_B = \mathfrak{i}^{-1}(\mathrm{M}_2(\mathfrak{o})), \quad \mathfrak{I} = \mathfrak{i}^{-1}(R_1),$$

where $\mathfrak{i}$ is the isomorphism given by (5.7). Let $\mathfrak{o}_{B_1}$ and $\mathfrak{o}_{B_2}$ be the unique maximal orders in B_1 and B_2, respectively. Then we have

$$\mathfrak{o}_{B_1} = \mathfrak{o} + \mathfrak{o}\mathbf{i} + \mathfrak{o}\mathbf{j}_1 + \mathfrak{o}\mathbf{i}\mathbf{j}_1, \quad \mathfrak{o}_{B_2} = \mathfrak{i}_2^{-1}(\mathfrak{o}_{B_1}),$$

where $\mathfrak{i}_2$ is the isomorphism given by (5.9). Put

$$\mathcal{K} = \mathfrak{o}_B^\times, \quad \mathcal{I} = \mathfrak{I}^\times, \quad \mathcal{K}_1 = \mathfrak{o}_{B_1}^\times, \quad \mathcal{K}_2 = \mathfrak{o}_{B_2}^\times.$$

Put $s = 1$. We take the complete polarization $\mathbb{V} = \mathbb{X}' \oplus \mathbb{Y}'$ as in §5.2.2 and identify $\mathbb{X}'$ with $V^\dagger \cong B_1$ as in §5.2.2.1. We define $\varphi' \in \mathcal{S}(\mathbb{X}')$ by $\varphi' = q^{\frac{1}{2}} \cdot \mathbb{I}_{\mathfrak{o}_{B_1}}$, i.e.,

$$\varphi'(x) = q^{\frac{1}{2}} \cdot \mathbb{I}_\mathfrak{o}(x_1)\mathbb{I}_\mathfrak{o}(x_2)\mathbb{I}_\mathfrak{o}(x_3)\mathbb{I}_\mathfrak{o}(x_4)$$

for $x = x_1\mathbf{v}_1 + \cdots + x_4\mathbf{v}_4$, where $\mathbf{v}_1, \ldots, \mathbf{v}_4$ is the basis of $\mathbb{X}'$ given by (5.8). Then we have

$$\omega_\psi(k, (k_1, k_2))\varphi' = \varphi'$$

for $k \in \mathcal{I}$, $k_1 \in \mathcal{K}_1$, $k_2 \in \mathcal{K}_2$ such that $\nu(k) = \nu(k_1)\nu(k_2)$.

5.6.4. The case (1d). We only consider the case $J_1 \in (F^\times)^2$; we switch the roles of B_1 and B_2 in the other case. In this case, we have:

- F is non-archimedean,
- ψ is of order zero,
- $2 \in \mathfrak{o}^\times$,
- $u \in \mathfrak{o}^\times \setminus (\mathfrak{o}^\times)^2$,
- $J_1 = t^2$ for some $t \in \mathfrak{o}^\times$,
- $J, J_2 \in \varpi\mathfrak{o}^\times$.

We define a maximal order $\mathfrak{o}_{B_1}$ in B_1 and an Iwahori subalgebra $\mathfrak{I}_1$ of B_1 by

$$\mathfrak{o}_{B_1} = \mathfrak{i}_1^{-1}(\mathrm{M}_2(\mathfrak{o})), \quad \mathfrak{I}_1 = \mathfrak{i}_1^{-1}\left(\left(\begin{smallmatrix} 1 & \\ & \varpi^{-1} \end{smallmatrix}\right) R_1 \left(\begin{smallmatrix} 1 & \\ & \varpi \end{smallmatrix}\right)\right),$$

where $\mathfrak{i}_1$ is the isomorphism given by (5.12). Let $\mathfrak{o}_B$ and $\mathfrak{o}_{B_2}$ be the unique maximal orders in B and B_2, respectively. Then we have

$$\mathfrak{o}_B = \mathfrak{o} + \mathfrak{o}\mathbf{i} + \mathfrak{o}\mathbf{j} + \mathfrak{o}\mathbf{i}\mathbf{j}, \quad \mathfrak{o}_{B_2} = \mathfrak{i}_2^{-1}(\mathfrak{o}_B),$$

where $\mathfrak{i}_2$ is the isomorphism given by (5.13). Put

$$\mathcal{K} = \mathfrak{o}_B^\times, \quad \mathcal{K}_1 = \mathfrak{o}_{B_1}^\times, \quad \mathcal{I}_1 = \mathfrak{I}_1^\times, \quad \mathcal{K}_2 = \mathfrak{o}_{B_2}^\times.$$

Put $s = 1$. We take the complete polarization $\mathbb{V} = \mathbb{X}' \oplus \mathbb{Y}'$ and identify $\mathbb{X}'$ with $W = B$ as in §5.2.3. We define $\varphi' \in \mathcal{S}(\mathbb{X}')$ by $\varphi' = q^{\frac{1}{2}} \cdot \mathbb{I}_{\mathfrak{o}_B}$, i.e.,

$$\varphi'(x) = q^{\frac{1}{2}} \cdot \mathbb{I}_\mathfrak{o}(x_1)\mathbb{I}_\mathfrak{o}(x_2)\mathbb{I}_\mathfrak{o}(x_3)\mathbb{I}_\mathfrak{o}(x_4)$$

for $x = x_1\mathbf{v}_1 + \cdots + x_4\mathbf{v}_4$, where $\mathbf{v}_1, \ldots, \mathbf{v}_4$ is the basis of $\mathbb{X}'$ given by (5.14). Then we have

$$\omega_\psi(k, (k_1, k_2))\varphi' = \varphi'$$

for $k \in \mathcal{K}$, $k_1 \in \mathcal{I}_1$, $k_2 \in \mathcal{K}_2$ such that $\nu(k) = \nu(k_1)\nu(k_2)$.

5.6.5. The case (ds).

5.6.5.1. *The case when B_1 and B_2 are split.* In this case, we have:

- $F = \mathbb{R}$,
- $\psi(x) = e^{2\pi\sqrt{-1}x}$,
- $u < 0$,
- $J = t^2$ for some $t \in F^\times$,
- $J_1 = s^2$ for some $s \in F^\times$,
- $J_2 > 0$.

Put $v = |u|^{\frac{1}{2}}$. We take the complete polarization $\mathbb{V} = \mathbb{X}' \oplus \mathbb{Y}'$ as in §5.2.2 and identify $\mathbb{X}'$ with $V^\dagger \cong B_1$ as in §5.2.2.1. For a non-negative integer k, we define $\varphi' = \varphi'_k \in \mathcal{S}(\mathbb{X}')$ by

$$\varphi'(x) = c_k^{-\frac{1}{2}} \cdot (x_2 - x_1 \mathbf{i})^k \cdot e^{-\frac{\pi}{2v}(x_2^2 - ux_1^2 + x_4^2 - ux_3^2)}$$

for $x = x_1\mathbf{v}_1 + \cdots + x_4\mathbf{v}_4$, where $\mathbf{v}_1, \ldots, \mathbf{v}_4$ is the basis of $\mathbb{X}'$ given by (5.8) and

$$c_k = \frac{k!|u|^{\frac{k}{2}+1}}{4\pi^k}.$$

LEMMA 5.6.1. *We have $\langle \varphi', \varphi' \rangle = 1$ and*

$$\omega_\psi(\alpha, (\alpha_1, \alpha_2))\varphi' = \chi_k(\alpha)^{-1}\chi_k(\alpha_1)\chi_k(\alpha_2)\varphi'$$

for $\alpha, \alpha_1, \alpha_2 \in E^\times$ such that $\nu(\alpha) = \nu(\alpha_1)\nu(\alpha_2)$.

PROOF. Recall that the Haar measure on $\mathbb{X}'$ is given by $dx = \frac{|u|}{4}\, dx_1 \cdots dx_4$. We have

$$\begin{aligned}
&\frac{|u|}{4} \int_{F^4} (x_2^2 - ux_1^2)^k e^{-\frac{\pi}{v}(x_2^2 - ux_1^2 + x_4^2 - ux_3^2)}\, dx_1 \cdots dx_4 \\
&= \frac{|u|}{4\pi^2} \left(\frac{v}{\pi}\right)^k \int_{F^4} (x_2^2 + x_1^2)^k e^{-(x_2^2 + x_1^2 + x_4^2 + x_3^2)}\, dx_1 \cdots dx_4 \\
&= \frac{|u|^{\frac{k}{2}+1}}{4\pi^{k+2}} \cdot (2\pi)^2 \int_0^\infty \int_0^\infty r_1^{2k} e^{-(r_1^2 + r_2^2)}\, r_1\, dr_1\, r_2\, dr_2 \\
&= \frac{|u|^{\frac{k}{2}+1}}{4\pi^k} \int_0^\infty \int_0^\infty r_1^k e^{-(r_1 + r_2)}\, dr_1\, dr_2 \\
&= \frac{|u|^{\frac{k}{2}+1}}{4\pi^k} \cdot \Gamma(k+1)
\end{aligned}$$

and hence $\langle \varphi', \varphi' \rangle = 1$. If we write $z_1 = x_2 + x_1\mathbf{i}$ and $z_2 = x_4 + x_3\mathbf{i}$, then

$$\varphi'(x) = c_k^{-\frac{1}{2}} \cdot (z_1^\rho)^k \cdot e^{-\frac{\pi}{2v}(z_1 z_1^\rho + z_2 z_2^\rho)},$$

and it is easy to see that

$$\omega_\psi(\nu^{\frac{1}{2}}, (\alpha_1, \alpha_2))\varphi' = \chi_k(\alpha_1)\chi_k(\alpha_2)\varphi'$$

for $\alpha_1, \alpha_2 \in E^\times$ and $\nu = \nu(\alpha_1)\nu(\alpha_2)$. On the other hand, we have

$$\begin{aligned}
\omega_\psi(\mathtt{H})\varphi'(x) &= \left(2 + x_1\frac{\partial}{\partial x_1} + \cdots + x_4\frac{\partial}{\partial x_4}\right)\varphi'(x), \\
\omega_\psi(\mathtt{X})\varphi'(x) &= \frac{\pi\sqrt{-1}}{2}(ux_1^2 - x_2^2 - ux_3^2 + x_4^2)\varphi'(x), \\
\omega_\psi(\mathtt{Y})\varphi'(x) &= -\frac{1}{2\pi\sqrt{-1}}\left(\frac{1}{u}\frac{\partial^2}{\partial x_1^2} - \frac{\partial^2}{\partial x_2^2} - \frac{1}{u}\frac{\partial^2}{\partial x_3^2} + \frac{\partial^2}{\partial x_4^2}\right)\varphi'(x),
\end{aligned}$$

where we identity $\mathrm{GU}(W) \cong B^\times$ with $\mathrm{GL}_2(F)$ via the isomorphism $\mathfrak{i}$ given by (5.7). Thus, noting that

$$\frac{\partial^2}{\partial z_1 \partial z_1^\rho} = \frac{1}{4}\left(\frac{\partial^2}{\partial x_2^2} - \frac{1}{u}\frac{\partial^2}{\partial x_1^2}\right), \quad \frac{\partial^2}{\partial z_2 \partial z_2^\rho} = \frac{1}{4}\left(\frac{\partial^2}{\partial x_4^2} - \frac{1}{u}\frac{\partial^2}{\partial x_3^2}\right),$$

we see that

$$\omega_\psi(v^{-1}\mathtt{X} - v\mathtt{Y})\varphi' = -\sqrt{-1}k\varphi'.$$

This implies that

$$\omega_\psi(\alpha, (1,1))\varphi' = \chi_k(\alpha)^{-1}\varphi'$$

for $\alpha \in E^1$ since

$$\begin{aligned}\mathfrak{i}(\alpha) &= \begin{pmatrix} a & b \\ bu & a \end{pmatrix} = \begin{pmatrix} 1 & \\ & v \end{pmatrix}\begin{pmatrix} \cos\theta & \sin\theta \\ -\sin\theta & \cos\theta \end{pmatrix}\begin{pmatrix} 1 & \\ & v^{-1} \end{pmatrix} \\ &= \exp((v^{-1}\mathtt{X} - v\mathtt{Y})\theta)\end{aligned}$$

if we write $\alpha = a + b\mathbf{i} = e^{\sqrt{-1}\theta}$. This completes the proof. □

5.6.5.2. *The case when B_1 and B_2 are ramified.* In this case, we have:

- $F = \mathbb{R}$,
- $\psi(x) = e^{2\pi\sqrt{-1}x}$,
- $u < 0$,
- $J = t^2$ for some $t \in F^\times$,
- $J_1 = -s^2$ for some $s \in F^\times$,
- $J_2 < 0$.

Put $v = |u|^{\frac{1}{2}}$. We take the complete polarization $\mathbb{V} = \mathbb{X}' \oplus \mathbb{Y}'$ as in §5.2.2 and identify $\mathbb{X}'$ with $V^\dagger \cong B_1$ as in §5.2.2.1. For a non-negative integer k, we define $\varphi' = \varphi'_k \in \mathcal{S}(\mathbb{X}')$ by

$$\varphi'(x) = c_k^{-\frac{1}{2}} \cdot (x_2 - x_1\mathbf{i})^k \cdot e^{-\frac{\pi}{2v}(x_2^2 - ux_1^2 + x_4^2 - ux_3^2)}$$

for $x = x_1\mathbf{v}_1 + \cdots + x_4\mathbf{v}_4$, where $\mathbf{v}_1, \ldots, \mathbf{v}_4$ is the basis of $\mathbb{X}'$ given by (5.8) and

$$c_k = \frac{k!|u|^{\frac{k}{2}+1}}{4\pi^k}.$$

LEMMA 5.6.2. *We have $\langle\varphi', \varphi'\rangle = 1$ and*

$$\omega_\psi(\alpha, (\alpha_1, \alpha_2))\varphi' = \chi_{k+2}(\alpha)^{-1}\chi_k(\alpha_1)\chi_k(\alpha_2)\varphi'$$

for $\alpha, \alpha_1, \alpha_2 \in E^\times$ such that $\nu(\alpha) = \nu(\alpha_1)\nu(\alpha_2)$.

PROOF. The proof is the same as that of Lemma 5.6.1 and we omit the details. Note that, in this case, we have

$$\begin{aligned}\omega_\psi(\mathtt{H})\varphi'(x) &= \left(2 + x_1\frac{\partial}{\partial x_1} + \cdots + x_4\frac{\partial}{\partial x_4}\right)\varphi'(x), \\ \omega_\psi(\mathtt{X})\varphi'(x) &= \frac{\pi\sqrt{-1}}{2}(ux_1^2 - x_2^2 + ux_3^2 - x_4^2)\varphi'(x), \\ \omega_\psi(\mathtt{Y})\varphi'(x) &= -\frac{1}{2\pi\sqrt{-1}}\left(\frac{1}{u}\frac{\partial^2}{\partial x_1^2} - \frac{\partial^2}{\partial x_2^2} + \frac{1}{u}\frac{\partial^2}{\partial x_3^2} - \frac{\partial}{\partial x_4^2}\right)\varphi'(x).\end{aligned}$$

□

5.6.6. The case (fd). We only consider the case $J_1 \in (F^\times)^2$; we switch the roles of B_1 and B_2 in the other case. In this case, we have:

- $F = \mathbb{R}$,
- $\psi(x) = e^{2\pi\sqrt{-1}x}$,
- $u < 0$,
- $J_1 = t^2$ for some $t \in F^\times$,
- $J = -s^2$ for some $s \in F^\times$,
- $J_2 < 0$.

Put $v = |u|^{\frac{1}{2}}$. We take the complete polarization $\mathbb{V} = \mathbb{X}' \oplus \mathbb{Y}'$ and identify $\mathbb{X}'$ with $W = B$ as in §5.2.3. For a non-negative integer k, we define $\varphi' = \varphi'_k \in \mathcal{S}(\mathbb{X}')$ by

$$\varphi'(x) = c_k^{-\frac{1}{2}} \cdot \left(x_1 - x_2\frac{\mathbf{i}}{u}\right)^k \cdot e^{-\frac{\pi v}{2}(x_1^2 - \frac{1}{u}x_2^2 + x_3^2 - \frac{1}{u}x_4^2)}$$

for $x = x_1\mathbf{v}_1 + \cdots + x_4\mathbf{v}_4$, where $\mathbf{v}_1, \ldots, \mathbf{v}_4$ is the basis of $\mathbb{X}'$ given by (5.14) and

$$c_k = \frac{k!}{\pi^k |u|^{\frac{k}{2}+1}}.$$

LEMMA 5.6.3. *We have* $\langle \varphi', \varphi' \rangle = 1$ *and*

$$\omega_\psi(\alpha, (\alpha_1, \alpha_2))\varphi' = \chi_k(\alpha)^{-1}\chi_{k+2}(\alpha_1)\chi_k(\alpha_2)\varphi'$$

for $\alpha, \alpha_1, \alpha_2 \in E^\times$ *such that* $\nu(\alpha) = \nu(\alpha_1)\nu(\alpha_2)$.

PROOF. Recall that the Haar measure on $\mathbb{X}'$ is given by $dx = \frac{1}{|u|}\,dx_1 \cdots dx_4$. We have

$$\begin{aligned}
&\frac{1}{|u|}\int_{F^4}\left(x_1^2 - \frac{1}{u}x_2^2\right)^k e^{-\pi v(x_1^2 - \frac{1}{u}x_2^2 + x_3^2 - \frac{1}{u}x_4^2)}\,dx_1 \cdots dx_4 \\
&= \frac{1}{\pi^2|u|}\left(\frac{1}{\pi v}\right)^k \int_{F^4} (x_1^2 + x_2^2)^k e^{-(x_1^2+x_2^2+x_3^2+x_4^2)}\,dx_1 \cdots dx_4 \\
&= \frac{1}{\pi^{k+2}|u|^{\frac{k}{2}+1}} \cdot (2\pi)^2 \int_0^\infty \int_0^\infty r_1^{2k} e^{-(r_1^2+r_2^2)}\, r_1\,dr_1\, r_2\,dr_2 \\
&= \frac{1}{\pi^k|u|^{\frac{k}{2}+1}} \int_0^\infty \int_0^\infty r_1^k e^{-(r_1+r_2)}\,dr_1\,dr_2 \\
&= \frac{1}{\pi^k|u|^{\frac{k}{2}+1}} \cdot \Gamma(k+1)
\end{aligned}$$

and hence $\langle \varphi', \varphi' \rangle = 1$. If we write $z_1 = x_1 + x_2\frac{\mathbf{i}}{u}$ and $z_2 = x_3 + x_4\frac{\mathbf{i}}{u}$, then

$$\varphi'(x) = c_k^{-\frac{1}{2}} \cdot (z_1^\rho)^k \cdot e^{-\frac{\pi v}{2}(z_1 z_1^\rho + z_2 z_2^\rho)},$$

and it is easy to see that

$$\omega_\psi(\alpha, (\nu^{\frac{1}{2}}, \alpha_2))\varphi' = \chi_k(\alpha)^{-1}\chi_k(\alpha_2)\varphi'$$

for $\alpha, \alpha_2 \in E^\times$ and $\nu = \nu(\alpha)\nu(\alpha_2)^{-1}$. On the other hand, we have

$$\omega_\psi(\mathtt{H})\varphi'(x) = -\left(2 + x_1\frac{\partial}{\partial x_1} + \cdots + x_4\frac{\partial}{\partial x_4}\right)\varphi'(x),$$

$$\omega_\psi(\mathtt{X})\varphi'(x) = \frac{1}{4\pi\sqrt{-1}}\left(\frac{\partial^2}{\partial x_1^2} - u\frac{\partial^2}{\partial x_2^2} + \frac{\partial^2}{\partial x_3^2} - u\frac{\partial}{\partial x_4^2}\right)\varphi'(x),$$

$$\omega_\psi(\mathtt{Y})\varphi'(x) = -\pi\sqrt{-1}\left(x_1^2 - \frac{1}{u}x_2^2 + x_3^2 - \frac{1}{u}x_4^2\right)\varphi'(x),$$

where we identity $\mathrm{GU}(V)^0 \cong (B_1^\times \times B_2^\times)/F^\times$ with a subgroup of $\mathrm{GL}_2(B)$ via the isomorphisms $\mathfrak{i}_1, \mathfrak{i}_2$ given by (5.12), (5.13). Thus, noting that

$$\frac{\partial^2}{\partial z_1 \partial z_1^\rho} = \frac{1}{4}\left(\frac{\partial^2}{\partial x_1^2} - u\frac{\partial^2}{\partial x_2^2}\right), \quad \frac{\partial^2}{\partial z_2 \partial z_2^\rho} = \frac{1}{4}\left(\frac{\partial^2}{\partial x_3^2} - u\frac{\partial^2}{\partial x_4^2}\right),$$

we see that

$$\omega_\psi(2v^{-1}\mathtt{X} - 2^{-1}v\mathtt{Y})\varphi' = \sqrt{-1}(k+2)\varphi'.$$

This implies that

$$\omega_\psi(1, (\alpha, 1))\varphi' = \chi_{k+2}(\alpha)\varphi'$$

for $\alpha \in E^1$ since

$$\begin{aligned}\mathfrak{i}_1(\alpha) &= \begin{pmatrix} a & 2b \\ \frac{bu}{2} & a \end{pmatrix} = \begin{pmatrix} 1 & \\ & 2^{-1}v \end{pmatrix}\begin{pmatrix} \cos\theta & \sin\theta \\ -\sin\theta & \cos\theta \end{pmatrix}\begin{pmatrix} 1 & \\ & 2v^{-1} \end{pmatrix} \\ &= \exp((2v^{-1}\mathtt{X} - 2^{-1}v\mathtt{Y})\theta)\end{aligned}$$

if we write $\alpha = a + b\mathbf{i} = e^{\sqrt{-1}\theta}$. This completes the proof. □

5.7. Schwartz functions on $\mathbb{X}$

Let $\varphi \in \mathcal{S}(\mathbb{X})$ be the partial Fourier transform of the Schwartz function $\varphi' \in \mathcal{S}(\mathbb{X}')$ given in §5.6. (We also write φ_μ and φ_k for the partial Fourier transforms of φ'_μ and φ'_k, respectively, to indicate the dependence on a unitary ramified character μ in the case (rps) and on a non-negative integer k in the cases (ds), (fd).) Then we have

$$\langle \varphi, \varphi \rangle = \langle \varphi', \varphi' \rangle = 1.$$

Also, since the partial Fourier transform is a $\mathrm{G}(\mathrm{U}(V)^0 \times \mathrm{U}(W))$-equivariant map, φ satisfies the same equivariance properties as φ'. In this section, we compute φ explicitly.

We need to introduce more notation. Put $\kappa_1 = 1$ and $\kappa_2 = -J_1$. We define a quadratic F-algebra K by

$$K = F + F\mathbf{j}.$$

We write

$$\begin{aligned} x &= \mathbf{e}_1 z_1 + \mathbf{e}_2 z_2 = x_1\mathbf{e}_1 + x_2\mathbf{e}_2 + x_3\mathbf{e}_3 + x_4\mathbf{e}_4 \in \mathbb{X}, \\ z_i &= \alpha_i + \beta_i\mathbf{j} \in K, \end{aligned}$$

so that

$$\alpha_1 = x_1, \quad \beta_1 = x_4, \quad \alpha_2 = x_2, \quad \beta_2 = \frac{1}{J_1}x_3.$$

Recall that the Weil index $\gamma_F(\psi)$ is an 8th root of unity such that

$$\int_F \phi(x)\psi(x^2)\,dx = \gamma_F(\psi)|2|^{-\frac{1}{2}}\int_F \hat{\phi}(x)\psi\left(-\frac{x^2}{4}\right)dx$$

for all $\phi \in \mathcal{S}(F)$, where $\hat{\phi}$ is the Fourier transform of ϕ with respect to ψ and dx is the self-dual Haar measure on F with respect to ψ. For any non-negative integer n, let $H_n(x)$ denote the Hermite polynomial defined by

$$H_n(x) = (-1)^n e^{x^2} \frac{d^n}{dx^n} \left(e^{-x^2} \right).$$

5.7.1. The case (ur).

5.7.1.1. *The case when E is split and F is unramified.* We use the notation of §5.6.1.1. By the partial Fourier transform given in §5.4.1, we have $\varphi(x) = |2|^{-1} \prod_{i=1}^4 \varphi_i(x_i)$, where

$$\varphi_i(x_i) = \left| \frac{2a_i}{b_i} \right| \psi \left(\frac{a_i}{2} x_i^2 \right) \int_{\mathfrak{o}} \psi \left(\frac{a_i}{b_i^2} (x_i')^2 - \frac{2a_i}{b_i} x_i x_i' \right) dx_i'.$$

LEMMA 5.7.1. *Assume that ψ is of order zero. Put*

$$I(a, b) = \int_{\mathfrak{o}} \psi(ax^2 + bx)\, dx$$

for $a, b \in F$.

(i) *We have*

$$I(a, b) = \begin{cases} \mathbb{I}_{\mathfrak{o}}(b) & \text{if } a \in \mathfrak{o}, \\ \psi(-\frac{b^2}{4a}) \gamma_F(a\psi) |2a|^{-\frac{1}{2}} \mathbb{I}_{\mathfrak{o}}(\frac{b}{2a}) & \text{if } 4a \notin \mathfrak{p}, \end{cases}$$

where $a\psi$ is the non-trivial character of F given by $(a\psi)(x) = \psi(ax)$.

(ii) *If $F = \mathbb{Q}_2$ and $2a \in \mathfrak{o}^\times$, then we have*

$$I(a, b) = \mathbb{I}_{\mathfrak{o}^\times}(\tfrac{b}{a}).$$

PROOF. If $a \in \mathfrak{o}$, then we have $I(a, b) = \hat{\mathbb{I}}_{\mathfrak{o}}(b) = \mathbb{I}_{\mathfrak{o}}(b)$. If $2a \notin \mathfrak{p}$, then we change the variable $x \mapsto x + \frac{y}{2a}$ with $y \in \mathfrak{o}$ to get

$$\begin{aligned} I(a, b) &= \int_{\mathfrak{o}} \psi \left(a \left(x + \frac{y}{2a} \right)^2 + b \left(x + \frac{y}{2a} \right) \right) dx \\ &= \psi \left(\frac{y^2}{4a} + \frac{by}{2a} \right) \int_{\mathfrak{o}} \psi(ax^2 + xy + bx)\, dx \\ &= \psi \left(\frac{y^2}{4a} + \frac{by}{2a} \right) I(a, b). \end{aligned}$$

Assume that $4a \notin \mathfrak{p}$. Then we have $I(a, b) = \psi(\frac{by}{2a}) I(a, b)$ for all $y \in \mathfrak{o}$, so that $I(a, b) = 0$ unless $\frac{b}{2a} \in \mathfrak{o}$. If $\frac{b}{2a} \in \mathfrak{o}$, then we have

$$I(a, b) = \int_{\mathfrak{o}} \psi \left(a \left(x + \frac{b}{2a} \right)^2 - \frac{b^2}{4a} \right) dx = \psi \left(-\frac{b^2}{4a} \right) \int_{\mathfrak{o}} \psi(ax^2)\, dx.$$

On the other hand, by definition, we have

$$\int_F \phi(x) \psi(ax^2)\, dx = \gamma_F(a\psi) |2a|^{-\frac{1}{2}} \int_F \hat{\phi}(x) \psi \left(-\frac{x^2}{4a} \right) dx$$

for all $\phi \in \mathcal{S}(F)$. Hence we have

$$\int_{\mathfrak{o}} \psi(ax^2)\, dx = \gamma_F(a\psi) |2a|^{-\frac{1}{2}} \int_{\mathfrak{o}} \psi \left(-\frac{x^2}{4a} \right) dx = \gamma_F(a\psi) |2a|^{-\frac{1}{2}}.$$

This proves (i).

Assume that $F = \mathbb{Q}_2$ and $2a \in \mathfrak{o}^\times$. As above, we have $I(a,b) = \psi(\frac{by}{2a})I(a,b)$ for all $y \in 2\mathfrak{o}$, so that $I(a,b) = 0$ unless $\frac{b}{a} \in \mathfrak{o}$. Also, we have

$$I(a,b) = \psi\left(-\frac{b^2}{4a}\right)\int_{\mathfrak{o}} \psi(ax^2)\,dx$$

if $\frac{b}{a} \in 2\mathfrak{o}$. Changing the variable $x \mapsto x+1$, we have

$$\int_{\mathfrak{o}} \psi(ax^2)\,dx = \int_{\mathfrak{o}} \psi(ax^2 + 2ax + a)\,dx = \psi(a)\int_{\mathfrak{o}} \psi(ax^2)\,dx.$$

Since $F = \mathbb{Q}_2$, ψ is of order zero, and $\psi(a)^2 = \psi(2a) = 1$, we must have $\psi(a) = -1$. Hence we have

$$\int_{\mathfrak{o}} \psi(ax^2)\,dx = 0,$$

so that $I(a,b) = 0$ if $\frac{b}{a} \in 2\mathfrak{o}$. Assume that $\frac{b}{a} \in \mathfrak{o}^\times$. Since $F = \mathbb{Q}_2$, we may write $a = y + \frac{1}{2}$ and $\frac{b}{2a} = z + \frac{1}{2}$ for some $y, z \in \mathfrak{o}$. Then we have

$$\begin{aligned} I(a,b) &= \int_{\mathfrak{o}} \psi\left(x^2\left(y+\frac{1}{2}\right) + 2x\left(y+\frac{1}{2}\right)\left(z+\frac{1}{2}\right)\right)dx \\ &= \int_{\mathfrak{o}} \psi\left(\frac{1}{2}x^2 + \frac{1}{2}x\right)dx \\ &= 1 \end{aligned}$$

since $\frac{1}{2}x(x+1) \in \mathfrak{o}$ for all $x \in \mathfrak{o}$. This proves (ii). $\square$

By Lemma 5.7.1, we have

$$I\left(\frac{a_i}{b_i^2}, -\frac{2a_i}{b_i}x_i\right) = \begin{cases} \mathbb{I}_{\mathfrak{o}}(\frac{2a_i}{b_i}x_i) & \text{if } \frac{a_i}{b_i^2} \in \mathfrak{o}, \\ \psi(-a_i x_i^2)\gamma_F(a_i\psi)\left|\frac{b_i^2}{2a_i}\right|^{\frac{1}{2}} \mathbb{I}_{\mathfrak{o}}(b_i x_i) & \text{if } \frac{4a_i}{b_i^2} \notin \mathfrak{p}, \end{cases}$$

so that

$$\begin{aligned} \varphi_1(x_1) &= |2| \cdot \psi\left(\frac{t}{2}x_1^2\right) \cdot \mathbb{I}_{\mathfrak{o}}(2x_1), \\ \varphi_2(x_2) &= |2J_1| \cdot \psi\left(-\frac{tJ_1}{2}x_2^2\right) \cdot \mathbb{I}_{\mathfrak{o}}(2J_1x_2), \\ \varphi_3(x_3) &= \gamma_F(-tJ_2\psi) \cdot |2J_2|^{\frac{1}{2}} \cdot \psi\left(\frac{tJ_2}{2}x_3^2\right) \cdot \mathbb{I}_{\mathfrak{o}}(2Jx_3), \\ \varphi_4(x_4) &= \gamma_F(tJ\psi) \cdot |2J|^{\frac{1}{2}} \cdot \psi\left(-\frac{tJ}{2}x_4^2\right) \cdot \mathbb{I}_{\mathfrak{o}}(2Jx_4). \end{aligned}$$

We have

$$\begin{aligned} \gamma_F(-tJ_2\psi) \cdot \gamma_F(tJ\psi) &= \gamma_F(-2tJ_2, \frac{1}{2}\psi) \cdot \gamma_F(2tJ, \frac{1}{2}\psi) \cdot \gamma_F(\frac{1}{2}\psi)^2 \\ &= \gamma_F(-J_1, \frac{1}{2}\psi) \cdot (-2tJ_2, 2tJ)_F \cdot \gamma_F(-1, \frac{1}{2}\psi)^{-1} \\ &= \gamma_F(J_1, \frac{1}{2}\psi) \cdot (2tJ_2, J_1)_F. \end{aligned}$$

Hence we have

$$\begin{aligned}\varphi(x) &= \gamma_F(J_1, \frac{1}{2}\psi) \cdot (2tJ_2, J_1)_F \cdot |2|^2 |J_1|^{\frac{3}{2}} |J_2| \\ &\quad \times \psi\left(\frac{t}{2}(x_1^2 - J_1 x_2^2 + J_2 x_3^2 - J x_4^2)\right) \cdot \mathbb{I}_{\mathfrak{o}}(2x_1)\mathbb{I}_{\mathfrak{o}}(2J_1 x_2)\mathbb{I}_{\mathfrak{o}}(2Jx_3)\mathbb{I}_{\mathfrak{o}}(2Jx_4) \\ &= \gamma_F(J_1, \frac{1}{2}\psi) \cdot (2tJ_2, J_1)_F \cdot |2|^2 |J_1|^{\frac{3}{2}} |J_2| \cdot \tilde{\varphi}_{\kappa_1}(z_1)\tilde{\varphi}_{\kappa_2}(z_2),\end{aligned}$$

where

$$\tilde{\varphi}_\kappa(z) = \psi\left(\frac{\kappa t}{2} \mathrm{N}_{K/F}(z)\right) \cdot \mathbb{I}_{\mathfrak{o}+\mathfrak{o}\frac{\mathbf{i}}{J}}(2\kappa z)$$

for $z \in K$.

5.7.1.2. *The case when E is split and F is ramified.* We use the notation of §5.6.1.2. Note that the inverse different $\mathfrak{d}^{-1}$ is equal to $\varpi^{-d}\mathfrak{o}$. By the partial Fourier transform given in §5.4.1, we have $\varphi(x) = q^d |2|^{-1} \prod_{i=1}^4 \varphi_i(x_i)$, where

$$\varphi_i(x_i) = \left|\frac{2a_i}{b_i}\right| \psi\left(\frac{a_i}{2}x_i^2\right) \int_{\mathfrak{o}} \psi\left(\frac{a_i}{b_i^2}(x_i')^2 - \frac{2a_i}{b_i}x_i x_i'\right) dx_i'.$$

In particular, we have

$$\varphi_i(x_i) \in \mathbb{Z}[q^{-\frac{1}{2}}, \mu_{p^\infty}]$$

for all $x_i \in F$, where p is the residual characteristic of F and μ_{p^∞} is the group of p-power roots of unity. If further $2 \in \mathfrak{o}^\times$, then we have

$$\varphi_i(x_i) = \psi\left(\frac{a_i}{2}x_i^2\right) \cdot \hat{\mathbb{I}}_{\mathfrak{o}}\left(-\frac{2a_i}{b_i}x_i\right) = q^{-\frac{d}{2}} \cdot \psi\left(\frac{a_i}{2}x_i^2\right) \cdot \mathbb{I}_{\mathfrak{d}^{-1}}(x_i)$$

and hence

$$\begin{aligned}\varphi(x) &= q^{-d} \cdot \psi\left(\frac{t}{2}(x_1^2 - J_1 x_2^2 - J_2 x_3^2 + J x_4^2)\right) \cdot \mathbb{I}_{\mathfrak{d}^{-1}}(x_1)\mathbb{I}_{\mathfrak{d}^{-1}}(x_2)\mathbb{I}_{\mathfrak{d}^{-1}}(x_3)\mathbb{I}_{\mathfrak{d}^{-1}}(x_4) \\ &= q^{-d} \cdot \tilde{\varphi}_{\kappa_1}(z_1)\tilde{\varphi}_{\kappa_2}(z_2),\end{aligned}$$

where

$$\tilde{\varphi}_\kappa(z) = \psi\left(\frac{\kappa t}{4} \mathrm{tr}_{K/F}(z^2)\right) \cdot \mathbb{I}_{\mathfrak{d}^{-1}+\mathfrak{d}^{-1}\mathbf{j}}(z)$$

for $z \in K$.

5.7.1.3. *The case when E is inert and $J \in (F^\times)^2$.* We use the notation of §5.6.1.3. By the partial Fourier transform given in §5.4.2.1, we have

$$\begin{aligned}\varphi(x) &= |J|^{\frac{1}{2}} \cdot \mathbb{I}_{\mathfrak{o}}(x_1 + tx_4)\hat{\mathbb{I}}_{\mathfrak{o}}\left(-\frac{1}{2}(x_1 - tx_4)\right) \\ &\quad \times \mathbb{I}_{\mathfrak{o}}\left(s\left(x_2 + \frac{t}{J_1}x_3\right)\right)\hat{\mathbb{I}}_{\mathfrak{o}}\left(\frac{J_1}{2s}\left(x_2 - \frac{t}{J_1}x_3\right)\right) \\ &= |J|^{\frac{1}{2}} \cdot \mathbb{I}_{\mathfrak{o}}(x_1)\mathbb{I}_{\mathfrak{o}}(sx_2)\mathbb{I}_{\mathfrak{o}}\left(\frac{st}{J_1}x_3\right)\mathbb{I}_{\mathfrak{o}}(tx_4) \\ &= |J|^{\frac{1}{2}} \cdot \mathbb{I}_{\mathfrak{o}+\mathfrak{o}\frac{\mathbf{i}}{t}}(z_1)\mathbb{I}_{\mathfrak{o}+\mathfrak{o}\frac{\mathbf{i}}{t}}(sz_2).\end{aligned}$$

5.7.1.4. *The case when E is inert, and $J_1 \in (F^\times)^2$ or $J_2 \in (F^\times)^2$.* We use the notation of §5.6.1.4. By the partial Fourier transform given in §5.4.3, we have

$$\begin{aligned}\varphi(x) &= |J|^{\frac{1}{2}} \cdot \mathbb{I}_{\mathfrak{o}}(x_1 + tx_2)\hat{\mathbb{I}}_{\mathfrak{o}}\left(-\frac{1}{2}(x_1 - tx_2)\right) \\ &\quad \times \mathbb{I}_{\mathfrak{o}}\left(s\left(x_4 + \frac{1}{t}x_3\right)\right)\hat{\mathbb{I}}_{\mathfrak{o}}\left(\frac{J}{2s}\left(x_4 - \frac{1}{t}x_3\right)\right) \\ &= |J|^{\frac{1}{2}} \cdot \mathbb{I}_{\mathfrak{o}}(x_1)\mathbb{I}_{\mathfrak{o}}(tx_2)\mathbb{I}_{\mathfrak{o}}\left(\frac{s}{t}x_3\right)\mathbb{I}_{\mathfrak{o}}(sx_4) \\ &= |J|^{\frac{1}{2}} \cdot \mathbb{I}_{\mathfrak{o}+\mathfrak{o}\frac{\mathbf{j}}{s}}(z_1)\mathbb{I}_{\mathfrak{o}+\mathfrak{o}\frac{\mathbf{j}}{s}}(tz_2).\end{aligned}$$

5.7.1.5. *The case when E is ramified.* We use the notation of §5.6.1.5. By the partial Fourier transform given in §5.4.2.2, we have

$$\begin{aligned}\varphi(x) &= q^{-\frac{1}{2}} \\ &\quad \times \mathbb{I}_{\mathfrak{o}}\left(\frac{1}{2}\left(x_1 + t_1x_2 + \frac{t}{t_1}x_3 + tx_4\right)\right)\hat{\mathbb{I}}_{\mathfrak{o}}\left(-\frac{1}{2}\left(x_1 - t_1x_2 + \frac{t}{t_1}x_3 - tx_4\right)\right) \\ &\quad \times \hat{\mathbb{I}}_{\mathfrak{o}}\left(-\frac{1}{2}\left(x_1 + t_1x_2 - \frac{t}{t_1}x_3 - tx_4\right)\right)\mathbb{I}_{\mathfrak{o}}\left(\frac{u}{2}\left(x_1 - t_1x_2 - \frac{t}{t_1}x_3 + tx_4\right)\right) \\ &= q^{-\frac{1}{2}} \cdot \mathbb{I}_{\mathfrak{o}}(x_1 - tx_4)\mathbb{I}_{\mathfrak{o}}\left(x_2 - \frac{t}{J_1}x_3\right) \\ &\quad \times \mathbb{I}_{\mathfrak{o}}\left(x_1 + t_1x_2 + \frac{t}{t_1}x_3 + tx_4\right)\mathbb{I}_{\varpi^{-1}\mathfrak{o}}\left(x_1 - t_1x_2 - \frac{t}{t_1}x_3 + tx_4\right) \\ &= q^{-\frac{1}{2}} \cdot \mathbb{I}_{\mathfrak{o}}(\alpha_1 - t\beta_1)\mathbb{I}_{\mathfrak{o}}(\alpha_2 - t\beta_2) \\ &\quad \times \mathbb{I}_{\mathfrak{o}}(\alpha_1 + t\beta_1 + t_1\alpha_2 + tt_1\beta_2)\mathbb{I}_{\varpi^{-1}\mathfrak{o}}(\alpha_1 + t\beta_1 - t_1\alpha_2 - tt_1\beta_2).\end{aligned}$$

5.7.2. The case (rps). We use the notation of §5.6.2. By the partial Fourier transform given in §5.4.1, we have $\varphi(x) = q^{\frac{n+1}{2}}(q-1)^{-\frac{1}{2}}\prod_{i=1}^4 \varphi_i(x_i)$, where

$$\varphi_i(x_i) = \psi\left(\frac{a_i}{2}x_i^2\right) \cdot \hat{\mathbb{I}}_{\mathfrak{o}}\left(-\frac{2a_i}{b_i}x_i\right) = \mathbb{I}_{\mathfrak{o}}(x_i)$$

for $i = 1, 2$,

$$\varphi_3(x_3) = \psi\left(\frac{a_3}{2}x_3^2\right) \cdot \hat{\mathbb{I}}_{\varpi^n\mathfrak{o}}\left(-\frac{2a_3}{b_3}x_3\right) = q^{-n} \cdot \psi\left(-\frac{tJ_2}{2}x_3^2\right) \cdot \mathbb{I}_{\varpi^{-n}\mathfrak{o}}(x_3),$$

and

$$\varphi_4(x_4) = \psi\left(\frac{a_4}{2}x_4^2\right) \cdot \widehat{\mathbb{I}_{\mathfrak{o}^\times}\mu}\left(-\frac{2a_4}{b_4}x_4\right) = \psi\left(\frac{tJ}{2}x_4^2\right) \cdot \widehat{\mathbb{I}_{\mathfrak{o}^\times}\mu}(x_4).$$

Since μ is of conductor q^n, we have

$$\widehat{\mathbb{I}_{\mathfrak{o}^\times}\mu} = q^{-n} \cdot \mathfrak{g}(\mu, \psi) \cdot \mathbb{I}_{\varpi^{-n}\mathfrak{o}^\times}\mu^{-1},$$

where

$$\mathfrak{g}(\mu, \psi) = \int_{\varpi^{-n}\mathfrak{o}^\times} \mu(y)\psi(y)\, dy.$$

Note that $|\mathfrak{g}(\mu,\psi)| = q^{\frac{n}{2}}$. Hence we have

$$\begin{aligned}\varphi(x) &= q^{-\frac{3}{2}n+\frac{1}{2}}(q-1)^{-\frac{1}{2}}\cdot\mathfrak{g}(\mu,\psi)\\ &\quad\times\psi\left(\frac{t}{2}(-J_2x_3^2+Jx_4^2)\right)\cdot\mathbb{I}_{\mathfrak{o}}(x_1)\mathbb{I}_{\mathfrak{o}}(x_2)\mathbb{I}_{\varpi^{-n}\mathfrak{o}}(x_3)\mathbb{I}_{\varpi^{-n}\mathfrak{o}^\times}(x_4)\mu(x_4)^{-1}\\ &= q^{-\frac{3}{2}n+\frac{1}{2}}(q-1)^{-\frac{1}{2}}\cdot\mathfrak{g}(\mu,\psi)\\ &\quad\times\psi\left(\frac{\kappa_1 tJ}{2}\beta_1^2\right)\cdot\mathbb{I}_{\mathfrak{o}}(\alpha_1)\mathbb{I}_{\varpi^{-n}\mathfrak{o}^\times}(\beta_1)\mu(\beta_1)^{-1}\\ &\quad\times\psi\left(\frac{\kappa_2 tJ}{2}\beta_2^2\right)\cdot\mathbb{I}_{\mathfrak{o}}(\alpha_2)\mathbb{I}_{\varpi^{-n}\mathfrak{o}}(\beta_2).\end{aligned}$$

5.7.3. The case (st).

5.7.3.1. *The case when B_1 and B_2 are split.* We use the notation of §5.6.3.1. By the partial Fourier transform given in §5.4.1, we have $\varphi(x) = q^{\frac{1}{2}}\prod_{i=1}^4\varphi_i(x_i)$, where

$$\varphi_i(x_i) = \psi\left(\frac{a_i}{2}x_i^2\right)\cdot\hat{\mathbb{I}}_{\mathfrak{o}}\left(-\frac{2a_i}{b_i}x_i\right) = \mathbb{I}_{\mathfrak{o}}(x_i)$$

for $i = 1,2,4$ and

$$\varphi_3(x_3) = \psi\left(\frac{a_3}{2}x_3^2\right)\cdot\hat{\mathbb{I}}_{\mathfrak{p}}\left(-\frac{2a_3}{b_3}x_3\right) = q^{-1}\cdot\psi\left(-\frac{tJ_2}{2}x_3^2\right)\cdot\mathbb{I}_{\varpi^{-1}\mathfrak{o}}(x_3).$$

Hence we have

$$\begin{aligned}\varphi(x) &= q^{-\frac{1}{2}}\cdot\psi\left(-\frac{tJ_2}{2}x_3^2\right)\cdot\mathbb{I}_{\mathfrak{o}}(x_1)\mathbb{I}_{\mathfrak{o}}(x_2)\mathbb{I}_{\varpi^{-1}\mathfrak{o}}(x_3)\mathbb{I}_{\mathfrak{o}}(x_4)\\ &= q^{-\frac{1}{2}}\cdot\mathbb{I}_{\mathfrak{o}}(\alpha_1)\mathbb{I}_{\mathfrak{o}}(\beta_1)\cdot\psi\left(\frac{\kappa_2 tJ}{2}\beta_2^2\right)\cdot\mathbb{I}_{\mathfrak{o}}(\alpha_2)\mathbb{I}_{\varpi^{-1}\mathfrak{o}}(\beta_2).\end{aligned}$$

5.7.3.2. *The case when B_1 and B_2 are ramified.* We use the notation of §5.6.3.2. By the partial Fourier transform given in §5.4.2.1, we have

$$\begin{aligned}\varphi(x) &= q^{-1}\cdot\mathbb{I}_{\mathfrak{o}}(x_1+tx_4)\hat{\mathbb{I}}_{\mathfrak{o}}\left(-\frac{1}{2}(x_1-tx_4)\right)\\ &\quad\times\mathbb{I}_{\mathfrak{o}}\left(s\left(x_2+\frac{t}{J_1}x_3\right)\right)\hat{\mathbb{I}}_{\mathfrak{o}}\left(\frac{J_1}{2s}\left(x_2-\frac{t}{J_1}x_3\right)\right)\\ &= q^{-1}\cdot\mathbb{I}_{\mathfrak{o}}(x_1)\mathbb{I}_{\varpi^{-1}\mathfrak{o}}(x_4)\mathbb{I}_{\mathfrak{o}}\left(x_2+\frac{t}{J_1}x_3\right)\mathbb{I}_{\varpi^{-1}\mathfrak{o}}\left(x_2-\frac{t}{J_1}x_3\right)\\ &= q^{-1}\cdot\mathbb{I}_{\mathfrak{o}}(\alpha_1)\mathbb{I}_{\varpi^{-1}\mathfrak{o}}(\beta_1)\mathbb{I}_{\mathfrak{o}}(\alpha_2+t\beta_2)\mathbb{I}_{\varpi^{-1}\mathfrak{o}}(\alpha_2-t\beta_2).\end{aligned}$$

5.7.4. The case (1d). We use the notation of §5.6.4. By the partial Fourier transform given in §5.4.3, we have

$$\begin{aligned}\varphi(x) &= q^{-\frac{1}{2}}\cdot\mathbb{I}_{\mathfrak{o}}(x_1+tx_2)\hat{\mathbb{I}}_{\mathfrak{o}}\left(-\frac{1}{2}(x_1-tx_2)\right)\\ &\quad\times\mathbb{I}_{\mathfrak{o}}\left(s\left(x_4+\frac{1}{t}x_3\right)\right)\hat{\mathbb{I}}_{\mathfrak{o}}\left(\frac{J}{2s}\left(x_4-\frac{1}{t}x_3\right)\right)\\ &= q^{-\frac{1}{2}}\cdot\mathbb{I}_{\mathfrak{o}}(x_1)\mathbb{I}_{\mathfrak{o}}(x_2)\mathbb{I}_{\mathfrak{o}}\left(x_4+\frac{1}{t}x_3\right)\mathbb{I}_{\varpi^{-1}\mathfrak{o}}\left(x_4-\frac{1}{t}x_3\right)\\ &= q^{-\frac{1}{2}}\cdot\mathbb{I}_{\mathfrak{o}}(\alpha_1)\mathbb{I}_{\mathfrak{o}}(\alpha_2)\mathbb{I}_{\mathfrak{o}}(\beta_1+t\beta_2)\mathbb{I}_{\varpi^{-1}\mathfrak{o}}(\beta_1-t\beta_2).\end{aligned}$$

5.7.5. The case (ds).

5.7.5.1. *The case when B_1 and B_2 are split.* We use the notation of §5.6.5.1. By the partial Fourier transform given in §5.4.2.1, we have

$$\varphi(x) = c_k^{-\frac{1}{2}} \left|\frac{uJ}{4}\right|^{\frac{1}{2}} \int_{-\infty}^{\infty} (x_2' - v\sqrt{-1}x_1')^k e^{-\frac{\pi}{2}(vx_1'^2+\frac{1}{v}x_2'^2)} e^{2\pi\sqrt{-1}x_2'y_2'}\, dx_2'$$
$$\times \int_{-\infty}^{\infty} e^{-\frac{\pi}{2}(vx_3'^2+\frac{1}{v}x_4'^2)} e^{2\pi\sqrt{-1}x_4'y_4'}\, dx_4'.$$

LEMMA 5.7.2. *Let k be a non-negative integer and v a positive real number. Put*

$$I(x,y) = \int_{-\infty}^{\infty} \left(x + \frac{\sqrt{-1}}{v}w\right)^k e^{-\frac{\pi}{2}(vx^2+\frac{1}{v}w^2)} e^{2\pi\sqrt{-1}wy}\, dw$$

for $x, y \in \mathbb{R}$. Then we have

$$I(x,y) = \frac{1}{\sqrt{2^{k-1}\pi^k v^{k-1}}} \cdot H_k\left(\sqrt{2\pi v}\left(\frac{1}{2}x - y\right)\right) \cdot e^{-\pi v(\frac{1}{2}x^2+2y^2)}.$$

PROOF. We have

$$I(x,y) = e^{-\frac{\pi v}{2}x^2} \sqrt{\frac{2v}{\pi}} \int_{-\infty}^{\infty} \left(x + \sqrt{\frac{2}{\pi v}}\sqrt{-1}w\right)^k e^{-w^2} e^{2\sqrt{2\pi v}\sqrt{-1}wy}\, dw$$
$$= e^{-\frac{\pi v}{2}x^2} \sqrt{\frac{2^{k+1}}{\pi^{k+1}v^{k-1}}} \int_{-\infty}^{\infty} \left(\sqrt{\frac{\pi v}{2}}x + \sqrt{-1}w\right)^k e^{-(w-\sqrt{2\pi v}\sqrt{-1}y)^2-2\pi v y^2}\, dw$$
$$= e^{-\frac{\pi v}{2}x^2-2\pi v y^2} \sqrt{\frac{2^{k+1}}{\pi^{k+1}v^{k-1}}} \int_{-\infty}^{\infty} \left(\sqrt{\frac{\pi v}{2}}x + \sqrt{-1}w - \sqrt{2\pi v}y\right)^k e^{-w^2}\, dw.$$

Hence the assertion follows from the integral representation of the Hermite polynomial:

$$H_k(x) = \frac{2^k}{\sqrt{\pi}} \int_{-\infty}^{\infty} (x + \sqrt{-1}w)^k e^{-w^2}\, dw. \qquad \square$$

By Lemma 5.7.2, we have

$$\varphi(x) = c_k^{-\frac{1}{2}} \left|\frac{uJ}{4}\right|^{\frac{1}{2}} \cdot \frac{(-v\sqrt{-1})^k}{\sqrt{2^{k-2}\pi^k v^{k-2}}}$$
$$\times H_k\left(\sqrt{2\pi v}\left(\frac{1}{2}x_1' - y_2'\right)\right) \cdot e^{-\pi v(\frac{1}{2}x_1'^2+2y_2'^2+\frac{1}{2}x_3'^2+2y_4'^2)}$$
$$= \frac{|uJ|^{\frac{1}{2}}(-\sqrt{-1})^k}{2^{\frac{k}{2}-1}\sqrt{k!}} \cdot H_k(\sqrt{2\pi v}x_1) \cdot e^{-\pi v(x_1^2+Jx_4^2+J_1x_2^2+\frac{J}{J_1}x_3^2)}$$
$$= \frac{|uJ|^{\frac{1}{2}}(-\sqrt{-1})^k}{2^{\frac{k}{2}-1}\sqrt{k!}} \cdot H_k(\sqrt{2\pi v}\alpha_1) \cdot e^{-\pi v(\alpha_1^2+J\beta_1^2)} \cdot e^{-\pi v J_1(\alpha_2^2+J\beta_2^2)}.$$

5.7.5.2. *The case when B_1 and B_2 are ramified.* We use the notation of §5.6.5.2. By the partial Fourier transform given in §5.4.2.1, we have

$$\varphi(x) = c_k^{-\frac{1}{2}} \left|\frac{uJ}{4}\right|^{\frac{1}{2}} \int_{-\infty}^{\infty} (x_2' - v\sqrt{-1}x_1')^k e^{-\frac{\pi}{2}(vx_1'^2+\frac{1}{v}x_2'^2)} e^{2\pi\sqrt{-1}x_2'y_2'}\, dx_2'$$
$$\times \int_{-\infty}^{\infty} e^{-\frac{\pi}{2}(vx_3'^2+\frac{1}{v}x_4'^2)} e^{2\pi\sqrt{-1}x_4'y_4'}\, dx_4'.$$

By Lemma 5.7.2, we have

$$\begin{aligned}\varphi(x) &= c_k^{-\frac{1}{2}} \left|\frac{uJ}{4}\right|^{\frac{1}{2}} \cdot \frac{(-v\sqrt{-1})^k}{\sqrt{2^{k-2}\pi^k v^{k-2}}} \\ &\quad \times H_k\left(\sqrt{2\pi v}\left(\frac{1}{2}x_1' - y_2'\right)\right) \cdot e^{-\pi v(\frac{1}{2}x_1'^2+2y_2'^2+\frac{1}{2}x_3'^2+2y_4'^2)} \\ &= \frac{|uJ|^{\frac{1}{2}}(-\sqrt{-1})^k}{2^{\frac{k}{2}-1}\sqrt{k!}} \cdot H_k(\sqrt{2\pi v}x_1) \cdot e^{-\pi v(x_1^2+Jx_4^2-J_1x_2^2-\frac{J}{J_1}x_3^2)} \\ &= \frac{|uJ|^{\frac{1}{2}}(-\sqrt{-1})^k}{2^{\frac{k}{2}-1}\sqrt{k!}} \cdot H_k(\sqrt{2\pi v}\alpha_1) \cdot e^{-\pi v(\alpha_1^2+J\beta_1^2)} \cdot e^{\pi v J_1(\alpha_2^2+J\beta_2^2)}.\end{aligned}$$

5.7.6. The case (fd). We use the notation of §5.6.6. By the partial Fourier transform given in §5.4.3, we have

$$\begin{aligned}\varphi(x) &= c_k^{-\frac{1}{2}} \left|\frac{J}{u}\right|^{\frac{1}{2}} \int_{-\infty}^{\infty} \left(x_1' + \frac{\sqrt{-1}}{v}x_2'\right)^k e^{-\frac{\pi}{2}(vx_1'^2+\frac{1}{v}x_2'^2)} e^{2\pi\sqrt{-1}x_2'y_2'}\, dx_2' \\ &\quad \times \int_{-\infty}^{\infty} e^{-\frac{\pi}{2}(vx_3'^2+\frac{1}{v}x_4'^2)} e^{2\pi\sqrt{-1}x_4'y_4'}\, dx_4'.\end{aligned}$$

By Lemma 5.7.2, we have

$$\begin{aligned}\varphi(x) &= c_k^{-\frac{1}{2}} \left|\frac{J}{u}\right|^{\frac{1}{2}} \cdot \frac{1}{\sqrt{2^{k-2}\pi^k v^{k-2}}} \\ &\quad \times H_k\left(\sqrt{2\pi v}\left(\frac{1}{2}x_1' - y_2'\right)\right) \cdot e^{-\pi v(\frac{1}{2}x_1'^2+2y_2'^2+\frac{1}{2}x_3'^2+2y_4'^2)} \\ &= \frac{|uJ|^{\frac{1}{2}}}{2^{\frac{k}{2}-1}\sqrt{k!}} \cdot H_k(\sqrt{2\pi v}x_1) \cdot e^{-\pi v(x_1^2+J_1x_2^2-Jx_4^2-\frac{J}{J_1}x_3^2)} \\ &= \frac{|uJ|^{\frac{1}{2}}}{2^{\frac{k}{2}-1}\sqrt{k!}} \cdot H_k(\sqrt{2\pi v}\alpha_1) \cdot e^{-\pi v(\alpha_1^2-J\beta_1^2)} \cdot e^{-\pi v J_1(\alpha_2^2-J\beta_2^2)}.\end{aligned}$$

CHAPTER 6

Explicit Form of the Rallis Inner Product Formula

In this chapter, we prove an explicit formula (Theorem 6.3.2) for the inner product of the theta lift $\theta_\varphi(f_B)$, where $\varphi \in \mathcal{S}(\mathbb{X}(\mathbb{A}))$ is the Schwartz function given in Chapter 5 and $f_B \in \pi_B$ is the automorphic form which will be given via the theory of new vectors (see §6.2). For this, we will explicate the Rallis inner product formula (Proposition 4.4.2), which essentially boils down to computing the integral

$$Z_v = \int_{B_v^1} \langle \omega_\psi(g_{1,v})\varphi_v, \varphi_v \rangle \langle \pi_{B,v}(g_{1,v}) f_{B,v}, f_{B,v} \rangle \, dg_{1,v}.$$

This computation is almost straightforward (see §6.6), except that it is not easy to directly compute $\langle \omega_\psi(g_{1,v})\varphi_v, \varphi_v \rangle$ since the Weil representation on $\mathcal{S}(\mathbb{X}_v)$ is complicated. However, we have defined φ_v as the image of a nice Schwartz function φ_v' under the partial Fourier transform $\mathcal{S}(\mathbb{X}_v') \to \mathcal{S}(\mathbb{X}_v)$. Hence, by the Plancherel formula, we have

$$\langle \omega_\psi(g_{1,v})\varphi_v, \varphi_v \rangle = \langle \omega_\psi(g_{1,v})\varphi_v', \varphi_v' \rangle.$$

Since the Weil representation on $\mathcal{S}(\mathbb{X}_v')$ has a simple description, we can easily compute the right-hand side (see §6.5).

6.1. Measures

In Chapter 4, for any connected reductive algebraic group G over a number field F, we have always taken the Tamagawa measure on $G(\mathbb{A})$, which is a product of Haar measures on G_v defined in terms of a non-zero invariant differential form of top degree on G over F. However, with respect to this Haar measure, the volume of a hyperspecial maximal compact subgroup of G_v is not necessarily 1 for almost all v. For our applications, it is more convenient to take the "standard" measure on $G(\mathbb{A})$, which is a product of Haar measures on G_v such that the volume of a maximal compact subgroup of G_v is 1 for all v. In this section, we give a precise definition of the standard measures on $\mathbb{A}^\times \backslash B^\times(\mathbb{A})$ and $B^1(\mathbb{A})$, where B is a quaternion algebra over F, and compare them with the Tamagawa measures.

Let F be a number field and ψ the standard additive character of $\mathbb{A}/F$. Let $D = D_F$ be the discriminant of F. We have $|D| = \prod_{v \in \Sigma_{\mathrm{fin}}} q_v^{d_v}$, where d_v is the non-negative integer such that ψ_v is trivial on $\varpi_v^{-d_v}\mathfrak{o}_v$ but non-trivial on $\varpi_v^{-d_v-1}\mathfrak{o}_v$. For each place v of F, let $\zeta_v(s)$ be the local zeta function of F_v defined by

$$\zeta_v(s) = \begin{cases} (1-q_v^{-s})^{-1} & \text{if } v \text{ is finite,} \\ \pi^{-\frac{s}{2}}\Gamma(\frac{s}{2}) & \text{if } v \text{ is real,} \\ 2(2\pi)^{-s}\Gamma(s) & \text{if } v \text{ is complex.} \end{cases}$$

Note that

$$\zeta_v(1) = \begin{cases} 1 & \text{if } v \text{ is real}, \\ \pi^{-1} & \text{if } v \text{ is complex}. \end{cases}$$

Let $\zeta_F(s) = \prod_{v \in \Sigma_{\mathrm{fin}}} \zeta_v(s)$ be the Dedekind zeta function of F. Put

$$\rho_F := \operatorname*{Res}_{s=1} \zeta_F(s) = \frac{2^{r_1}(2\pi)^{r_2} hR}{|D|^{\frac{1}{2}} w},$$

where r_1 is the number of the real places of F, r_2 is the number of the complex places of F, $h = h_F$ is the class number of F, $R = R_F$ is the regulator of F, and $w = w_F$ is the number of roots of unity in F. For any connected reductive algebraic group G over F, let $\tau(G)$ denote the Tamagawa number of G.

From now on, we assume that F is totally real.

6.1.1. Measures on $\mathbb{A}^\times$. For each place v of F, we define a Haar measure $d^\times x_v^{\mathrm{Tam}}$ on $F_v^\times$ by

$$d^\times x_v^{\mathrm{Tam}} := \zeta_v(1) \cdot \frac{dx_v}{|x_v|},$$

where dx_v is the self-dual Haar measure on F_v with respect to ψ_v. Note that:

- $\mathrm{vol}(\mathfrak{o}_v, dx_v) = q_v^{-\frac{d_v}{2}}$ if v is finite,
- dx_v is the Lebesgue measure if v is real.

Then the Tamagawa measure on $\mathbb{A}^\times$ is given by

$$d^\times x^{\mathrm{Tam}} := \rho_F^{-1} \cdot \prod_v d^\times x_v^{\mathrm{Tam}}.$$

We have $\tau(\mathbb{G}_m) = 1$.

On the other hand, we define the standard measure on $\mathbb{A}^\times$ as a product measure $d^\times x := \prod_v d^\times x_v$, where

- $d^\times x_v$ is the Haar measure on $F_v^\times$ such that $\mathrm{vol}(\mathfrak{o}_v^\times, d^\times x_v) = 1$ if v is finite,
- $d^\times x_v = \frac{dx_v}{|x_v|}$ if v is real.

We have

$$d^\times x^{\mathrm{Tam}} = |D|^{-\frac{1}{2}} \rho_F^{-1} \cdot d^\times x. \tag{6.1}$$

6.1.2. Measures on $B^\times(\mathbb{A})$. For each place v of F, we define a Haar measure $d^\times \boldsymbol{\alpha}_v^{\mathrm{Tam}}$ on $B_v^\times$ by

$$d^\times \boldsymbol{\alpha}_v^{\mathrm{Tam}} := \zeta_v(1) \cdot \frac{d\boldsymbol{\alpha}_v}{|\nu(\boldsymbol{\alpha}_v)|^2},$$

where $d\boldsymbol{\alpha}_v$ is the self-dual Haar measure on B_v with respect to the pairing

$$(\boldsymbol{\alpha}_v, \boldsymbol{\beta}_v) \mapsto \psi_v(\mathrm{tr}_{B_v/F_v}(\boldsymbol{\alpha}_v \boldsymbol{\beta}_v)).$$

Then the Tamagawa measure on $B^\times(\mathbb{A})$ is given by

$$d^\times \boldsymbol{\alpha}^{\mathrm{Tam}} := \rho_F^{-1} \cdot \prod_v d^\times \boldsymbol{\alpha}_v^{\mathrm{Tam}}.$$

Also, the Tamagawa measure on $(B^\times/\mathbb{G}_m)(\mathbb{A}) = B^\times(\mathbb{A})/\mathbb{A}^\times$ is given by the quotient measure $d^\times \boldsymbol{\alpha}^{\mathrm{Tam}}/d^\times x^{\mathrm{Tam}}$. We have $\tau(B^\times/\mathbb{G}_m) = 2$.

On the other hand, we define the standard measure on $B^\times(\mathbb{A})$ as a product measure $d^\times \boldsymbol{\alpha} := \prod_v d^\times \boldsymbol{\alpha}_v$, where $d^\times \boldsymbol{\alpha}_v$ is given as follows:

- For $v \in \Sigma_{\mathrm{fin}} \smallsetminus \Sigma_{B,\mathrm{fin}}$, fix an isomorphism $i_v : B_v \to \mathrm{M}_2(F_v)$ of quaternion F_v-algebras and let $d^\times \boldsymbol{\alpha}_v$ be the Haar measure on $B_v^\times$ such that
$$\mathrm{vol}(i_v^{-1}(\mathrm{GL}_2(\mathfrak{o}_v)), d^\times \boldsymbol{\alpha}_v) = 1.$$
Since i_v is unique up to inner automorphisms, $d^\times \boldsymbol{\alpha}_v$ is independent of the choice of i_v.
- For $v \in \Sigma_{B,\mathrm{fin}}$, let $d^\times \boldsymbol{\alpha}_v$ be the Haar measure on $B_v^\times$ such that $\mathrm{vol}(\mathfrak{o}_{B_v}^\times, d^\times \boldsymbol{\alpha}_v) = 1$, where $\mathfrak{o}_{B_v}$ is the unique maximal order in B_v.
- For $v \in \Sigma_\infty \smallsetminus \Sigma_{B,\infty}$, fix an isomorphism $i_v : B_v \to \mathrm{M}_2(F_v)$ of quaternion F_v-algebras and define a Haar measure $d^\times \boldsymbol{\alpha}_v$ on $B_v^\times$ by
$$d^\times \boldsymbol{\alpha}_v = \frac{dx_v\, dy_v}{|y_v|^2} \frac{dz_v}{z_v} d\kappa_v$$
for $\boldsymbol{\alpha}_v = i_v^{-1}\left(\left(\begin{smallmatrix} 1 & x_v \\ & 1 \end{smallmatrix}\right)\left(\begin{smallmatrix} y_v & \\ & 1 \end{smallmatrix}\right) z_v \kappa_v\right)$ with $x_v \in \mathbb{R}$, $y_v \in \mathbb{R}^\times$, $z_v \in \mathbb{R}_+^\times$, $\kappa_v \in \mathrm{SO}(2)$, where dx_v, dy_v, dz_v are the Lebesgue measures and $d\kappa_v$ is the Haar measure on $\mathrm{SO}(2)$ such that $\mathrm{vol}(\mathrm{SO}(2), d\kappa_v) = 1$. Since i_v is unique up to inner automorphisms, $d^\times \boldsymbol{\alpha}_v$ is independent of the choice of i_v.
- For $v \in \Sigma_{B,\infty}$, let $d^\times \boldsymbol{\alpha}_v$ be the Haar measure on $B_v^\times$ such that
$$\mathrm{vol}(B_v^\times / F_v^\times, d^\times \boldsymbol{\alpha}_v / d^\times x_v) = 1.$$

Also, we define the standard measure on $B^\times(\mathbb{A})/\mathbb{A}^\times$ as the quotient measure $d^\times \boldsymbol{\alpha}/d^\times x$.

LEMMA 6.1.1. *We have*
$$\begin{aligned} d^\times \boldsymbol{\alpha}^{\mathrm{Tam}} &= (2\pi)^{|\Sigma_\infty \smallsetminus \Sigma_{B,\infty}|} \cdot (4\pi^2)^{|\Sigma_{B,\infty}|} \\ &\quad \times \prod_{v \in \Sigma_{B,\mathrm{fin}}} (q_v - 1)^{-1} \cdot |D|^{-2} \cdot \rho_F^{-1} \cdot \zeta_F(2)^{-1} \cdot d^\times \boldsymbol{\alpha}. \end{aligned}$$

PROOF. For each place v of F, let C_v be the constant such that $d^\times \boldsymbol{\alpha}_v^{\mathrm{Tam}} = C_v \cdot d^\times \boldsymbol{\alpha}_v$. If $v \in \Sigma_{\mathrm{fin}} \smallsetminus \Sigma_{B,\mathrm{fin}}$, we identify B_v with $\mathrm{M}_2(F_v)$. Then we have $\mathrm{vol}(\mathrm{M}_2(\mathfrak{o}_v), d\boldsymbol{\alpha}_v) = q_v^{-2d_v}$ and hence
$$\begin{aligned} C_v &= \mathrm{vol}(\mathrm{GL}_2(\mathfrak{o}_v), d^\times \boldsymbol{\alpha}_v^{\mathrm{Tam}}) \\ &= \zeta_v(1) \cdot |\mathrm{GL}_2(\mathbb{F}_{q_v})| \cdot \mathrm{vol}(1 + \mathrm{M}_2(\mathfrak{p}_v), d\boldsymbol{\alpha}_v) \\ &= q_v^{-2d_v} \cdot \zeta_v(2)^{-1}. \end{aligned}$$

If $v \in \Sigma_{B,\mathrm{fin}}$, then we have $\mathrm{vol}(\mathfrak{o}_{B_v}, d\boldsymbol{\alpha}_v) = q_v^{-2d_v - 1}$ and hence
$$\begin{aligned} C_v &= \mathrm{vol}(\mathfrak{o}_{B_v}^\times, d^\times \boldsymbol{\alpha}_v^{\mathrm{Tam}}) \\ &= \zeta_v(1) \cdot |\mathbb{F}_{q_v^2}^\times| \cdot \mathrm{vol}(1 + \mathfrak{p}_{B_v}, d\boldsymbol{\alpha}_v) \\ &= q_v^{-2d_v} \cdot (q_v - 1)^{-1} \cdot \zeta_v(2)^{-1}. \end{aligned}$$

If $v \in \Sigma_\infty \smallsetminus \Sigma_{B,\infty}$, we identify B_v with $\mathrm{M}_2(\mathbb{R})$. Then $d^\times \boldsymbol{\alpha}_v^{\mathrm{Tam}}$ arises from the gauge form on $\mathrm{GL}_2(\mathbb{R})$ determined (up to sign) by the lattice $\mathrm{M}_2(\mathbb{Z})$ in $\mathrm{Lie}\,\mathrm{GL}_2(\mathbb{R}) = \mathrm{M}_2(\mathbb{R})$. Also, the measures $\frac{dx_v\, dy_v}{|y_v|^2}$, $\frac{dz_v}{z_v}$, $d\kappa_v$ in the definition of $d^\times \boldsymbol{\alpha}_v$ arise from the (left invariant) gauge forms determined by the lattices

$$\mathbb{Z}\begin{pmatrix}1 & 0\\ 0 & 0\end{pmatrix}+\mathbb{Z}\begin{pmatrix}0 & 1\\ 0 & 0\end{pmatrix}, \quad \mathbb{Z}\begin{pmatrix}1 & 0\\ 0 & 1\end{pmatrix}, \quad 2\pi\mathbb{Z}\begin{pmatrix}0 & 1\\ -1 & 0\end{pmatrix},$$

respectively. Hence we have

$$C_v = 2\pi.$$

If $v \in \Sigma_{B,\infty}$, we identify B_v with

$$\mathbb{H} := \left\{ \begin{pmatrix}\alpha & \beta\\ -\bar{\beta} & \bar{\alpha}\end{pmatrix} \,\middle|\, \alpha, \beta \in \mathbb{C} \right\}.$$

Then $d^\times \boldsymbol{\alpha}_v^{\mathrm{Tam}}$ arises from the gauge form on $\mathbb{H}^\times$ determined by the lattice spanned by

$$\frac{1}{\sqrt{2}}\begin{pmatrix}1 & 0\\ 0 & 1\end{pmatrix}, \quad \frac{1}{\sqrt{2}}\begin{pmatrix}\sqrt{-1} & 0\\ 0 & -\sqrt{-1}\end{pmatrix}, \quad \frac{1}{\sqrt{2}}\begin{pmatrix}0 & 1\\ -1 & 0\end{pmatrix}, \quad \frac{1}{\sqrt{2}}\begin{pmatrix}0 & \sqrt{-1}\\ \sqrt{-1} & 0\end{pmatrix}$$

in $\mathrm{Lie}\,\mathbb{H}^\times = \mathbb{H}$. Let $d^\times \dot{\boldsymbol{\alpha}}_v$ be the Haar measure on $\mathbb{H}^\times/\mathbb{R}^\times$ which arises from the gauge form determined by the lattice spanned by

$$\frac{1}{2}\begin{pmatrix}\sqrt{-1} & 0\\ 0 & -\sqrt{-1}\end{pmatrix}, \quad \begin{pmatrix}0 & 1\\ -1 & 0\end{pmatrix}, \quad \begin{pmatrix}0 & \sqrt{-1}\\ \sqrt{-1} & 0\end{pmatrix},$$

so that we have $d^\times \boldsymbol{\alpha}_v^{\mathrm{Tam}}/d^\times x_v = 2 \cdot d^\times \dot{\boldsymbol{\alpha}}_v$. Note that this lattice is an integral lattice in $\mathrm{Lie}\,\mathbb{H}^\times/\mathbb{R}^\times$ constructed in [**51**]. It follows from [**51**] that $\mathrm{vol}(\mathbb{H}^\times/\mathbb{R}^\times, d^\times \dot{\boldsymbol{\alpha}}_v) = 2\pi^2$ and hence

$$C_v = \mathrm{vol}(\mathbb{H}^\times/\mathbb{R}^\times, d^\times \boldsymbol{\alpha}_v^{\mathrm{Tam}}/d^\times x_v) = 4\pi^2.$$

This completes the proof. □

EXAMPLE 6.1.2 (Eichler's mass formula [**14**], [**74**], [**81**]). Suppose that B is totally definite. Put $B_\infty = B \otimes_{\mathbb{Q}} \mathbb{R}$ and fix a maximal compact subgroup $\mathcal{K}$ of $B^\times(\mathbb{A}_{\mathrm{fin}})$. We can write

$$B^\times(\mathbb{A}) = \bigsqcup_{i=1}^{n} \mathbb{A}^\times B^\times(F) B_\infty^\times \boldsymbol{\alpha}_i \mathcal{K},$$

where $\{\boldsymbol{\alpha}_i \in B^\times(\mathbb{A}_{\mathrm{fin}}) \,|\, 1 \le i \le n\}$ is a (finite) set of representatives for

$$\mathbb{A}^\times B^\times(F) B_\infty^\times \backslash B^\times(\mathbb{A})/\mathcal{K}.$$

Put

$$\mathfrak{mass} := \mathrm{vol}\left(\mathbb{A}^\times B^\times(F)\backslash B^\times(\mathbb{A}), \frac{d^\times \boldsymbol{\alpha}}{d^\times x}\right) = \sum_{i=1}^{n} \frac{1}{|\Gamma_i|},$$

where $\Gamma_i = F^\times \backslash (B^\times(F) \cap \mathbb{A}_{\mathrm{fin}}^\times \boldsymbol{\alpha}_i \mathcal{K} \boldsymbol{\alpha}_i^{-1})$. Then it follows from (6.1) and Lemma 6.1.1 that

$$\begin{aligned}
\mathfrak{mass} &= \tau(B^\times/\mathbb{G}_m) \cdot (4\pi^2)^{-d} \cdot \prod_{v \in \Sigma_{B,\mathrm{fin}}} (q_v - 1) \cdot |D|^{\frac{3}{2}} \cdot \zeta_F(2) \\
&= (-1)^d \cdot 2^{-d+1} \cdot \prod_{v \in \Sigma_{B,\mathrm{fin}}} (q_v - 1) \cdot \zeta_F(-1),
\end{aligned}$$

where $d = [F : \mathbb{Q}]$.

Finally, we compare the standard measure on $\mathbb{A}^\times\backslash B^\times(\mathbb{A})$ with the measure on the Shimura variety. Let $(G,X)=(\mathrm{Res}_{F/\mathbb{Q}}(B^\times),X_B)$ be the Shimura datum given in §1.2. If $v\in\Sigma_\infty\smallsetminus\Sigma_{B,\infty}$, we identify B_v with $\mathrm{M}_2(\mathbb{R})$. As explained in §1.2, this gives rise to an identification

$$X=\prod_{v\in\Sigma_\infty\smallsetminus\Sigma_{B,\infty}}\mathfrak{h}^\pm. \tag{6.2}$$

Fix an open compact subgroup K of $B^\times(\mathbb{A}_{\mathrm{fin}})$ such that

$$\hat{\mathfrak{o}}^\times\subset K,$$

where $\hat{\mathfrak{o}}^\times:=\prod_{v\in\Sigma_{\mathrm{fin}}}\mathfrak{o}_v^\times\subset\mathbb{A}_{\mathrm{fin}}^\times$. Let $\mathrm{Sh}_K(G,X)$ be the associated Shimura variety:

$$\mathrm{Sh}_K(G,X)=B^\times(F)\backslash X\times B^\times(\mathbb{A}_{\mathrm{fin}})/K.$$

Since $\mathfrak{h}^\pm=\mathrm{GL}_2(\mathbb{R})/\mathbb{R}_+^\times\cdot\mathrm{SO}(2)$, we have a natural surjective map

$$p:B^\times(\mathbb{A})\longrightarrow\mathrm{Sh}_K(G,X).$$

Recall that in Definition 1.2.8, we have taken the measure on $\mathrm{Sh}_K(G,X)$ given as follows:

- On X, we take the product over $v\in\Sigma_\infty\smallsetminus\Sigma_{B,\infty}$ of the $\mathrm{GL}_2(\mathbb{R})$-invariant measure
$$\frac{dx_v\,dy_v}{|y_v|^2}$$
for $x_v+\sqrt{-1}y_v\in\mathfrak{h}^\pm$, where dx_v, dy_v are the Lebesgue measures. This measure is independent of the choice of identification (6.2).
- On $B^\times(\mathbb{A}_{\mathrm{fin}})/K$, we take the counting measure.
- If B is not totally definite, then $\mathfrak{o}^\times\backslash B^\times(F)$ acts on $X\times B^\times(\mathbb{A}_{\mathrm{fin}})/K$ properly discontinuously, and we take a natural measure $d\mu_x$ on $\mathrm{Sh}_K(G,X)$ induced by the product of the above measures.
- If B is totally definite, then $\mathrm{Sh}_K(G,X)$ is a finite set, and for any $x\in\mathrm{Sh}_K(G,X)$, its stabilizer Γ_x in $\mathfrak{o}^\times\backslash B^\times(F)$ is a finite group. We take a measure $d\mu_x$ on $\mathrm{Sh}_K(G,X)$ given by
$$\int_{\mathrm{Sh}_K(G,X)}\phi(x)\,d\mu_x=\sum_{x\in\mathrm{Sh}_K(G,X)}|\Gamma_x|^{-1}\phi(x).$$

LEMMA 6.1.3. *Let ϕ be an integrable function on $\mathrm{Sh}_K(G,X)$ such that $\phi(x\cdot z)=\phi(x)$ for all $x\in\mathrm{Sh}_K(G,X)$ and $z\in\mathbb{A}^\times$. Then we have*

$$\begin{aligned}&\int_{\mathrm{Sh}_K(G,X)}\phi(x)\,d\mu_x\\&=2^{|\Sigma_\infty\smallsetminus\Sigma_{B,\infty}|}\cdot[K_0:K]\cdot h_F\cdot\int_{\mathbb{A}^\times B^\times(F)\backslash B^\times(\mathbb{A})}p^*\phi(\boldsymbol{\alpha})\,d^\times\boldsymbol{\alpha},\end{aligned}\tag{6.3}$$

where K_0 is any maximal compact subgroup of $B^\times(\mathbb{A}_{\mathrm{fin}})$ containing K.

PROOF. Put $F_\infty=F\otimes_{\mathbb{Q}}\mathbb{R}$ and $B_\infty=B\otimes_{\mathbb{Q}}\mathbb{R}$. We can write

$$B^\times(\mathbb{A})=\bigsqcup_{i=1}^{n}B^\times(F)B_\infty^\times\boldsymbol{\alpha}_iK,$$

where $\{\boldsymbol{\alpha}_i\in B^\times(\mathbb{A}_{\mathrm{fin}})\,|\,1\le i\le n\}$ is a (finite) set of representatives for

$$B^\times(F)B_\infty^\times\backslash B^\times(\mathbb{A})/K.$$

Then we have

$$\mathrm{Sh}_K(G,X)=\bigsqcup_{i=1}^{n}\Gamma_i\backslash X,$$

where $\Gamma_i=\mathfrak{o}^\times\backslash(B^\times(F)\cap\boldsymbol{\alpha}_iK\boldsymbol{\alpha}_i^{-1})$. For each i, we have a natural commutative diagram

$$\begin{array}{ccc} B_\infty^\times\boldsymbol{\alpha}_iK & \xrightarrow{\quad p_i\quad} & X \\ \downarrow & & \downarrow \\ B^\times(F)B_\infty^\times\boldsymbol{\alpha}_iK & \xrightarrow{\quad p\quad} & \Gamma_i\backslash X \end{array}\ .$$

First we assume that B is not totally definite. Since both sides of (6.3) are proportional, we may assume that for each i, the restriction of ϕ to $\Gamma_i\backslash X$ is of the form

$$\phi(x)=\sum_{\gamma\in\Gamma_i}\varphi_i(\gamma x)$$

for some continuous compactly supported function φ_i on X. Then, noting that Γ_i acts on X faithfully, we have

$$\int_{\Gamma_i\backslash X}\phi(x)\,d\mu_x=\int_X\varphi_i(x)\,d\mu_x,$$

where the measure $d\mu_x$ on X on the right-hand side is as defined above. By the definition of the standard measure, we have

$$\int_X\varphi_i(x)\,d\mu_x=2^{|\Sigma_\infty\smallsetminus\Sigma_{B,\infty}|}\cdot\mathrm{vol}(K)^{-1}\cdot\int_{B_\infty^\times\boldsymbol{\alpha}_iK/F_\infty^\times\hat{\mathfrak{o}}^\times}p_i^*\varphi_i(\boldsymbol{\alpha})\,d^\times\boldsymbol{\alpha}.$$

(Here the factor 2 arises from $[\mathbb{R}^\times:\mathbb{R}_+^\times]$.) Since

$$p^*\phi(\boldsymbol{\alpha})=\phi(p(\boldsymbol{\alpha}))=\sum_{\gamma\in\Gamma_i}\varphi_i(\gamma p_i(\boldsymbol{\alpha}))=\sum_{\gamma\in\Gamma_i}\varphi_i(p_i(\gamma\boldsymbol{\alpha}))=\sum_{\gamma\in\Gamma_i}p_i^*\varphi_i(\gamma\boldsymbol{\alpha})$$

for $\boldsymbol{\alpha}\in B_\infty^\times\boldsymbol{\alpha}_iK$, we have

$$\int_{B_\infty^\times\boldsymbol{\alpha}_iK/F_\infty^\times\hat{\mathfrak{o}}^\times}p_i^*\varphi_i(\boldsymbol{\alpha})\,d^\times\boldsymbol{\alpha}=\int_{\Gamma_i\backslash B_\infty^\times\boldsymbol{\alpha}_iK/F_\infty^\times\hat{\mathfrak{o}}^\times}p^*\phi(\boldsymbol{\alpha})\,d^\times\boldsymbol{\alpha}.$$

Thus, noting that

$$\Gamma_i\backslash B_\infty^\times\boldsymbol{\alpha}_iK/F_\infty^\times\hat{\mathfrak{o}}^\times=B^\times(F)\backslash B^\times(F)B_\infty^\times\boldsymbol{\alpha}_iK/F_\infty^\times\hat{\mathfrak{o}}^\times,$$

we have

$$\begin{aligned}&\int_{\Gamma_i\backslash X}\phi(x)\,d\mu_x\\&\quad=2^{|\Sigma_\infty\smallsetminus\Sigma_{B,\infty}|}\cdot\mathrm{vol}(K)^{-1}\cdot\int_{B^\times(F)\backslash B^\times(F)B_\infty^\times\boldsymbol{\alpha}_iK/F_\infty^\times\hat{\mathfrak{o}}^\times}p^*\phi(\boldsymbol{\alpha})\,d^\times\boldsymbol{\alpha}.\end{aligned}$$

Summing over i, we obtain

$$\begin{aligned}
&\int_{\mathrm{Sh}_K(G,X)} \phi(x)\, d\mu_x \\
&= 2^{|\Sigma_\infty \smallsetminus \Sigma_{B,\infty}|} \cdot \mathrm{vol}(K)^{-1} \cdot \int_{B^\times(F)\backslash B^\times(\mathbb{A})/F_\infty^\times \hat{\mathfrak{o}}^\times} p^*\phi(\boldsymbol{\alpha})\, d^\times \boldsymbol{\alpha} \\
&= 2^{|\Sigma_\infty \smallsetminus \Sigma_{B,\infty}|} \cdot \mathrm{vol}(K)^{-1} \cdot \mathrm{vol}(F^\times \backslash \mathbb{A}^\times / F_\infty^\times \hat{\mathfrak{o}}^\times) \cdot \int_{\mathbb{A}^\times B^\times(F)\backslash B^\times(\mathbb{A})} p^*\phi(\boldsymbol{\alpha})\, d^\times \boldsymbol{\alpha}.
\end{aligned}$$

On the other hand, we have $\mathrm{vol}(K) = [K_0 : K]^{-1}$ for any maximal compact subgroup K_0 of $B^\times(\mathbb{A}_{\mathrm{fin}})$ containing K, and $\mathrm{vol}(F^\times \backslash \mathbb{A}^\times / F_\infty^\times \hat{\mathfrak{o}}^\times) = h_F$ since the standard measure on $\mathbb{A}^\times / F_\infty^\times \hat{\mathfrak{o}}^\times$ is the counting measure. This proves (6.3).

Next we assume that B is totally definite. Since

$$\mathrm{vol}(B^\times(F)\backslash B^\times(F) B_\infty^\times \boldsymbol{\alpha}_i K / F_\infty^\times \hat{\mathfrak{o}}^\times) = |\Gamma_i|^{-1} \cdot \mathrm{vol}(K),$$

we have

$$\begin{aligned}
\int_{B^\times(F)\backslash B^\times(\mathbb{A})/F_\infty^\times \hat{\mathfrak{o}}^\times} p^*\phi(\boldsymbol{\alpha})\, d^\times \boldsymbol{\alpha} &= \mathrm{vol}(K) \cdot \sum_{i=1}^n |\Gamma_i|^{-1} p^*\phi(\boldsymbol{\alpha}_i) \\
&= \mathrm{vol}(K) \cdot \int_{\mathrm{Sh}_K(G,X)} \phi(x)\, d\mu_x.
\end{aligned}$$

The rest of the proof is the same as before. □

6.1.3. Measures on $B^1(\mathbb{A})$. We recall the exact sequence

$$1 \longrightarrow B^1 \longrightarrow B^\times \xrightarrow{\ \nu\ } \mathbb{G}_m \longrightarrow 1$$

of algebraic groups over F. For each place v of F, this induces an exact sequence

$$1 \longrightarrow B_v^1 \longrightarrow B_v^\times \xrightarrow{\ \nu\ } F_v^\times.$$

We define a Haar measure dg_v^{Tam} on B_v^1 by requiring that

$$\int_{B_v^\times} \phi(\boldsymbol{\alpha}_v)\, d^\times \boldsymbol{\alpha}_v^{\mathrm{Tam}} = \int_{\nu(B_v^\times)} \dot{\phi}(x_v)\, d^\times x_v^{\mathrm{Tam}}$$

for all $\phi \in L^1(B_v^\times)$, where

$$\dot{\phi}(\nu(\boldsymbol{\alpha}_v)) := \int_{B_v^1} \phi(g_v \boldsymbol{\alpha}_v)\, dg_v^{\mathrm{Tam}}.$$

Note that $\nu(B_v^\times) = F_v^\times$ unless $v \in \Sigma_{B,\infty}$, in which case we have $\nu(B_v^\times) = \mathbb{R}_+^\times$. Then the Tamagawa measure on $B^1(\mathbb{A})$ is given by

$$dg^{\mathrm{Tam}} := \prod_v dg_v^{\mathrm{Tam}}.$$

We have $\tau(B^1) = 1$.

On the other hand, we define the standard measure on $B^1(\mathbb{A})$ as a product measure $dg := \prod_v dg_v$, where dg_v is given as follows:

- For $v \in \Sigma_{\mathrm{fin}} \smallsetminus \Sigma_{B,\mathrm{fin}}$, fix an isomorphism $i_v : B_v \to \mathrm{M}_2(F_v)$ of quaternion F_v-algebras, which is unique up to inner automorphisms by elements of $\mathrm{GL}_2(F_v)$, and let dg_v be the Haar measure on B_v^1 such that $\mathrm{vol}(i_v^{-1}(\mathrm{SL}_2(\mathfrak{o}_v)), dg_v) = 1$.

Noting that there are exactly 2 conjugacy classes of maximal compact subgroups of $\mathrm{SL}_2(F_v)$, i.e., those of $\mathrm{SL}_2(\mathfrak{o}_v)$ and $\begin{pmatrix} \varpi_v & \\ & 1 \end{pmatrix} \mathrm{SL}_2(\mathfrak{o}_v) \begin{pmatrix} \varpi_v^{-1} & \\ & 1 \end{pmatrix}$, we have

$$\mathrm{vol}(i_v^{-1}(h_v \mathrm{SL}_2(\mathfrak{o}_v) h_v^{-1}), dg_v) = \mathrm{vol}(i_v^{-1}(\mathrm{SL}_2(\mathfrak{o}_v)), dg_v)$$

for $h_v \in \mathrm{GL}_2(F_v)$. Hence dg_v is independent of the choice of i_v.

- For $v \in \Sigma_{B,\mathrm{fin}}$, let dg_v be the Haar measure on B_v^1 such that $\mathrm{vol}(B_v^1, dg_v) = 1$.
- For $v \in \Sigma_\infty \smallsetminus \Sigma_{B,\infty}$, fix an isomorphism $i_v : B_v \to \mathrm{M}_2(F_v)$ of quaternion F_v-algebras, which is unique up to inner automorphisms by elements of $\mathrm{GL}_2(F_v)$, and define a Haar measure dg_v on B_v^1 by
$$dg_v = \frac{dx_v\, dy_v}{y_v^2}\, d\kappa_v$$
for $g_v = i_v^{-1}\left(\begin{pmatrix} 1 & x_v \\ & 1 \end{pmatrix} \begin{pmatrix} \sqrt{y_v} & \\ & \sqrt{y_v}^{-1} \end{pmatrix} \kappa_v \right)$ with $x_v \in \mathbb{R}$, $y_v \in \mathbb{R}_+^\times$, $\kappa_v \in \mathrm{SO}(2)$, where dx_v, dy_v are the Lebesgue measures and $d\kappa_v$ is the Haar measure on $\mathrm{SO}(2)$ such that $\mathrm{vol}(\mathrm{SO}(2), d\kappa_v) = 1$. This measure dg_v does not change if we replace i_v by $\mathrm{Ad}(h_v) \circ i_v$ for $h_v \in \mathrm{SL}_2(F_v)$. If we replace i_v by $\mathrm{Ad}\begin{pmatrix} -1 & \\ & 1 \end{pmatrix} \circ i_v$, then dg_v becomes $\frac{dx_v\, dy_v}{y_v^2}\, d\kappa_v$ for $g_v = i_v^{-1}\left(\begin{pmatrix} 1 & -x_v \\ & 1 \end{pmatrix} \begin{pmatrix} \sqrt{y_v} & \\ & \sqrt{y_v}^{-1} \end{pmatrix} \kappa_v^{-1} \right)$, which is in fact equal to the original dg_v. Hence dg_v is independent of the choice of i_v.
- For $v \in \Sigma_{B,\infty}$, let dg_v be the Haar measure on B_v^1 such that $\mathrm{vol}(B_v^1, dg_v) = 1$.

LEMMA 6.1.4. *We have*

$$\begin{aligned} dg^{\mathrm{Tam}} &= \pi^{|\Sigma_\infty \smallsetminus \Sigma_{B,\infty}|} \cdot (4\pi^2)^{|\Sigma_{B,\infty}|} \\ &\quad \times \prod_{v \in \Sigma_{B,\mathrm{fin}}} (q_v - 1)^{-1} \cdot |D|^{-\frac{3}{2}} \cdot \zeta_F(2)^{-1} \cdot dg. \end{aligned}$$

PROOF. For each place v of F, let C_v be the constant such that $dg_v^{\mathrm{Tam}} = C_v \cdot dg_v$. If $v \in \Sigma_{\mathrm{fin}} \smallsetminus \Sigma_{B,\mathrm{fin}}$, we identify B_v with $\mathrm{M}_2(F_v)$. As in the proof of Lemma 6.1.1, we have

$$\begin{aligned} C_v &= \mathrm{vol}(\mathrm{SL}_2(\mathfrak{o}_v), dg_v^{\mathrm{Tam}}) \\ &= \frac{\mathrm{vol}(\mathrm{GL}_2(\mathfrak{o}_v), d^\times \boldsymbol{\alpha}_v^{\mathrm{Tam}})}{\mathrm{vol}(\mathfrak{o}_v^\times, d^\times x_v^{\mathrm{Tam}})} \\ &= q_v^{-\frac{3d_v}{2}} \cdot \zeta_v(2)^{-1}. \end{aligned}$$

If $v \in \Sigma_{B,\mathrm{fin}}$, then as in the proof of Lemma 6.1.1, we have

$$\begin{aligned} C_v &= \mathrm{vol}(B_v^1, dg_v^{\mathrm{Tam}}) \\ &= \frac{\mathrm{vol}(\mathfrak{o}_{B_v}^\times, d^\times \boldsymbol{\alpha}_v^{\mathrm{Tam}})}{\mathrm{vol}(\mathfrak{o}_v^\times, d^\times x_v^{\mathrm{Tam}})} \\ &= q_v^{-\frac{3d_v}{2}} \cdot (q_v - 1)^{-1} \cdot \zeta_v(2)^{-1}. \end{aligned}$$

If $v \in \Sigma_\infty \smallsetminus \Sigma_{B,\infty}$, we identify B_v with $\mathrm{M}_2(\mathbb{R})$. For $\boldsymbol{\alpha}_v \in \mathrm{GL}_2(\mathbb{R})^+$, we write $\boldsymbol{\alpha}_v = z_v \cdot g_v$ with $z_v \in \mathbb{R}_+^\times$ and $g_v \in \mathrm{SL}_2(\mathbb{R})$. Then we have

$$d^\times \boldsymbol{\alpha}_v^{\mathrm{Tam}} = 2 \cdot d^\times z_v\, dg_v^{\mathrm{Tam}}, \quad d^\times \boldsymbol{\alpha}_v = d^\times z_v\, dg_v$$

on $\mathrm{GL}_2(\mathbb{R})^+$. Since $d^\times \boldsymbol{\alpha}_v^{\mathrm{Tam}} = 2\pi \cdot d^\times \boldsymbol{\alpha}_v$ as in the proof of Lemma 6.1.1, we have

$$C_v = \frac{1}{2} \cdot 2\pi = \pi.$$

If $v \in \Sigma_{B,\infty}$, then we have $d^\times \boldsymbol{\alpha}_v^{\mathrm{Tam}} = 2 \cdot d^\times z_v \, dg_v^{\mathrm{Tam}}$ for $\boldsymbol{\alpha}_v = z_v \cdot g_v$ with $z_v \in \mathbb{R}_+^\times$ and $g_v \in B_v^1$. Hence we have

$$\begin{aligned} C_v &= \mathrm{vol}(B_v^1, dg_v^{\mathrm{Tam}}) \\ &= \frac{1}{2} \cdot \mathrm{vol}(B_v^\times / \mathbb{R}_+^\times, d^\times \boldsymbol{\alpha}_v^{\mathrm{Tam}} / d^\times z_v) \\ &= \mathrm{vol}(B_v^\times / \mathbb{R}^\times, d^\times \boldsymbol{\alpha}_v^{\mathrm{Tam}} / d^\times z_v) \\ &= 4\pi^2 \end{aligned}$$

as in the proof of Lemma 6.1.1. This completes the proof. □

EXAMPLE 6.1.5 (Siegel's formula [**72**]). Suppose that $B = \mathrm{M}_2(F)$. Put

$$\mathfrak{vol} := \mathrm{vol}\left(\mathrm{SL}_2(\mathfrak{o}) \backslash \mathfrak{h}^d, \prod_{v \in \Sigma_\infty} \frac{dx_v \, dy_v}{y_v^2} \right),$$

where $d = [F : \mathbb{Q}]$. Since

$$\mathrm{SL}_2(\mathfrak{o}) \backslash \mathfrak{h}^d \cong \mathrm{SL}_2(F) \backslash \mathrm{SL}_2(\mathbb{A}) / K,$$

where $K = \prod_{v \in \Sigma_\infty} \mathrm{SO}(2) \times \prod_{v \in \Sigma_{\mathrm{fin}}} \mathrm{SL}_2(\mathfrak{o}_v)$, we have

$$\begin{aligned} \mathrm{vol}(\mathrm{SL}_2(F) \backslash \mathrm{SL}_2(\mathbb{A}), dg) &= \mathfrak{vol} \cdot \mathrm{vol}\left(\{\pm 1\} \backslash K, \prod_{v \in \Sigma_\infty} d\kappa_v \cdot \prod_{v \in \Sigma_{\mathrm{fin}}} dg_v \right) \\ &= \mathfrak{vol} \cdot \frac{1}{2}. \end{aligned}$$

On the other hand, it follows from Lemma 6.1.4 that

$$\begin{aligned} \mathrm{vol}(\mathrm{SL}_2(F) \backslash \mathrm{SL}_2(\mathbb{A}), dg) &= \tau(B^1) \cdot \pi^{-d} \cdot |D|^{\frac{3}{2}} \cdot \zeta_F(2) \\ &= (-2\pi)^d \cdot \zeta_F(-1). \end{aligned}$$

Hence we have

$$\mathfrak{vol} = (-1)^d \cdot 2^{d+1} \cdot \pi^d \cdot \zeta_F(-1).$$

6.2. New vectors

In this section, we define a 1-dimensional subspace of new vectors in the space of an irreducible representation of $B_v^\times$. For the moment, we fix a place v of F and suppress the subscript v from the notation. We only consider representations π of $B^\times$ listed below:

- If F is non-archimedean and B is split, then

(ur) $\pi = \mathrm{Ind}(\chi \otimes \mu)$ is a principal series representation, where χ and μ are unitary unramified; or

(rps) $\pi = \mathrm{Ind}(\chi \otimes \mu)$ is a principal series representation, where χ is unitary unramified and μ is unitary ramified of conductor q^n; or

(st) $\pi = \mathrm{St} \otimes \chi$ is a twist of the Steinberg representation, where χ is unitary unramified.

- If F is non-archimedean and B is ramified, then

(1d) $\pi = \chi \circ \nu$ is a 1-dimensional representation, where χ is unitary unramified.

- If $F = \mathbb{R}$ and B is split, then

(ds) $\pi = \mathrm{DS}_k$ is the irreducible unitary (limit of) discrete series representation of weight k.

- If $F = \mathbb{R}$ and B is ramified, then

(fd) $\pi = \mathrm{Sym}^k$ is the irreducible unitary $(k+1)$-dimensional representation.

If F is non-archimedean, we define a compact subgroup K_n of $\mathrm{GL}_2(F)$ by

$$K_n = \left\{ \begin{pmatrix} a & b \\ c & d \end{pmatrix} \in \mathrm{GL}_2(\mathfrak{o}) \,\middle|\, c \in \varpi^n \mathfrak{o} \right\}.$$

Note that $I := K_1$ is an Iwahori subgroup of $\mathrm{GL}_2(F)$. If $F = \mathbb{R}$, we define a character χ_k of $\mathbb{C}^\times$ by

$$\chi_k(\alpha) = \left(\frac{\alpha}{\sqrt{\alpha\alpha^\rho}} \right)^k.$$

6.2.1. The case (ur). Fix an isomorphism $i : B \to \mathrm{M}_2(F)$. This determines a maximal compact subgroup $\mathcal{K} = i^{-1}(\mathrm{GL}_2(\mathfrak{o}))$ of $B^\times$. We say that $f \in \pi$ is a new vector with respect to $\mathcal{K}$ if

$$\pi(k)f = f$$

for all $k \in \mathcal{K}$.

6.2.2. The case (rps). Fix an isomorphism $i : B \to \mathrm{M}_2(F)$. This determines a compact subgroup $\mathcal{K}_n = i^{-1}(K_n)$ of $B^\times$. We define a character $\boldsymbol{\mu}$ of $\mathcal{K}_n$ by $\boldsymbol{\mu}(k) = \mu(d)$ for $k = i^{-1}\left(\begin{smallmatrix} a & b \\ c & d \end{smallmatrix}\right)$. We say that $f \in \pi$ is a new vector with respect to $(\mathcal{K}_n, \boldsymbol{\mu})$ if

$$\pi(k)f = \boldsymbol{\mu}(k)f$$

for all $k \in \mathcal{K}_n$.

6.2.3. The case (st). Fix an isomorphism $i : B \to \mathrm{M}_2(F)$. This determines an Iwahori subgroup $\mathcal{I} = i^{-1}(I)$ of $B^\times$. We say that $f \in \pi$ is a new vector with respect to $\mathcal{I}$ if

$$\pi(k)f = f$$

for all $k \in \mathcal{I}$.

6.2.4. The case (1d). Let $\mathcal{K} = \mathfrak{o}_B^\times$ be the unique maximal compact subgroup of $B^\times$. Then we have

$$\pi(k)f = f$$

for all $k \in \mathcal{K}$ and $f \in \pi$. For uniformity, we call any $f \in \pi$ a new vector with respect to $\mathcal{K}$.

6.2.5. The cases (ds), (fd). Fix an embedding $h : \mathbb{C}^\times \hookrightarrow B^\times$. We say that $f \in \pi$ is a new vector with respect to h if

$$\pi(h(z))f = \chi_k(z)f$$

for all $z \in \mathbb{C}^\times$.

6.3. An explicit Rallis inner product formula

Suppose that F is global. Let $\pi \cong \otimes_v \pi_v$ be an irreducible unitary cuspidal automorphic representation of $\mathrm{GL}_2(\mathbb{A})$ such that for $v \in \Sigma_{\mathrm{fin}}$,

- $\pi_v = \mathrm{Ind}(\chi_v \otimes \mu_v)$, where χ_v and μ_v are unitary unramified; or
- $\pi_v = \mathrm{Ind}(\chi_v \otimes \mu_v)$, where χ_v is unitary unramified and μ_v is unitary ramified of conductor $q_v^{n_v}$; or
- $\pi_v = \mathrm{St} \otimes \chi_v$, where χ_v is unitary unramified,

and for $v \in \Sigma_\infty$,

- $\pi_v = \mathrm{DS}_{k_v}$, where $k_v \geq 1$.

We assume that π_v is unramified for all finite places v of F such that F_v is ramified or of residual characteristic 2. Put $\Sigma_\pi = \{v \mid \pi_v \text{ is a discrete series}\}$, $\Sigma_{\pi,\mathrm{fin}} = \Sigma_\pi \cap \Sigma_{\mathrm{fin}}$, and

$$\Sigma'_{\pi,\mathrm{fin}} := \{v \in \Sigma_{\mathrm{fin}} \mid \pi_v \text{ is a ramified principal series}\}.$$

We consider a non-zero vector $f = \otimes_v f_v \in \pi$ such that:

- for $v \in \Sigma_{\mathrm{fin}} \smallsetminus (\Sigma_{\pi,\mathrm{fin}} \cup \Sigma'_{\pi,\mathrm{fin}})$, f_v is a new vector with respect to $\mathrm{GL}_2(\mathfrak{o}_v)$;
- for $v \in \Sigma_{\pi,\mathrm{fin}}$, f_v is a new vector with respect to the Iwahori subgroup I of $\mathrm{GL}_2(F_v)$ given in §6.2;
- for $v \in \Sigma'_{\pi,\mathrm{fin}}$, f_v is a new vector with respect to $(K_{n_v}, \boldsymbol{\mu}_v)$, where K_{n_v} is the compact subgroup of $\mathrm{GL}_2(F_v)$ given in §6.2 and $\boldsymbol{\mu}_v$ is the character of K_{n_v} defined by $\boldsymbol{\mu}_v\left(\begin{smallmatrix} a & b \\ c & d \end{smallmatrix}\right) = \mu_v(d)$;
- for $v \in \Sigma_\infty$, f_v is a new vector with respect to the embedding $h_v : \mathbb{C}^\times \hookrightarrow \mathrm{GL}_2(\mathbb{R})$ defined by $h_v(a + b\sqrt{-1}) = \left(\begin{smallmatrix} a & b \\ -b & a \end{smallmatrix}\right)$.

We normalize such a vector f, which is unique up to scalars, so that

$$W_f \begin{pmatrix} \delta^{-1} & \\ & 1 \end{pmatrix} = e^{-2\pi d},$$

where W_f is the Whittaker function of f defined by

$$W_f(g) = \int_{F \backslash \mathbb{A}} f\left(\begin{pmatrix} 1 & x \\ & 1 \end{pmatrix} g \right) \overline{\psi(x)} \, dx$$

with the Tamagawa measure dx on $\mathbb{A}$, $\delta = (\varpi_v^{d_v}) \in \mathbb{A}_{\mathrm{fin}}^\times$, and $d = [F : \mathbb{Q}]$ (see also Lemmas 6.4.1, 6.4.3, 6.4.5 and (6.5) below). Let $\langle f, f \rangle$ be the Petersson norm of f defined by

$$\langle f, f \rangle = \int_{\mathbb{A}^\times \mathrm{GL}_2(F) \backslash \mathrm{GL}_2(\mathbb{A})} |f(g)|^2 \, dg,$$

where dg is the standard measure on $\mathbb{A}^\times \backslash \mathrm{GL}_2(\mathbb{A})$. In §6.4 below, we will prove:

PROPOSITION 6.3.1. *We have*

$$\langle f, f \rangle = 2 \cdot \prod_{v \in \Sigma_\infty} \frac{(k_v - 1)!}{2^{2k_v+1} \pi^{k_v+1}} \cdot \prod_{v \in \Sigma_{\pi,\mathrm{fin}} \cup \Sigma'_{\pi,\mathrm{fin}}} \frac{q_v}{q_v + 1} \cdot |D| \cdot L(1, \pi, \mathrm{ad}),$$

where $L(s, \pi, \mathrm{ad}) = \prod_{v \in \Sigma_{\mathrm{fin}}} L(s, \pi_v, \mathrm{ad})$ *is the adjoint L-function of* π.

Let B, B_1, B_2 be quaternion algebras over F such that $B = B_1 \cdot B_2$ in the Brauer group. We assume that $\Sigma_B \neq \varnothing$ and $\Sigma_B \cup \Sigma_{B_1} \cup \Sigma_{B_2} \subset \Sigma_\pi$, i.e., B is division and the Jacquet–Langlands transfers π_B, π_{B_1}, π_{B_2} of π to $B^\times(\mathbb{A})$, $B_1^\times(\mathbb{A})$, $B_2^\times(\mathbb{A})$ exist. Now, we choose a totally imaginary quadratic extension E of F such that E embeds into B, B_1, B_2, and write $E = F + F\mathbf{i}$, $B = E + E\mathbf{j}$, $B_1 = E + E\mathbf{j}_1$, $B_2 = E + E\mathbf{j}_2$.

We also impose the ramification conditions on u, J, J_1, J_2 in §5.5. We consider non-zero vectors $f_B = \otimes_v f_{B,v} \in \pi_B$, $f_{B_1} = \otimes_v f_{B_1,v} \in \pi_{B_1}$, $f_{B_2} = \otimes_v f_{B_2,v} \in \pi_{B_2}$ such that:

- for $v \in \Sigma_{\mathrm{fin}} \smallsetminus (\Sigma_{\pi,\mathrm{fin}} \cup \Sigma'_{\pi,\mathrm{fin}})$, $f_{B,v}$, $f_{B_1,v}$, $f_{B_2,v}$ are new vectors with respect to $\mathcal{K}$, $\mathcal{K}_1$, $\mathcal{K}_2$, respectively, given in §5.6.1;
- for $v \in \Sigma_{\pi,\mathrm{fin}} \smallsetminus (\Sigma_{B,\mathrm{fin}} \cup \Sigma_{B_1,\mathrm{fin}} \cup \Sigma_{B_2,\mathrm{fin}})$, $f_{B,v}$, $f_{B_1,v}$, $f_{B_2,v}$ are new vectors with respect to $\mathcal{I}$, $\mathcal{I}_1$, $\mathcal{I}_2$, respectively, given in §5.6.3.1;
- for $v \in \Sigma_{B_1,\mathrm{fin}} \cap \Sigma_{B_2,\mathrm{fin}}$, $f_{B,v}$, $f_{B_1,v}$, $f_{B_2,v}$ are new vectors with respect to $\mathcal{I}$, $\mathcal{K}_1$, $\mathcal{K}_2$, respectively, given in §5.6.3.2;
- for $v \in \Sigma_{B,\mathrm{fin}} \cap \Sigma_{B_2,\mathrm{fin}}$, $f_{B,v}$, $f_{B_1,v}$, $f_{B_2,v}$ are new vectors with respect to $\mathcal{K}$, $\mathcal{I}_1$, $\mathcal{K}_2$, respectively, given in §5.6.4; we switch the roles of B_1 and B_2 for $v \in \Sigma_{B,\mathrm{fin}} \cap \Sigma_{B_1,\mathrm{fin}}$;
- for $v \in \Sigma'_{\pi,\mathrm{fin}}$, $f_{B,v}$, $f_{B_1,v}$, $f_{B_2,v}$ are new vectors with respect to $(\mathcal{K}_{n_v}, \boldsymbol{\mu}_v)$, $(\mathcal{K}_{1,n_v}, \boldsymbol{\mu}_v)$, $(\mathcal{K}_{2,n_v}, \boldsymbol{\mu}_v^{-1} \cdot \mu_v \circ \nu)$, respectively, given in §5.6.2;
- for $v \in \Sigma_\infty$, $f_{B,v}$, $f_{B_1,v}$, $f_{B_2,v}$ are new vectors with respect to the embeddings $\mathbb{C}^\times \cong E_v^\times \hookrightarrow B_v^\times$, $\mathbb{C}^\times \cong E_v^\times \hookrightarrow B_{1,v}^\times$, $\mathbb{C}^\times \cong E_v^\times \hookrightarrow B_{2,v}^\times$, respectively, given in §§5.6.5, 5.6.6.

We fix such vectors f_B, f_{B_1}, f_{B_2}, which are unique up to scalars. We emphasize that the 1-dimensional subspaces of π_B, π_{B_1}, π_{B_2} spanned by f_B, f_{B_1}, f_{B_2}, respectively, depend on the choice of E, $\mathbf{i}$, $\mathbf{j}$, $\mathbf{j}_1$, $\mathbf{j}_2$. Let $\langle f_B, f_B \rangle$ be the Petersson norm of f_B defined by

$$\langle f_B, f_B \rangle = \int_{\mathbb{A}^\times B^\times(F)\backslash B^\times(\mathbb{A})} |f_B(g)|^2 \, dg,$$

where dg is the standard measure on $\mathbb{A}^\times \backslash B^\times(\mathbb{A})$. We define the Petersson norms $\langle f_{B_1}, f_{B_1} \rangle$ and $\langle f_{B_2}, f_{B_2} \rangle$ similarly.

Let $\varphi = \otimes_v \varphi_v \in \mathcal{S}(\mathbb{X}(\mathbb{A}))$ be the Schwartz function given in §5.7, where

- $\varphi_v = \varphi_{\mu_v}$ for $v \in \Sigma'_{\pi,\mathrm{fin}}$;
- $\varphi_v = \varphi_{k_v}$ for $v \in \Sigma_\infty \smallsetminus (\Sigma_{B,\infty} \cup \Sigma_{B_1,\infty} \cup \Sigma_{B_2,\infty})$;
- $\varphi_v = \varphi_{k_v - 2}$ for $v \in \Sigma_{B_1,\infty} \cap \Sigma_{B_2,\infty}$;
- $\varphi_v = \varphi_{k_v - 2}$ for $v \in \Sigma_{B,\infty} \cap \Sigma_{B_2,\infty}$; we switch the roles of B_1 and B_2 for $v \in \Sigma_{B,\infty} \cap \Sigma_{B_1,\infty}$.

In §4.2, we have defined the theta lift $\theta_\varphi(f_B)$, but for our purposes, we slightly modify its definition: on the right-hand side of (4.1), we take the standard measure on $B^1(\mathbb{A})$ rather than the Tamagawa measure on $B^1(\mathbb{A})$. We regard $\theta_\varphi(f_B)$ as an automorphic form on $B_1^\times(\mathbb{A}) \times B_2^\times(\mathbb{A})$. Then it follows from the equivariance properties of φ that there exists a constant $\alpha(B_1, B_2) \in \mathbb{C}$ (once we fix f_B, f_{B_1}, f_{B_2}) such that

$$\theta_\varphi(f_B) = \alpha(B_1, B_2) \cdot (f_{B_1} \times f_{B_2}). \tag{6.4}$$

Now we state an explicit Rallis inner product formula.

THEOREM 6.3.2. *We have*

$$|\alpha(B_1, B_2)|^2 \cdot \langle f_{B_1}, f_{B_1} \rangle \cdot \langle f_{B_2}, f_{B_2} \rangle = C \cdot \langle f, f \rangle \cdot \langle f_B, f_B \rangle,$$

where $C = |D|^2 \cdot \prod_v C_v$ *with*

$$C_v = \begin{cases} 1 & \text{if } v \in \Sigma_{\mathrm{fin}} \smallsetminus (\Sigma_{\pi,\mathrm{fin}} \cup \Sigma'_{\pi,\mathrm{fin}}), \\ \dfrac{q_v}{(q_v+1)^2} & \text{if } v \in \Sigma_{\pi,\mathrm{fin}} \smallsetminus (\Sigma_{B,\mathrm{fin}} \cup \Sigma_{B_1,\mathrm{fin}} \cup \Sigma_{B_2,\mathrm{fin}}), \\ q_v & \text{if } v \in \Sigma_{B_1,\mathrm{fin}} \cap \Sigma_{B_2,\mathrm{fin}}, \\ q_v & \text{if } v \in \Sigma_{B,\mathrm{fin}}, \\ \dfrac{1}{q_v^{n_v-3}(q_v-1)(q_v+1)^2} & \text{if } v \in \Sigma'_{\pi,\mathrm{fin}}, \\ \dfrac{2^{2k_v+2}\pi^{k_v}}{k_v!} & \text{if } v \in \Sigma_\infty \smallsetminus (\Sigma_{B,\infty} \cup \Sigma_{B_1,\infty} \cup \Sigma_{B_2,\infty}), \\ \dfrac{2^{2k_v}\pi^{k_v-2}}{(k_v-1)^2 \cdot (k_v-2)!} & \text{if } v \in \Sigma_{B_1,\infty} \cap \Sigma_{B_2,\infty}, \\ \dfrac{2^{2k_v-2}\pi^{k_v-2}}{(k_v-1)^2 \cdot (k_v-2)!} & \text{if } v \in \Sigma_{B,\infty}. \end{cases}$$

PROOF. By Proposition 4.4.2, we have

$$C_{B_1} \cdot C_{B_2} \cdot (C_B^1)^2 \cdot |\alpha(B_1,B_2)|^2 \cdot \langle f_{B_1}, f_{B_1}\rangle \cdot \langle f_{B_2}, f_{B_2}\rangle \\ = 2 \cdot C_B \cdot C_B^1 \cdot \frac{L^S(1,\pi,\mathrm{ad})}{\zeta_F^S(2)^2} \cdot \langle f_B, f_B\rangle \cdot \prod_{v \in S} Z_v$$

for a sufficiently large finite set S of places of F, where

- C_B is the constant such that $dg^{\mathrm{Tam}} = C_B \cdot dg$, where dg^{Tam} is the Tamagawa measure on $\mathbb{A}^\times \backslash B^\times(\mathbb{A})$ and dg is the standard measure on $\mathbb{A}^\times \backslash B^\times(\mathbb{A})$; we define C_{B_1} and C_{B_2} similarly;
- C_B^1 is the constant such that $dg_1^{\mathrm{Tam}} = C_B^1 \cdot dg_1$, where dg_1^{Tam} is the Tamagawa measure on $B^1(\mathbb{A})$ and dg is the standard measure on $B^1(\mathbb{A})$;
- Z_v is the integral defined by

$$Z_v = \int_{B_v^1} \langle \omega_\psi(g_{1,v})\varphi_v, \varphi_v\rangle \langle \pi_{B,v}(g_{1,v}) f_{B,v}, f_{B,v}\rangle \, dg_{1,v}$$

(cf. (4.2)), where
· the hermitian inner product $\langle \cdot, \cdot \rangle$ on $\mathcal{S}(\mathbb{X}_v)$ is normalized as in §5.3;
· the invariant hermitian inner product $\langle \cdot, \cdot \rangle$ on $\pi_{B,v}$ is normalized so that

$$\langle f_{B,v}, f_{B,v}\rangle = 1;$$

· $dg_{1,v}$ is the standard measure on B_v^1.

Hence, by (6.1) and Lemmas 6.1.1, 6.1.4, we have

$$|\alpha(B_1,B_2)|^2 \cdot \langle f_{B_1}, f_{B_1}\rangle \cdot \langle f_{B_2}, f_{B_2}\rangle \\ = C' \cdot L(1,\pi,\mathrm{ad}) \cdot \langle f_B, f_B\rangle \cdot \prod_{v \in S_{\mathrm{fin}}} \frac{\zeta_v(2)^2}{L(1,\pi_v,\mathrm{ad})} \cdot \prod_{v \in S} Z_v,$$

where $S_{\mathrm{fin}} = S \cap \Sigma_{\mathrm{fin}}$ and

$$\begin{aligned} C' &= \frac{2 \cdot C_B}{C_{B_1} \cdot C_{B_2} \cdot C_B^1 \cdot \zeta_F(2)^2} \\ &= 2 \cdot 2^{|\Sigma_\infty \smallsetminus \Sigma_{B,\infty}|} \cdot (2\pi)^{-|\Sigma_\infty \smallsetminus \Sigma_{B_1,\infty}| - |\Sigma_\infty \smallsetminus \Sigma_{B_2,\infty}|} \cdot (4\pi^2)^{-|\Sigma_{B_1,\infty}| - |\Sigma_{B_2,\infty}|} \\ &\quad \times \prod_{v \in \Sigma_{B_1,\mathrm{fin}}} (q_v - 1) \cdot \prod_{v \in \Sigma_{B_2,\mathrm{fin}}} (q_v - 1) \cdot |D|^3 \\ &= 2 \cdot |D|^3 \cdot \prod_v C'_v \end{aligned}$$

with

$$C'_v = \begin{cases} 1 & \text{if } v \in \Sigma_{\mathrm{fin}} \smallsetminus (\Sigma_{B,\mathrm{fin}} \cup \Sigma_{B_1,\mathrm{fin}} \cup \Sigma_{B_2,\mathrm{fin}}), \\ (q_v - 1)^2 & \text{if } v \in \Sigma_{B_1,\mathrm{fin}} \cap \Sigma_{B_2,\mathrm{fin}}, \\ q_v - 1 & \text{if } v \in \Sigma_{B,\mathrm{fin}}, \\ (2\pi^2)^{-1} & \text{if } v \in \Sigma_\infty \smallsetminus (\Sigma_{B,\infty} \cup \Sigma_{B_1,\infty} \cup \Sigma_{B_2,\infty}), \\ (8\pi^4)^{-1} & \text{if } v \in \Sigma_{B_1,\infty} \cap \Sigma_{B_2,\infty}, \\ (8\pi^3)^{-1} & \text{if } v \in \Sigma_{B,\infty}. \end{cases}$$

Moreover, by Proposition 6.3.1, we have

$$\begin{aligned} &|\alpha(B_1, B_2)|^2 \cdot \langle f_{B_1}, f_{B_1} \rangle \cdot \langle f_{B_2}, f_{B_2} \rangle \\ &\qquad = C'' \cdot \langle f, f \rangle \cdot \langle f_B, f_B \rangle \cdot \prod_{v \in S_{\mathrm{fin}}} \frac{\zeta_v(2)^2}{L(1, \pi_v, \mathrm{ad})} \cdot \prod_{v \in S} Z_v, \end{aligned}$$

where $C'' = |D|^2 \cdot \prod_v C''_v$ with

$$C''_v = \begin{cases} 1 & \text{if } v \in \Sigma_{\mathrm{fin}} \smallsetminus (\Sigma_{\pi,\mathrm{fin}} \cup \Sigma'_{\pi,\mathrm{fin}}), \\ \dfrac{q_v + 1}{q_v} & \text{if } v \in \Sigma_{\pi,\mathrm{fin}} \smallsetminus (\Sigma_{B,\mathrm{fin}} \cup \Sigma_{B_1,\mathrm{fin}} \cup \Sigma_{B_2,\mathrm{fin}}), \\ \dfrac{(q_v - 1)^2 (q_v + 1)}{q_v} & \text{if } v \in \Sigma_{B_1,\mathrm{fin}} \cap \Sigma_{B_2,\mathrm{fin}}, \\ \dfrac{(q_v - 1)(q_v + 1)}{q_v} & \text{if } v \in \Sigma_{B,\mathrm{fin}}, \\ \dfrac{q_v + 1}{q_v} & \text{if } v \in \Sigma'_{\pi,\mathrm{fin}}, \\ \dfrac{2^{2k_v} \pi^{k_v - 1}}{(k_v - 1)!} & \text{if } v \in \Sigma_\infty \smallsetminus (\Sigma_{B,\infty} \cup \Sigma_{B_1,\infty} \cup \Sigma_{B_2,\infty}), \\ \dfrac{2^{2k_v - 2} \pi^{k_v - 3}}{(k_v - 1)!} & \text{if } v \in \Sigma_{B_1,\infty} \cap \Sigma_{B_2,\infty}, \\ \dfrac{2^{2k_v - 2} \pi^{k_v - 2}}{(k_v - 1)!} & \text{if } v \in \Sigma_{B,\infty}. \end{cases}$$

Now Theorem 6.3.2 follows from this and Lemmas 6.6.1, 6.6.2, 6.6.3, 6.6.4, 6.6.5, 6.6.6 in §6.6 below, where we compute the integral Z_v explicitly. □

6.4. Computation of $\langle f, f \rangle$

Proposition 6.3.1 follows from a standard computation of the Rankin–Selberg integral, but we give the details of the proof for the convenience of the reader. We retain the notation of §6.3. Put

$$\mathbf{n}(x) = \begin{pmatrix} 1 & x \\ & 1 \end{pmatrix}, \quad \mathbf{t}(y) = \begin{pmatrix} y & \\ & 1 \end{pmatrix}, \quad w = \begin{pmatrix} & -1 \\ 1 & \end{pmatrix}.$$

By [**36**, §4], [**47**, §2.2], [**82**, Proposition 3.1], we have

$$C \cdot \langle f, f\rangle = \frac{2}{\rho_F} \cdot \frac{\mathrm{Res}_{s=1} L^S(s, \pi \times \pi^\vee)}{\zeta_F^S(2)} \cdot |D|^{-\frac{1}{2}} \cdot \prod_{v \in S} \|W_v\|^2$$

for a sufficiently large finite set S of places of F, where

- C is the constant such that $dg^{\mathrm{Tam}} = C \cdot dg$, where dg^{Tam} is the Tamagawa measure on $\mathbb{A}^\times \backslash \mathrm{GL}_2(\mathbb{A})$ and dg is the standard measure on $\mathbb{A}^\times \backslash \mathrm{GL}_2(\mathbb{A})$;
- the Whittaker function W_f of f is decomposed as a product $W_f = \prod_v W_v$, where W_v is the Whittaker function of π_v with respect to ψ_v normalized so that
  - · $W_v(\mathbf{t}(\varpi_v^{-d_v})) = 1$ for $v \in \Sigma_{\mathrm{fin}} \setminus (\Sigma_{\pi,\mathrm{fin}} \cup \Sigma'_{\pi,\mathrm{fin}})$;
  - · $W_v(1) = 1$ for $v \in \Sigma_{\pi,\mathrm{fin}}$;
  - · $W_v(1) = 1$ for $v \in \Sigma'_{\pi,\mathrm{fin}}$;
  - · $W_v(1) = e^{-2\pi}$ for $v \in \Sigma_\infty$;
- $\|W_v\|^2$ is the integral defined by

$$\|W_v\|^2 = \int_{F_v^\times} |W_v(\mathbf{t}(y_v))|^2 \, d^\times y_v,$$

  where $d^\times y_v$ is the standard measure on $F_v^\times$.

We remark that:

- the volume of $F^\times \backslash \mathbb{A}^1$ given in [**82**, Proposition 3.1] is equal to ρ_F,
- $\prod_v d^\times y_v^{\mathrm{Tam}} = |D|^{-\frac{1}{2}} \cdot \prod_v d^\times y_v$.

Hence, by (6.1) and Lemma 6.1.1, we have

$$\langle f, f\rangle = 2 \cdot (2\pi)^{-[F:\mathbb{Q}]} \cdot |D| \cdot L(1, \pi, \mathrm{ad}) \cdot \prod_{v \in S_{\mathrm{fin}}} \frac{\zeta_v(2)}{\zeta_v(1) \cdot L(1, \pi_v, \mathrm{ad})} \cdot \prod_{v \in S} \|W_v\|^2,$$

where $S_{\mathrm{fin}} = S \cap \Sigma_{\mathrm{fin}}$. Now Proposition 6.3.1 follows from this and Lemmas 6.4.2, 6.4.4, 6.4.6, 6.4.7 below, where we compute the integral $\|W_v\|^2$ explicitly.

For the rest of this section, we fix a place v of F and suppress the subscript v from the notation.

6.4.1. The case $v \in \Sigma_{\mathrm{fin}} \setminus (\Sigma_{\pi,\mathrm{fin}} \cup \Sigma'_{\pi,\mathrm{fin}})$. Recall that $\pi = \mathrm{Ind}(\chi \otimes \mu)$, where χ and μ are unitary unramified. Put $\alpha = \chi(\varpi)$ and $\beta = \mu(\varpi)$. We have

$$L(s, \pi, \mathrm{ad}) = \frac{1}{(1 - q^{-s})(1 - \alpha\beta^{-1}q^{-s})(1 - \alpha^{-1}\beta q^{-s})}.$$

LEMMA 6.4.1. *We have*

$$W(\mathbf{t}(\varpi^{i-d})) = \begin{cases} q^{-\frac{i}{2}} \cdot \dfrac{\alpha^{i+1} - \beta^{i+1}}{\alpha - \beta} & \text{if } i \geq 0, \\ 0 & \text{if } i < 0. \end{cases}$$

PROOF. Recall that d is the non-negative integer such that ψ is trivial on $\varpi^{-d}\mathfrak{o}$ but non-trivial on $\varpi^{-d-1}\mathfrak{o}$. We define a non-trivial character ψ^0 of F of order zero by $\psi^0(x) = \psi(\varpi^{-d}x)$. Let W^0 be the Whittaker function of π with respect to ψ^0 such that

- $W^0(gk) = W^0(g)$ for all $g \in \mathrm{GL}_2(F)$ and $k \in \mathrm{GL}_2(\mathfrak{o})$,

- $W^0(1) = 1$.

Then we have $W(g) = W^0(\mathbf{t}(\varpi^d))g)$, so that the assertion follows from the formula of Casselman–Shalika [**8**]. □

LEMMA 6.4.2. *We have*

$$\|W\|^2 = \frac{\zeta(1) \cdot L(1, \pi, \mathrm{ad})}{\zeta(2)}.$$

PROOF. By Lemma 6.4.1, we have

$$\begin{aligned}
\|W\|^2 &= \sum_{i=0}^{\infty} |W(\mathbf{t}(\varpi^{i-d}))|^2 \\
&= \frac{1}{(\alpha-\beta)(\alpha^{-1}-\beta^{-1})} \cdot \sum_{i=0}^{\infty} q^{-i}(\alpha^{i+1}-\beta^{i+1})(\alpha^{-i-1}-\beta^{-i-1}) \\
&= \frac{1}{(\alpha-\beta)(\alpha^{-1}-\beta^{-1})} \\
&\quad \times \left(\frac{1}{1-q^{-1}} - \frac{\alpha\beta^{-1}}{1-\alpha\beta^{-1}q^{-1}} - \frac{\alpha^{-1}\beta}{1-\alpha^{-1}\beta q^{-1}} + \frac{1}{1-q^{-1}} \right) \\
&= \frac{1+q^{-1}}{(1-q^{-1})(1-\alpha\beta^{-1}q^{-1})(1-\alpha^{-1}\beta q^{-1})}.
\end{aligned}$$

□

6.4.2. The case $v \in \Sigma_{\pi,\mathrm{fin}}$. Recall that $\pi = \mathrm{St} \otimes \chi$, where χ is unitary unramified. Put $\alpha = \chi(\varpi)$. We have $L(s, \pi, \mathrm{ad}) = \zeta(s+1)$.

LEMMA 6.4.3. *We have*

$$W(\mathbf{t}(\varpi^i)) = \begin{cases} q^{-i} \cdot \alpha^i & \text{if } i \geq 0, \\ 0 & \text{if } i < 0. \end{cases}$$

PROOF. We may assume that $\chi = 1$. We recall the exact sequence

$$0 \longrightarrow \mathrm{St} \longrightarrow \mathrm{Ind}(|\cdot|^{\frac{1}{2}} \otimes |\cdot|^{-\frac{1}{2}}) \stackrel{M}{\longrightarrow} \mathbf{1} \longrightarrow 0,$$

where $M : \mathrm{Ind}(|\cdot|^{\frac{1}{2}} \otimes |\cdot|^{-\frac{1}{2}}) \to \mathrm{Ind}(|\cdot|^{-\frac{1}{2}} \otimes |\cdot|^{\frac{1}{2}})$ is the intertwining operator defined by

$$M(f)(g) = \int_F f(w\mathbf{n}(x)g)\, dx$$

with the Haar measure dx on F such that $\mathrm{vol}(\mathfrak{o}, dx) = 1$. In particular, we have

$$\mathrm{St} = \{ f \in \mathrm{Ind}(|\cdot|^{\frac{1}{2}} \otimes |\cdot|^{-\frac{1}{2}}) \mid M(f)(1) = 0 \}.$$

Also, we have

$$\dim_{\mathbb{C}} \mathrm{St}^I = 1, \quad \dim_{\mathbb{C}} \mathrm{Ind}(|\cdot|^{\frac{1}{2}} \otimes |\cdot|^{-\frac{1}{2}})^I = 2.$$

Let f_1, f_w be the basis of $\mathrm{Ind}(|\cdot|^{\frac{1}{2}} \otimes |\cdot|^{-\frac{1}{2}})^I$ determined by

$$f_1|_{\mathrm{GL}_2(\mathfrak{o})} = \mathbb{I}_I, \quad f_w|_{\mathrm{GL}_2(\mathfrak{o})} = \mathbb{I}_{IwI}.$$

Then $f_1 - q^{-1} f_w$ is a basis of St^I. Indeed, noting that

$$w\mathbf{n}(x) = \begin{pmatrix} & -1 \\ 1 & x \end{pmatrix} = \begin{pmatrix} x^{-1} & -1 \\ & x \end{pmatrix} \begin{pmatrix} 1 & \\ x^{-1} & 1 \end{pmatrix},$$

we have

$$M(f_1)(1) = \sum_{j=1}^{\infty} \int_{\varpi^{-j}\mathfrak{o}^\times} |x|^{-2}\, dx = \sum_{j=1}^{\infty} q^{-j}(1-q^{-1}) = q^{-1}$$

and

$$M(f_w)(1) = \int_{\mathfrak{o}} dx = 1.$$

We consider the Jacquet integral

$$\mathcal{W}_k(g) := \int_F f_k(w\mathbf{n}(x)g)\overline{\psi(x)}\, dx$$

for $k = 1, w$, where we recall that ψ is assumed to be of order zero. We have

$$\begin{aligned} \mathcal{W}_k(\mathbf{t}(y)) &= |y|^{-1} \int_F f_k(w\mathbf{n}(xy^{-1}))\overline{\psi(x)}\, dx \\ &= \int_F f_k(w\mathbf{n}(x))\overline{\psi(xy)}\, dx. \end{aligned}$$

If $k = 1$, then we have

$$\mathcal{W}_1(\mathbf{t}(y)) = \sum_{j=1}^{\infty} \int_{\varpi^{-j}\mathfrak{o}^\times} |x|^{-2}\overline{\psi(xy)}\, dx = \sum_{j=1}^{\infty} q^{-j} \cdot \hat{\mathbb{I}}_{\mathfrak{o}^\times}(\varpi^{-j}y).$$

Since

$$\hat{\mathbb{I}}_{\mathfrak{o}^\times}(x) = \begin{cases} 1-q^{-1} & \text{if } x \in \mathfrak{o}, \\ -q^{-1} & \text{if } x \in \varpi^{-1}\mathfrak{o}^\times, \\ 0 & \text{otherwise,} \end{cases}$$

we have

$$\begin{aligned} &\mathcal{W}_1(\mathbf{t}(\varpi^i)) \\ &= \begin{cases} \sum_{j=1}^{i} q^{-j}(1-q^{-1}) + q^{-(i+1)}\cdot(-q^{-1}) = q^{-1} - q^{-i-1} - q^{-i-2} & \text{if } i > 0, \\ q^{-1}\cdot(-q^{-1}) = -q^{-2} & \text{if } i = 0, \\ 0 & \text{if } i < 0. \end{cases} \end{aligned}$$

If $k = w$, then we have

$$\mathcal{W}_w(\mathbf{t}(y)) = \int_{\mathfrak{o}} \overline{\psi(xy)}\, dx = \mathbb{I}_{\mathfrak{o}}(y).$$

Hence, if we put $\mathcal{W} = \mathcal{W}_1 - q^{-1}\mathcal{W}_w$, then we have

$$\mathcal{W}(\mathbf{t}(\varpi^i)) = \begin{cases} -q^{-i-1}(1+q^{-1}) & \text{if } i \geq 0, \\ 0 & \text{if } i < 0. \end{cases}$$

Thus $W = \mathcal{W}(1)^{-1} \cdot \mathcal{W}$ and the assertion follows. □

LEMMA 6.4.4. *We have*

$$\|W\|^2 = L(1, \pi, \mathrm{ad}).$$

PROOF. By Lemma 6.4.3, we have

$$\|W\|^2 = \sum_{i=0}^{\infty} |W(\mathbf{t}(\varpi^i))|^2 = \sum_{i=0}^{\infty} q^{-2i} = \frac{1}{1-q^{-2}}.$$ □

6.4.3. The case $v \in \Sigma'_{\pi,\mathrm{fin}}$. Recall that $\pi = \mathrm{Ind}(\chi \otimes \mu)$, where χ is unitary unramified and μ is unitary ramified of conductor q^n. Put $\alpha = \chi(\varpi)$. We have $L(s,\pi,\mathrm{ad}) = \zeta(s)$.

LEMMA 6.4.5. *We have*

$$W(\mathbf{t}(\varpi^i)) = \begin{cases} q^{-\frac{i}{2}} \cdot \alpha^i & \text{if } i \geq 0, \\ 0 & \text{if } i < 0. \end{cases}$$

PROOF. Let $f \in \mathrm{Ind}(\chi \otimes \mu)$ be the new vector with respect to $(K_n, \boldsymbol{\mu})$ determined by

$$f|_{\mathrm{GL}_2(\mathfrak{o})} = \mathbb{I}_{K_n}\boldsymbol{\mu}.$$

We consider the Jacquet integral

$$\mathcal{W}(g) := \int_F f(w\mathbf{n}(x)g)\overline{\psi(x)}\,dx,$$

where we recall that ψ is assumed to be of order zero. We have

$$\begin{aligned} \mathcal{W}(\mathbf{t}(y)) &= \mu(y)|y|^{-\frac{1}{2}} \int_F f(w\mathbf{n}(xy^{-1}))\overline{\psi(x)}\,dx \\ &= \mu(y)|y|^{\frac{1}{2}} \int_F f(w\mathbf{n}(x))\overline{\psi(xy)}\,dx. \end{aligned}$$

Noting that

$$w\mathbf{n}(x) = \begin{pmatrix} & -1 \\ 1 & x \end{pmatrix} = \begin{pmatrix} x^{-1} & -1 \\ & x \end{pmatrix}\begin{pmatrix} 1 & \\ x^{-1} & 1 \end{pmatrix},$$

we have

$$\begin{aligned} \int_F f(w\mathbf{n}(x))\overline{\psi(xy)}\,dx &= \sum_{j=n}^{\infty} \int_{\varpi^{-j}\mathfrak{o}^\times} \chi(x)^{-1}\mu(x)|x|^{-1}\overline{\psi(xy)}\,dx \\ &= \sum_{j=n}^{\infty} \alpha^j \mu(\varpi)^{-j} \cdot \widehat{\mathbb{I}_{\mathfrak{o}^\times}\mu}(\varpi^{-j}y). \end{aligned}$$

Since $\widehat{\mathbb{I}_{\mathfrak{o}^\times}\mu} = q^{-n} \cdot \mathfrak{g}(\mu,\psi) \cdot \mathbb{I}_{\varpi^{-n}\mathfrak{o}^\times}\mu^{-1}$, where

$$\mathfrak{g}(\mu,\psi) = \int_{\varpi^{-n}\mathfrak{o}^\times} \mu(x)\psi(x)\,dx,$$

we have

$$\begin{aligned} \mathcal{W}(\mathbf{t}(\varpi^i)) &= \mu(\varpi)^i q^{-\frac{i}{2}} \cdot \alpha^{i+n}\mu(\varpi)^{-(i+n)} \cdot q^{-n} \cdot \mathfrak{g}(\mu,\psi) \cdot \mu(\varpi)^n \\ &= q^{-\frac{i}{2}-n} \cdot \alpha^{i+n} \cdot \mathfrak{g}(\mu,\psi) \end{aligned}$$

if $i \geq 0$, and $\mathcal{W}(\mathbf{t}(\varpi^i)) = 0$ if $i < 0$. Thus $W = \mathcal{W}(1)^{-1} \cdot \mathcal{W}$ and the assertion follows. □

LEMMA 6.4.6. *We have*

$$\|W\|^2 = L(1,\pi,\mathrm{ad}).$$

PROOF. By Lemma 6.4.5, we have

$$\|W\|^2 = \sum_{i=0}^{\infty} |W(\mathbf{t}(\varpi^i))|^2 = \sum_{i=0}^{\infty} q^{-i} = \frac{1}{1-q^{-1}}.$$ □

6.4.4. The case $v \in \Sigma_\infty$. Recall that $\pi = \mathrm{DS}_k$ and $\psi(x) = e^{2\pi\sqrt{-1}x}$. It is known that

$$W(\mathbf{t}(y)) = \begin{cases} y^{\frac{k}{2}} e^{-2\pi y} & \text{if } y > 0, \\ 0 & \text{if } y < 0. \end{cases} \tag{6.5}$$

LEMMA 6.4.7. *We have*

$$\|W\|^2 = \frac{(k-1)!}{(4\pi)^k}.$$

PROOF. By (6.5), we have

$$\|W\|^2 = \int_0^\infty |W(\mathbf{t}(y))|^2 \frac{dy}{y} = \int_0^\infty y^{k-1} e^{-4\pi y}\, dy = \frac{\Gamma(k)}{(4\pi)^k}. \qquad \square$$

6.5. Matrix coefficients of the Weil representation

Suppose that F is local. In this section, we compute the function

$$\Phi(g) := \langle \omega_\psi(g)\varphi, \varphi \rangle$$

on $\mathrm{U}(W) \cong B^1$ explicitly, where $\varphi \in \mathcal{S}(\mathbb{X})$ is the Schwartz function given in §5.7. Since φ is the partial Fourier transform of the Schwartz function $\varphi' \in \mathcal{S}(\mathbb{X}')$ given in §5.6, we have

$$\Phi(g) = \langle \omega_\psi(g)\varphi', \varphi' \rangle.$$

Put

$$\mathbf{m}(a) = \begin{pmatrix} a & \\ & a^{-1} \end{pmatrix}, \quad \mathbf{n}(b) = \begin{pmatrix} 1 & b \\ & 1 \end{pmatrix}$$

for $a \in F^\times$, $b \in F$.

6.5.1. The case (ur). We identify $B^\times$ with $\mathrm{GL}_2(F)$ via:

- the isomorphism $\mathfrak{i}$ given by (5.1) if E is split and F is unramified;
- the isomorphism $\mathrm{Ad}\left(\begin{smallmatrix} 1 & \\ & \varpi^{-d} \end{smallmatrix}\right) \circ \mathfrak{i}$, where $\mathfrak{i}$ is the isomorphism given by (5.1) and $\varpi^{-d}\mathfrak{o}$ is the inverse different, if E is split and F is ramified;
- the isomorphism $\mathfrak{i}$ given by (5.7) if E is inert and $J \in (F^\times)^2$;
- any fixed isomorphism $\mathfrak{i} : B \to \mathrm{M}_2(F)$ such that $\mathfrak{i}(\mathfrak{o}_B) = \mathrm{M}_2(\mathfrak{o})$, where $\mathfrak{o}_B$ is the maximal order in B given in §5.6.1.4, if E is inert, and $J_1 \in (F^\times)^2$ or $J_2 \in (F^\times)^2$;
- the isomorphism $\mathfrak{i}$ given by (5.7) if E is ramified.

Under this identification, we have $\mathcal{K} = \mathrm{GL}_2(\mathfrak{o})$, where $\mathcal{K}$ is the maximal compact subgroup of $B^\times$ given in §5.6.1.

LEMMA 6.5.1. *We have $\Phi(\mathbf{m}(a)) = |a|^2$ for $a \in \mathfrak{o} \smallsetminus \{0\}$.*

PROOF. Put

$$\phi(a) := \int_F \mathbb{I}_{\mathfrak{o}}(ax)\mathbb{I}_{\mathfrak{o}}(x)\, dx = q^{-\frac{d}{2}} \times \begin{cases} 1 & \text{if } a \in \mathfrak{o}, \\ |a|^{-1} & \text{otherwise}, \end{cases}$$

where dx is the self-dual Haar measure on F with respect to ψ. Note that $d = 0$ unless E is split and F is ramified.

Assume that E is split and F is unramified. We use the notation of §5.6.1.1. Then the Weil representation ω_ψ on $\mathcal{S}(\mathbb{X}')$ is given in §5.3.1. We have

$$\begin{aligned}\Phi(\mathbf{m}(a)) &= |a|^2 \cdot \int_{\mathbb{X}'} \varphi'(ax)\overline{\varphi'(x)}\, dx \\ &= |a|^2 \cdot \prod_{i=1}^{4} \int_F \mathbb{I}_{\mathfrak{o}}(ax_i)\mathbb{I}_{\mathfrak{o}}(x_i)\, dx_i \\ &= |a|^2 \cdot \phi(a)^4.\end{aligned}$$

This yields the desired identity.

Assume that E is split and F is ramified. We use the notation of §5.6.1.2. Then the Weil representation ω_ψ on $\mathcal{S}(\mathbb{X}')$ is given in §5.3.1, where $B^\times$ is identified with $\mathrm{GL}_2(F)$ via $\mathfrak{i}$ rather than $\mathrm{Ad}\left(\begin{smallmatrix} 1 & \\ & \varpi^{-d}\end{smallmatrix}\right) \circ \mathfrak{i}$. We have

$$\begin{aligned}\Phi(\mathbf{m}(a)) &= |a|^2 \cdot \int_{\mathbb{X}'} \varphi'(ax)\overline{\varphi'(x)}\, dx \\ &= q^{2d} \cdot |a|^2 \cdot \prod_{i=1}^{4} \int_F \mathbb{I}_{\mathfrak{o}}(ax_i)\mathbb{I}_{\mathfrak{o}}(x_i)\, dx_i \\ &= q^{2d} \cdot |a|^2 \cdot \phi(a)^4.\end{aligned}$$

This yields the desired identity.

Assume that E is inert and $J \in (F^\times)^2$. We use the notation of §5.6.1.3. Then the Weil representation ω_ψ on $\mathcal{S}(\mathbb{X}')$ is given in §5.3.2. We have

$$\begin{aligned}\Phi(\mathbf{m}(a)) &= |a|^2 \cdot \int_{\mathbb{X}'} \varphi'(ax)\overline{\varphi'(x)}\, dx \\ &= |a|^2 \cdot \prod_{i=1}^{4} \int_F \mathbb{I}_{\mathfrak{o}}(ax_i)\mathbb{I}_{\mathfrak{o}}(x_i)\, dx_i \\ &= |a|^2 \cdot \phi(a)^4.\end{aligned}$$

This yields the desired identity.

Assume that E is inert and $J_1 \in (F^\times)^2$; the case when E is inert and $J_2 \in (F^\times)^2$ is similar. We use the notation of §5.6.1.4. Then the Weil representation ω_ψ on $\mathcal{S}(\mathbb{X}')$ is given in §5.3.3. We have

$$\begin{aligned}\Phi(\mathbf{m}(a)) &= \int_{\mathbb{X}'} \varphi'(x\mathbf{m}(a))\overline{\varphi'(x)}\, dx \\ &= \int_{\mathrm{M}_2(F)} \varphi'(x\mathbf{m}(a))\overline{\varphi'(x)}\, dx \\ &= \phi(a)^2 \cdot \phi(a^{-1})^2,\end{aligned}$$

where we identify $\mathbb{X}' \cong W$ with $\mathrm{M}_2(F)$ via the fixed isomorphism $\mathfrak{i}$ and normalize the Haar measure on $\mathrm{M}_2(F)$ so that $\mathrm{vol}(\mathrm{M}_2(\mathfrak{o})) = 1$. This yields the desired identity.

Assume that E is ramified. We use the notation of §5.6.1.5. Then the Weil representation ω_ψ on $\mathcal{S}(\mathbb{X}')$ is given in §5.3.2. We have

$$\begin{aligned}\Phi(\mathbf{m}(a)) &= |a|^2 \cdot \int_{\mathbb{X}'} \varphi'(ax)\overline{\varphi'(x)}\, dx \\ &= |a|^2 \cdot \prod_{i=1}^{4} \int_F \mathbb{I}_{\mathfrak{o}}(ax_i)\mathbb{I}_{\mathfrak{o}}(x_i)\, dx_i \\ &= |a|^2 \cdot \phi(a)^4.\end{aligned}$$

This yields the desired identity. □

6.5.2. The case (rps). We identify $B^\times$ with $\mathrm{GL}_2(F)$ via the isomorphism $\mathfrak{i}$ given by (5.1). Under this identification, we have $\mathcal{K} = \mathrm{GL}_2(\mathfrak{o})$, where $\mathcal{K}$ is the maximal compact subgroup of $B^\times$ given in §5.6.2. We write $\Phi = \Phi_\mu$ to indicate the dependence of $\varphi = \varphi_\mu$ on a unitary ramified character μ of conductor q^n.

LEMMA 6.5.2. *We have*

$$\Phi_\mu(\mathbf{n}(b)\mathbf{m}(a)) = \begin{cases} \mu(a) & \text{if } a \in \mathfrak{o}^\times \text{ and } b \in \mathfrak{o}, \\ 0 & \text{otherwise.} \end{cases}$$

PROOF. We use the notation of §5.6.2. Then the Weil representation ω_ψ on $\mathcal{S}(\mathbb{X}')$ is given in §5.3.1. We have

$$\begin{aligned}&\Phi_\mu(\mathbf{n}(b)\mathbf{m}(a)) = |a|^2 \cdot \int_{\mathbb{X}'} \varphi'(ax)\overline{\varphi'(x)}\psi\left(\frac{1}{2}b\langle x, x\rangle^\dagger\right) dx \\ &= q^{n+1}(q-1)^{-1} \cdot \mu(a) \cdot |a|^2 \\ &\quad \times \int_{F^4} \mathbb{I}_{\mathfrak{o}}(ax_1)\mathbb{I}_{\mathfrak{o}}(x_1)\mathbb{I}_{\mathfrak{o}}(ax_2)\mathbb{I}_{\mathfrak{o}}(x_2)\mathbb{I}_{\varpi^n\mathfrak{o}}(ax_3)\mathbb{I}_{\varpi^n\mathfrak{o}}(x_3)\mathbb{I}_{\mathfrak{o}^\times}(ax_4)\mathbb{I}_{\mathfrak{o}^\times}(x_4) \\ &\qquad\qquad \times \psi(b(x_1x_4 - x_2x_3))\, dx_1\, dx_2\, dx_3\, dx_4.\end{aligned}$$

Since $\mathbb{I}_{\mathfrak{o}^\times}(ax_4)\mathbb{I}_{\mathfrak{o}^\times}(x_4) = \mathbb{I}_{\mathfrak{o}^\times}(a)\mathbb{I}_{\mathfrak{o}^\times}(x_4)$, the above integral is zero unless $a \in \mathfrak{o}^\times$, in which case it is equal to

$$\begin{aligned}&\int_{F^4} \mathbb{I}_{\mathfrak{o}}(x_1)\mathbb{I}_{\mathfrak{o}}(x_2)\mathbb{I}_{\varpi^n\mathfrak{o}}(x_3)\mathbb{I}_{\mathfrak{o}^\times}(x_4)\psi(b(x_1x_4 - x_2x_3))\, dx_1\, dx_2\, dx_3\, dx_4 \\ &= \int_F \mathbb{I}_{\mathfrak{o}}(bx_3)\mathbb{I}_{\varpi^n\mathfrak{o}}(x_3)\, dx_3 \cdot \int_F \mathbb{I}_{\mathfrak{o}}(bx_4)\mathbb{I}_{\mathfrak{o}^\times}(x_4)\, dx_4 \\ &= \begin{cases} q^{-n}(1-q^{-1}) & \text{if } b \in \mathfrak{o}, \\ 0 & \text{otherwise.} \end{cases}\end{aligned}$$

This yields the lemma. □

6.5.3. The case (st). We identify $B^\times$ with $\mathrm{GL}_2(F)$ via:

- the isomorphism $\mathfrak{i}$ given by (5.1) if B_1 and B_2 are split;
- the isomorphism $\mathfrak{i}$ given by (5.7) if B_1 and B_2 are ramified.

Under this identification, we have $\mathcal{K} = \mathrm{GL}_2(\mathfrak{o})$, where $\mathcal{K}$ is the maximal compact subgroup of $B^\times$ given in §5.6.3. Put

$$w = \begin{pmatrix} & -1 \\ 1 & \end{pmatrix}.$$

LEMMA 6.5.3. *We have*

$$\Phi(\mathbf{m}(\varpi^i)) = \begin{cases} q^{-2i} & \text{if } i \geq 0, \\ q^{2i} & \text{if } i \leq 0, \end{cases}$$

$$\Phi(\mathbf{m}(\varpi^i)w) = \begin{cases} \gamma_{B_1} \cdot q^{-2i-1} & \text{if } i \geq 0, \\ \gamma_{B_1} \cdot q^{2i+1} & \text{if } i < 0, \end{cases}$$

where

$$\gamma_{B_1} = \begin{cases} 1 & \text{if } B_1 \text{ is split}, \\ -1 & \text{if } B_1 \text{ is ramified}. \end{cases}$$

PROOF. For convenience, we write $\mathfrak{p}^i = \varpi^i \mathfrak{o}$ for $i \in \mathbb{Z}$. Put

$$\phi(j,k) := \int_F \mathbb{I}_{\mathfrak{p}^j}(x)\mathbb{I}_{\mathfrak{p}^k}(x)\, dx = \begin{cases} q^{-j} & \text{if } j \geq k, \\ q^{-k} & \text{if } j \leq k \end{cases}$$

for $j, k \in \mathbb{Z}$, where dx is the self-dual Haar measure on F with respect to ψ.

Assume that B_1 and B_2 are split. We use the notation of §5.6.3.1. Then the Weil representation ω_ψ on $\mathcal{S}(\mathbb{X}')$ is given in §5.3.1. We have

$$\varphi'(x) = q^{\frac{1}{2}} \cdot \mathbb{I}_{\mathfrak{o}}(x_1)\mathbb{I}_{\mathfrak{o}}(x_2)\mathbb{I}_{\mathfrak{p}}(x_3)\mathbb{I}_{\mathfrak{o}}(x_4),$$
$$\omega_\psi(w)\varphi'(x) = q^{-\frac{1}{2}} \cdot \mathbb{I}_{\mathfrak{o}}(x_1)\mathbb{I}_{\mathfrak{p}^{-1}}(x_2)\mathbb{I}_{\mathfrak{o}}(x_3)\mathbb{I}_{\mathfrak{o}}(x_4),$$

so that

$$\begin{aligned} \Phi(\mathbf{m}(\varpi^i)) &= q^{-2i} \cdot \int_{\mathbb{X}'} \varphi'(\varpi^i x)\overline{\varphi'(x)}\, dx \\ &= q^{-2i+1} \cdot \phi(-i,0)^3 \cdot \phi(-i+1,1), \\ \Phi(\mathbf{m}(\varpi^i)w) &= q^{-2i} \cdot \int_{\mathbb{X}'} \omega_\psi(w)\varphi'(\varpi^i x)\overline{\varphi'(x)}\, dx \\ &= q^{-2i} \cdot \phi(-i,0)^2 \cdot \phi(-i-1,0) \cdot \phi(-i,1). \end{aligned}$$

This yields the desired identity.

Assume that B_1 and B_2 are ramified. We use the notation of §5.6.3.2. Then the Weil representation ω_ψ on $\mathcal{S}(\mathbb{X}')$ is given in §5.3.2. We have

$$\varphi'(x) = q^{\frac{1}{2}} \cdot \mathbb{I}_{\mathfrak{o}}(x_1)\mathbb{I}_{\mathfrak{o}}(x_2)\mathbb{I}_{\mathfrak{o}}(x_3)\mathbb{I}_{\mathfrak{o}}(x_4),$$
$$\omega_\psi(w)\varphi'(x) = -q^{-\frac{1}{2}} \cdot \mathbb{I}_{\mathfrak{o}}(x_1)\mathbb{I}_{\mathfrak{o}}(x_2)\mathbb{I}_{\mathfrak{p}^{-1}}(x_3)\mathbb{I}_{\mathfrak{p}^{-1}}(x_4),$$

so that

$$\begin{aligned} \Phi(\mathbf{m}(\varpi^i)) &= q^{-2i} \cdot \int_{\mathbb{X}'} \varphi'(\varpi^i x)\overline{\varphi'(x)}\, dx \\ &= q^{-2i} \cdot \phi(-i,0)^4, \\ \Phi(\mathbf{m}(\varpi^i)w) &= q^{-2i} \cdot \int_{\mathbb{X}'} \omega_\psi(w)\varphi'(\varpi^i x)\overline{\varphi'(x)}\, dx \\ &= -q^{-2i-1} \cdot \phi(-i,0)^2 \cdot \phi(-i-1,0)^2. \end{aligned}$$

This yields the desired identity. □

6.5.4. The case (1d). Let $\mathcal{K} = \mathfrak{o}_B^\times$ be the unique maximal compact subgroup of $B^\times$. We have $B^1 \subset \mathcal{K}$.

LEMMA 6.5.4. *We have $\Phi(g) = 1$ for $g \in B^1$.*

PROOF. We use the notation of §5.6.4. We have $\omega_\psi(g)\varphi' = \varphi'$ and hence $\Phi(g) = \langle \varphi', \varphi' \rangle = 1$ for all $g \in B^1$. □

6.5.5. The case (ds). We identify $B^\times$ with $\mathrm{GL}_2(F)$ via the isomorphism $\mathfrak{i}$ given by (5.7). We write $\Phi = \Phi_k$ to indicate the dependence of $\varphi = \varphi_k$ on a non-negative integer k.

LEMMA 6.5.5. *We have*

$$\Phi_k(\mathbf{m}(a)) = \left(\frac{a + a^{-1}}{2}\right)^{-k-2}$$

for $a > 0$.

PROOF. Assume that B_1 and B_2 are split. We use the notation of §5.6.5.1. Then the Weil representation ω_ψ on $\mathcal{S}(\mathbb{X}')$ is given in §5.3.2. We have

$$\begin{aligned}
\Phi_k(\mathbf{m}(a)) &= a^2 \cdot \int_{\mathbb{X}'} \varphi'(ax)\overline{\varphi'(x)}\, dx \\
&= \frac{|u|}{4} \cdot c_k^{-1} \cdot a^{k+2} \cdot \int_{F^4} (x_2^2 - ux_1^2)^k \cdot e^{-\frac{\pi}{2v}(a^2+1)(x_2^2 - ux_1^2 + x_4^2 - ux_3^2)}\, dx_1 \cdots dx_4 \\
&= \frac{|u|}{4} \cdot c_k^{-1} \cdot a^{k+2} \cdot v^{-2} \cdot \left(\frac{\pi}{2v}(a^2+1)\right)^{-k-2} \cdot \phi(k) \cdot \phi(0),
\end{aligned}$$

where

$$\phi(k) := \int_{-\infty}^{\infty} \int_{-\infty}^{\infty} (x^2 + y^2)^k e^{-(x^2+y^2)}\, dx\, dy$$

with the Lebesgue measures dx, dy on $\mathbb{R}$. Since

$$\phi(k) = \int_0^{2\pi} \int_0^{\infty} r^{2k} e^{-r^2} r\, dr\, d\theta = \pi \cdot \int_0^{\infty} r^k e^{-r}\, dr = \pi \cdot k!,$$

we have

$$\begin{aligned}
\Phi_k(\mathbf{m}(a)) &= \frac{|u|}{4} \cdot \frac{4\pi^k}{k!|u|^{\frac{k}{2}+1}} \cdot a^{k+2} \cdot v^{-2} \cdot \left(\frac{\pi}{2v}(a^2+1)\right)^{-k-2} \cdot \pi^2 \cdot k! \\
&= \left(\frac{a + a^{-1}}{2}\right)^{-k-2}.
\end{aligned}$$

Assume that B_1 and B_2 are ramified. We use the notation of §5.6.5.2. Then the Weil representation ω_ψ on $\mathcal{S}(\mathbb{X}')$ is given in §5.3.2 and the computation is the same as in the case when B_1 and B_2 are split. □

6.5.6. The case (fd). We identify $\mathbb{C}^\times$ with a subgroup of $B^\times$ via the isomorphism $E \cong \mathbb{C}$ such that $\mathbf{i}/\sqrt{-1} > 0$ and the fixed embedding $E \hookrightarrow B$. Let ϕ_k be the matrix coefficient of Sym^k such that

- $\phi_k(\alpha g \beta) = \chi_k(\alpha)\chi_k(\beta)\phi_k(g)$ for $\alpha, \beta \in \mathbb{C}^\times$ and $g \in B^\times$,
- $\phi_k(1) = 1$.

We write $\Phi = \Phi_k$ to indicate the dependence of $\varphi = \varphi_k$ on a non-negative integer k.

LEMMA 6.5.6. *We have* $\Phi_k(g) = \overline{\phi_k(g)}$ *for* $g \in B^1$.

PROOF. We use the notation of §5.6.6. Then the Weil representation ω_ψ on $\mathcal{S}(\mathbb{X}')$ is given in §5.3.3. If we write $x = z_1 + z_2 \frac{\mathbf{j}}{s} \in \mathbb{X}' \cong W = B$ with $z_1, z_2 \in E$, then

$$\varphi'(x) = c_k^{-\frac{1}{2}} \cdot (z_1^\rho)^k \cdot e^{-\frac{\pi v}{2}(z_1 z_1^\rho + z_2 z_2^\rho)}.$$

Let $\mathcal{S}_k$ be the subspace of $\mathcal{S}(\mathbb{X}')$ generated by $\omega_\psi(g)\varphi'$ for all $g \in B^1$. Since

$$\omega_\psi(g)\varphi'(z_1, z_2) = \varphi'(z_1\alpha_1 - z_2\alpha_2^\rho, z_1\alpha_2 + z_2\alpha_1^\rho)$$

for $g = \alpha_1 + \alpha_2 \frac{\mathbf{j}}{s} \in B^1$ with $\alpha_1, \alpha_2 \in E$, $\mathcal{S}_k$ is generated by

$$(z_1^\rho)^i \cdot (z_2^\rho)^{k-i} \cdot e^{-\frac{\pi v}{2}(z_1 z_1 \rho + z_2 z_2^\rho)}$$

for all $0 \le i \le k$. Moreover, the representation of B^1 on $\mathcal{S}_k$ is isomorphic to the unique irreducible $(k+1)$-dimensional representation $\mathrm{Sym}^k|_{B^1}$, so that Φ_k is a matrix coefficient of $\mathrm{Sym}^k|_{B^1}$. On the other hand, by Lemma 5.6.3, we have $\Phi_k(\alpha g \beta) = \chi_k(\alpha)^{-1}\chi_k(\beta)^{-1}\Phi_k(g)$ for $\alpha, \beta \in \mathbb{C}^1$ and $\Phi_k(1) = \langle \varphi', \varphi' \rangle = 1$. Hence we must have $\Phi_k = \bar{\phi}_k|_{B^1}$. □

6.6. Computation of Z_v

To finish the proof of Theorem 6.3.2, it remains to compute the integral Z_v. We fix a place v of F and suppress the subscript v from the notation. Recall that

$$Z = \int_{B^1} \Phi(g)\Psi(g)\, dg,$$

where

- Φ is the function on B^1 given in §6.5;
- Ψ is the function on B^1 defined by

  $$\Psi(g) = \langle \pi_B(g) f_B, f_B \rangle,$$

  where $f_B \in \pi_B$ is the new vector as in §6.3 and $\langle \cdot, \cdot \rangle$ is the invariant hermitian inner product on π_B normalized so that $\langle f_B, f_B \rangle = 1$;
- dg is the standard measure on B^1.

6.6.1. The case (ur). In this case, $\pi_B = \mathrm{Ind}(\chi \otimes \mu)$, where χ and μ are unitary unramified. We have

$$L(s, \pi, \mathrm{ad}) = \frac{1}{(1 - q^{-s})(1 - \gamma q^{-s})(1 - \gamma^{-1} q^{-s})},$$

where $\gamma = \chi(\varpi) \cdot \mu(\varpi)^{-1}$.

LEMMA 6.6.1. *We have*

$$Z = \frac{L(1, \pi, \mathrm{ad})}{\zeta(2)^2}.$$

PROOF. We retain the notation of §§5.6.1, 6.5.1. Put $\mathcal{K}' = \mathrm{SL}_2(\mathfrak{o})$. Then we have

$$Z = \sum_{i=0}^{\infty} \Phi(\mathbf{m}(\varpi^i))\Psi(\mathbf{m}(\varpi^i)) \operatorname{vol}(\mathcal{K}'\mathbf{m}(\varpi^i)\mathcal{K}').$$

By Macdonald's formula [**50**], [**7**], we have

$$\Psi(\mathbf{m}(\varpi^i)) = \frac{q^{-i}}{1 + q^{-1}} \cdot \left(\gamma^i \cdot \frac{1 - \gamma^{-1} q^{-1}}{1 - \gamma^{-1}} + \gamma^{-i} \cdot \frac{1 - \gamma q^{-1}}{1 - \gamma} \right).$$

Also, we see that

$$\mathrm{vol}(\mathcal{K}'\mathbf{m}(\varpi^i)\mathcal{K}') = \begin{cases} 1 & \text{if } i = 0, \\ q^{2i}(1+q^{-1}) & \text{if } i \geq 1. \end{cases}$$

Combining these with Lemma 6.5.1, we obtain

$$\begin{aligned} Z &= 1 + \sum_{i=1}^{\infty} q^{-i} \cdot \left(\gamma^i \cdot \frac{1-\gamma^{-1}q^{-1}}{1-\gamma^{-1}} + \gamma^{-i} \cdot \frac{1-\gamma q^{-1}}{1-\gamma} \right) \\ &= 1 + \frac{\gamma q^{-1}}{1-\gamma q^{-1}} \cdot \frac{1-\gamma^{-1}q^{-1}}{1-\gamma^{-1}} + \frac{\gamma^{-1}q^{-1}}{1-\gamma^{-1}q^{-1}} \cdot \frac{1-\gamma q^{-1}}{1-\gamma} \\ &= \frac{(1+q^{-1})(1-q^{-2})}{(1-\gamma q^{-1})(1-\gamma^{-1}q^{-1})}. \end{aligned}$$

□

6.6.2. The case (rps). In this case, $\pi_B = \mathrm{Ind}(\chi \otimes \mu)$ and $\Phi = \Phi_\mu$, where χ is unitary unramified and μ is unitary ramified of conductor q^n. We have $L(s, \pi, \mathrm{ad}) = \zeta(s)$.

LEMMA 6.6.2. *We have*

$$Z = \frac{1}{q^{n-4}(q-1)(q+1)^3} \cdot \frac{L(1, \pi, \mathrm{ad})}{\zeta(2)^2}.$$

PROOF. Following [**46**, Chapter VIII], we shall compute Z explicitly. We retain the notation of §§5.6.2, 6.5.2. Put $\mathcal{K}' = \mathrm{SL}_2(\mathfrak{o})$ and $\mathcal{K}'_n = \mathcal{K}_n \cap \mathrm{SL}_2(\mathfrak{o})$. We take the invariant hermitian inner product $\langle \cdot, \cdot \rangle$ on π_B defined by

$$\langle f_1, f_2 \rangle = \int_{\mathcal{K}} f_1(k) \overline{f_2(k)} \, dk,$$

where dk is the Haar measure on $\mathcal{K}$ such that $\mathrm{vol}(\mathcal{K}) = 1$. Then f_B is determined by

$$f_B|_{\mathcal{K}} = \mathrm{vol}(\mathcal{K}_n)^{-\frac{1}{2}} \cdot \mathbb{I}_{\mathcal{K}_n} \boldsymbol{\mu}.$$

We can define a new vector $\tilde{f}_B \in \pi_B$ with respect to $(\mathcal{K}_n, \boldsymbol{\mu})$ by

$$\tilde{f}_B(h) = \int_{B^1} \Phi(g) f_B(hg) \, dg$$

for $h \in B^\times$. Since $\tilde{f}_B = \frac{\tilde{f}_B(1)}{f_B(1)} \cdot f_B$ and $\langle f_B, f_B \rangle = 1$, we have

$$Z = \langle \tilde{f}_B, f_B \rangle = \mathrm{vol}(\mathcal{K}_n)^{\frac{1}{2}} \cdot \tilde{f}_B(1).$$

We have

$$\begin{aligned} \tilde{f}_B(1) &= \int_{B^1} \Phi(g) f_B(g) \, dg \\ &= \int_{\mathcal{K}'} \int_{F^\times} \int_F \Phi(\mathbf{n}(b)\mathbf{m}(a)k) \cdot f_B(\mathbf{n}(b)\mathbf{m}(a)k) \cdot |a|^{-2} \, db \, da \, dk \\ &= \mathrm{vol}(\mathcal{K}_n)^{-\frac{1}{2}} \cdot \int_{\mathcal{K}'_n} \int_{F^\times} \int_F \Phi(\mathbf{n}(b)\mathbf{m}(a)) \boldsymbol{\mu}(k)^{-1} \cdot \chi(a)\mu(a)^{-1}|a| \boldsymbol{\mu}(k) \\ &\qquad \times |a|^{-2} \, db \, da \, dk \\ &= \mathrm{vol}(\mathcal{K}_n)^{-\frac{1}{2}} \cdot \mathrm{vol}(\mathcal{K}'_n) \cdot \int_{F^\times} \int_F \Phi(\mathbf{n}(b)\mathbf{m}(a)) \cdot \chi(a)\mu(a)^{-1}|a|^{-1} \, db \, da, \end{aligned}$$

where

- db is the Haar measure on F such that $\mathrm{vol}(\mathfrak{o}) = 1$;
- da is the Haar measure on $F^\times$ such that $\mathrm{vol}(\mathfrak{o}^\times) = 1$;
- dk is the Haar measure on $\mathcal{K}'$ such that $\mathrm{vol}(\mathcal{K}') = 1$.

By Lemma 6.5.2, we have

$$\int_{F^\times}\int_F \Phi(\mathbf{n}(b)\mathbf{m}(a)) \cdot \chi(a)\mu(a)^{-1}|a|^{-1}\, db\, da = \int_{\mathfrak{o}^\times}\int_{\mathfrak{o}} \chi(a)|a|^{-1}\, db\, da = 1.$$

Hence we have

$$Z = \mathrm{vol}(\mathcal{K}'_n) = \frac{1}{q^{n-1}(q+1)}.$$

□

6.6.3. The case (st). In this case, $\pi_B = \mathrm{St}\otimes\chi$, where χ is unitary unramified. We have $L(s, \pi, \mathrm{ad}) = \zeta(s+1)$.

LEMMA 6.6.3.

(i) *If B_1 and B_2 are split, then we have*

$$Z = \frac{q^2}{(q+1)^3} \cdot \frac{L(1, \pi, \mathrm{ad})}{\zeta(2)^2}.$$

(ii) *If B_1 and B_2 are ramified, then we have*

$$Z = \frac{q^2}{(q-1)^2(q+1)} \cdot \frac{L(1, \pi, \mathrm{ad})}{\zeta(2)^2}.$$

PROOF. We retain the notation of §§5.6.3, 6.5.3. Put $\mathcal{I}' = \mathcal{I}\cap\mathrm{SL}_2(\mathfrak{o})$. Let $\tilde{W} = N(T_0)/T_0$ be the extended affine Weyl group of $\mathrm{GL}_2(F)$, where $T_0 = \{\left(\begin{smallmatrix} a & \\ & d\end{smallmatrix}\right) \mid a, d \in \mathfrak{o}^\times\}$ and $N(T_0)$ is the normalizer of T_0 in $\mathrm{GL}_2(F)$. Then we have

$$\mathrm{GL}_2(F) = \bigsqcup_{\tilde{w}\in\tilde{W}} \mathcal{I}\tilde{w}\mathcal{I}.$$

We can write $\tilde{W} = \Omega \ltimes W_a$ with $\Omega = \langle\omega\rangle$ and $W_a = \langle w_1, w_2\rangle$, where

$$\omega = \begin{pmatrix} & 1 \\ \varpi & \end{pmatrix}, \quad w_1 = \begin{pmatrix} & 1 \\ 1 & \end{pmatrix}, \quad w_2 = \begin{pmatrix} & \varpi^{-1} \\ \varpi & \end{pmatrix}.$$

Noting that $w_1^2 = w_2^2 = 1$ and $w_1 w_2 = \mathbf{m}(\varpi)$, we have

$$\mathrm{SL}_2(F) = \bigsqcup_{j=0}^{1}\bigsqcup_{i=-\infty}^{\infty} \mathcal{I}'\mathbf{m}(\varpi^i)w^j\mathcal{I}'$$

and hence

$$Z = \sum_{j=0}^{1}\sum_{i=-\infty}^{\infty} \Phi(\mathbf{m}(\varpi^i)w^j)\Psi(\mathbf{m}(\varpi^i)w^j)\,\mathrm{vol}(\mathcal{I}'\mathbf{m}(\varpi^i)w^j\mathcal{I}').$$

Let ℓ be the length function on $\tilde{W}$, so that $\ell(\omega) = 0$ and $\ell(w_1) = \ell(w_2) = 1$. By [**18**, §7], we have

$$\Psi(\omega^k\tilde{w}) = (-\chi(\varpi))^k \cdot (-q)^{-\ell(\tilde{w})}$$

for $k \in \mathbb{Z}$ and $\tilde{w} \in W_a$. Also, we see that $|\mathcal{I}\tilde{w}\mathcal{I}/\mathcal{I}| = q^{\ell(\tilde{w})}$ for $\tilde{w} \in \tilde{W}$. Hence we have

$$\begin{aligned}\Psi(\mathbf{m}(\varpi^i)w^j)\operatorname{vol}(\mathcal{I}'\mathbf{m}(\varpi^i)w^j\mathcal{I}') &= (-1)^{\ell(\mathbf{m}(\varpi^i)w^j)}\cdot\operatorname{vol}(\mathcal{I}') \\ &= \frac{1}{q+1}\times\begin{cases}1 & \text{if } j=0,\\ -1 & \text{if } j=1,\end{cases}\end{aligned}$$

so that

$$Z = \frac{1}{q+1}\cdot\left(\sum_{i=-\infty}^{\infty}\Phi(\mathbf{m}(\varpi^i)) - \sum_{i=-\infty}^{\infty}\Phi(\mathbf{m}(\varpi^i)w)\right).$$

Combining this with Lemma 6.5.3, we obtain

$$\begin{aligned}Z &= \frac{1}{q+1}\cdot\left(\sum_{i=0}^{\infty}q^{-2i} + \sum_{i=1}^{\infty}q^{-2i} - \sum_{i=0}^{\infty}q^{-2i-1} - \sum_{i=1}^{\infty}q^{-2i+1}\right)\\ &= \frac{1}{q+1}\cdot\frac{1+q^{-2}-q^{-1}-q^{-1}}{1-q^{-2}}\\ &= \frac{q-1}{(q+1)^2}\end{aligned}$$

if B_1 and B_2 are split, and

$$\begin{aligned}Z &= \frac{1}{q+1}\cdot\left(\sum_{i=0}^{\infty}q^{-2i} + \sum_{i=1}^{\infty}q^{-2i} + \sum_{i=0}^{\infty}q^{-2i-1} + \sum_{i=1}^{\infty}q^{-2i+1}\right)\\ &= \frac{1}{q+1}\cdot\frac{1+q^{-2}+q^{-1}+q^{-1}}{1-q^{-2}}\\ &= \frac{1}{q-1}\end{aligned}$$

if B_1 and B_2 are ramified. □

6.6.4. The case (1d). In this case, $\pi_B = \chi\circ\nu$, where χ is unitary unramified. We have $L(s,\pi,\mathrm{ad}) = \zeta(s+1)$.

LEMMA 6.6.4. *We have*

$$Z = \frac{q^2}{(q-1)(q+1)}\cdot\frac{L(1,\pi,\mathrm{ad})}{\zeta(2)^2}.$$

PROOF. We retain the notation of §§5.6.4, 6.5.4. Then by Lemma 6.5.4, we have

$$Z = \int_{B^1} dg = 1.$$ □

6.6.5. The case (ds). In this case, $\pi_B = \mathrm{DS}_k$ and $\Phi = \Phi_l$, where

$$l = \begin{cases}k & \text{if } B_1 \text{ and } B_2 \text{ are split},\\ k-2 & \text{if } B_1 \text{ and } B_2 \text{ are ramified}.\end{cases}$$

LEMMA 6.6.5.

(i) *If B_1 and B_2 are split, then we have*

$$Z = \frac{4\pi}{k}.$$

(ii) *If B_1 and B_2 are ramified, then we have*

$$Z = \frac{4\pi}{k-1}.$$

PROOF. We retain the notation of §§5.6.5, 6.5.5. In particular, we identify $\mathbb{C}^\times$ with a subgroup of $B^\times$. Then we have

$$\begin{aligned} Z &= 4\pi \cdot \int_{\mathbb{C}^1} \int_0^\infty \int_{\mathbb{C}^1} \Phi(\kappa_1 \mathbf{m}(e^t)\kappa_2)\Psi(\kappa_1 \mathbf{m}(e^t)\kappa_2) \sinh(2t)\, d\kappa_1\, dt\, d\kappa_2 \\ &= 4\pi \cdot \int_0^\infty \Phi(\mathbf{m}(e^t))\Psi(\mathbf{m}(e^t)) \sinh(2t)\, dt, \end{aligned}$$

where

- dt is the Lebesgue measure;
- $d\kappa_1$ and $d\kappa_2$ are the Haar measures on $\mathbb{C}^1$ such that $\mathrm{vol}(\mathbb{C}^1) = 1$.

It is known that

$$\Psi(\mathbf{m}(e^t)) = \cosh(t)^{-k}.$$

Combining these with Lemma 6.5.5, we obtain

$$\begin{aligned} Z &= 4\pi \cdot \int_0^\infty \cosh(t)^{-k-l-2} \sinh(2t)\, dt \\ &= 8\pi \cdot \int_0^\infty \cosh(t)^{-k-l-1} \sinh(t)\, dt \\ &= 8\pi \cdot \int_1^\infty t^{-k-l-1}\, dt \\ &= \frac{8\pi}{k+l}. \end{aligned}$$

□

6.6.6. The case (fd). In this case, $\pi_B = \mathrm{Sym}^{k-2}$ and $\Phi = \Phi_{k-2}$.

LEMMA 6.6.6. *We have*

$$Z = \frac{1}{k-1}.$$

PROOF. We retain the notation of §§5.6.6, 6.5.6. Then by Lemma 6.5.6 and the Schur orthogonality relations, we have

$$Z = \int_{B^1} |\phi_{k-2}(g)|^2\, dg = \frac{1}{k-1}.$$

□

CHAPTER 7

The Main Conjecture on the Arithmetic of Theta Lifts

7.1. On the choices of u, J_1 and J_2

We suppose now that we are given a totally real number field F and two quaternion algebras B_1 and B_2 over F. Let us define for convenience:

$$\mathfrak{d}_{B_1 \smallsetminus B_2} = \prod_{\mathfrak{q} \mid \mathfrak{d}_{B_1},\, \mathfrak{q} \nmid \mathfrak{d}_{B_2}} \mathfrak{q}, \qquad \mathfrak{d}_{B_2 \smallsetminus B_1} = \prod_{\mathfrak{q} \mid \mathfrak{d}_{B_2},\, \mathfrak{q} \nmid \mathfrak{d}_{B_1}} \mathfrak{q},$$

$$\mathfrak{d}_{B_1 \cup B_2} = \prod_{\mathfrak{q} \mid \mathfrak{d}_{B_1} \mathfrak{d}_{B_2}} \mathfrak{q}, \qquad \mathfrak{d}_{B_1 \cap B_2} = \prod_{\mathfrak{q} \mid (\mathfrak{d}_{B_1}, \mathfrak{d}_{B_2})} \mathfrak{q}$$

and

$$\Sigma_{B_1 \smallsetminus B_2} = \Sigma_{B_1} \smallsetminus \Sigma_{B_2}, \qquad \Sigma_{B_2 \smallsetminus B_1} = \Sigma_{B_2} \smallsetminus \Sigma_{B_1},$$

$$\Sigma_{B_1 \cup B_2} = \Sigma_{B_1} \cup \Sigma_{B_2}, \qquad \Sigma_{B_1 \cap B_2} = \Sigma_{B_1} \cap \Sigma_{B_2}.$$

For the constructions so far (especially the constructions of splittings), the only condition needed is:

(7.1) At every place v of F, at least one of u, J_1, J_2, J is a square.

However, to formulate the main conjecture we will need to make a more careful choice. In this section, we show that we can make such a choice that satisfies a number of useful auxiliary conditions. For purposes of clarity, we first make the following definition.

DEFINITION 7.1.1. Let F, B_1 and B_2 be given. Suppose that we are also given an ideal $\mathfrak{f}$ of $\mathcal{O}_F$ that is coprime to $\mathfrak{d}_{B_1 \cup B_2}$. A triple (u, J_1, J_2) of elements in $F^\times$ will be said to be well adapted to $(F, B_1, B_2, \mathfrak{f})$ if:

(i) u, J_1, J_2 lie in $\mathcal{O}_F$.
(ii) At every place v of F, at least one of u, J_1, J_2, J is a square (where $J := J_1 J_2$).
(iii) $u \ll 0$, so that $E := F + F\mathbf{i}$, $\mathbf{i}^2 = u$ is a CM field.
(iv) u is a unit at any prime $\mathfrak{q}$ of F that is unramified in E.
(v) If $\mathfrak{q}$ is a prime of F dividing 2, then $E_\mathfrak{q}$ is the unique unramified quadratic extension of $F_\mathfrak{q}$ if $\mathfrak{q} \mid \mathfrak{d}_{B_1 \cup B_2}$ and $E_\mathfrak{q}/F_\mathfrak{q}$ is split otherwise.
(vi) (a) $B_1 \simeq E + E\mathbf{j}_1$, with $\mathbf{j}_1^2 = J_1$ and $\mathbf{i}\mathbf{j}_1 = -\mathbf{j}_1\mathbf{i}$.
(b) $B_2 \simeq E + E\mathbf{j}_2$, with $\mathbf{j}_2^2 = J_2$ and $\mathbf{i}\mathbf{j}_2 = -\mathbf{j}_2\mathbf{i}$.
(c) $B \simeq E + E\mathbf{j}$ with $\mathbf{j}^2 = J = J_1 J_2$ and $\mathbf{i}\mathbf{j} = -\mathbf{j}\mathbf{i}$, where $B \simeq B_1 \cdot B_2$ in the Brauer group of F. (This is actually a formal consequence of the previous two subparts.)
(vii) (a) If $\mathfrak{q} \mid \mathfrak{d}_{B_1 \smallsetminus B_2}$, then J_1 is a uniformizer at $\mathfrak{q}$ and J_2 is the square of a unit.
(b) If $\mathfrak{q} \mid \mathfrak{d}_{B_2 \smallsetminus B_1}$, then J_2 is a uniformizer at $\mathfrak{q}$ and J_1 is the square of a unit.

(c) If $\mathfrak{q} \mid \mathfrak{d}_{B_1 \cap B_2}$, then J_1 and J_2 are both uniformizers at $\mathfrak{q}$ such that J_1/J_2 is the square of a unit.

(viii) u, J_1, J_2 and J are squares of units at the primes of $\mathcal{O}_F$ dividing $\mathfrak{f}$.

A *well adapted coordinate system* for $(F, B_1, B_2, \mathfrak{f})$ will consist of a well adapted choice of (u, J_1, J_2) together with a choice of isomorphisms as in (vi) above. (If $\mathfrak{f} = (1)$ we may simply drop it from the notation.)

REMARK 7.1.2. If $\mathfrak{f}' \mid \mathfrak{f}$, then a well adapted coordinate system for $(F, B_1, B_2, \mathfrak{f})$ gives rise naturally to a well adapted coordinate system for $(F, B_1, B_2, \mathfrak{f}')$.

REMARK 7.1.3. In our application later in this chapter, we will assume that $\mathfrak{d}_{B_1 \cup B_2}$ is prime to $2\mathfrak{D}_{F/\mathbb{Q}}$ (where $\mathfrak{D}_{F/\mathbb{Q}}$ denotes the different of $F/\mathbb{Q}$) and that $2\mathfrak{D}_{F/\mathbb{Q}} \mid \mathfrak{f}$. Note that with this assumption, for any prime $\mathfrak{q}$ of $\mathcal{O}_F$ above 2, condition (viii) implies that the element u must be a square of a unit at such a $\mathfrak{q}$ and hence $E_\mathfrak{q}/F_\mathfrak{q}$ is split. Thus the conditions (viii) and (v) both apply at such a $\mathfrak{q}$ but are consistent. Then choosing a well adapted coordinate system for $(F, B_1, B_2, \mathfrak{f})$ ensures that the list of possibilities in the table of §5.5 covers all possible ramification scenarios at any place v of F.

The following key proposition shows that we can pick a well adapted coordinate system that is especially convenient for studying ℓ-integrality questions.

PROPOSITION 7.1.4. *Suppose that ℓ is a rational prime that is coprime to $\mathfrak{d}_{B_1 \cup B_2}$ and $\{\mathfrak{f}_1, \cdots, \mathfrak{f}_n\}$ is a collection of primes of $\mathcal{O}_F$ (possibly empty) that are coprime to $\ell\mathfrak{d}_{B_1 \cup B_2}$. Then we can find elements $u, J_1, J_2 \in F$ such that the following hold:*

(i) *u, J_1, J_2 lie in $\mathcal{O}_F$.*

(ii) *At every place v of F, at least one of u, J_1, J_2, J is a square.*

(iii) *$u \ll 0$, so that $E := F + F\mathbf{i}$, $\mathbf{i}^2 = u$ is a CM field.*

(iv) *u is a unit at any prime $\mathfrak{q}$ that is unramified in E.*

(v) *If $\mathfrak{q}$ is a prime of F dividing 2, then $E_\mathfrak{q}$ is the unique unramified quadratic extension of $F_\mathfrak{q}$ if $\mathfrak{q} \mid \mathfrak{d}_{B_1 \cup B_2}$ and $E_\mathfrak{q}/F_\mathfrak{q}$ is split otherwise.*

(vi) (a) *$B_1 \simeq E + E\mathbf{j}_1$, with $\mathbf{j}_1^2 = J_1$ and $\mathbf{i}\mathbf{j}_1 = -\mathbf{j}_1\mathbf{i}$.*

(b) *$B_2 \simeq E + E\mathbf{j}_2$, with $\mathbf{j}_2^2 = J_2$ and $\mathbf{i}\mathbf{j}_2 = -\mathbf{j}_2\mathbf{i}$.*

(c) *$B \simeq E + E\mathbf{j}$ with $\mathbf{j}^2 = J = J_1 J_2$ and $\mathbf{i}\mathbf{j} = -\mathbf{j}\mathbf{i}$.*

(vii) (a) *If $\mathfrak{q} \mid \mathfrak{d}_{B_1 \smallsetminus B_2}$, then J_1 is a uniformizer at $\mathfrak{q}$ and J_2 is the square of a unit.*

(b) *If $\mathfrak{q} \mid \mathfrak{d}_{B_2 \smallsetminus B_1}$, then J_2 is a uniformizer at $\mathfrak{q}$ and J_1 is the square of a unit.*

(c) *If $\mathfrak{q} \mid \mathfrak{d}_{B_1 \cap B_2}$, then J_1 and J_2 are both uniformizers at $\mathfrak{q}$ such that J_1/J_2 is the square of a unit.*

(viii) *u, J_1, J_2 and J are squares of units at the primes in $\{\mathfrak{f}_1, \ldots, \mathfrak{f}_n\}$ and at all primes $\mathfrak{l}$ of F above ℓ.*

In other words, there exists a well adapted coordinate system for $(F, B_1, B_2, (\ell)\mathfrak{f})$ with $\mathfrak{f} := \prod_i \mathfrak{f}_i$.

Let K denote the quadratic extension of F given by

$$K = F + F\mathbf{j}. \tag{7.2}$$

Note that the condition (viii) above implies that both E and K are split at the primes in $\{\mathfrak{l} \mid \ell\} \cup \{\mathfrak{f}_1, \ldots, \mathfrak{f}_n\}$.

Prop. 7.1.4 will suffice for the current paper. The following enhancement of it will be useful in [**31**], [**32**].

PROPOSITION 7.1.5. *Let ℓ and $\mathfrak{f}_1, \dots, \mathfrak{f}_n$ be as in the previous proposition. Suppose that the prime ℓ satisfies the following conditions:*

- *ℓ is unramified in F.*
- *$\ell > 5$ and for any $\mathfrak{q} \mid \mathfrak{d}_{B_1 \smallsetminus B_2} \cdot \mathfrak{d}_{B_2 \smallsetminus B_1}$, we have*

$$\mathrm{N}\mathfrak{q} \not\equiv 0, \pm 1 \pmod{\ell}.$$

Then we can choose u, J_1, J_2 such that in addition to (i) *through* (viii) *above, we have:*

(ix) *If E or K is ramified at a prime $\mathfrak{p}$, then* $\mathrm{N}\mathfrak{p} \not\equiv 0, \pm 1 \pmod{\ell}$.

The following Lemmas 7.1.6 and 7.1.7 will be useful in the proofs of Prop. 7.1.4 and Prop. 7.1.5 respectively.

LEMMA 7.1.6. *Let F be a number field, Ξ_f a finite subset of Σ_{fin} and $\Xi_\infty \subseteq \Sigma_\infty$ a set of real infinite places. Let I be an ideal in $\mathcal{O}_F$ prime to the primes in Ξ_f. Then there exists a prime ideal $\mathfrak{q} \subset \mathcal{O}_F$ such that $I \cdot \mathfrak{q} = (\alpha)$ is principal with α satisfying:*

(a) *α is a square of a unit at the primes in Ξ_f.*

(b) *$\sigma_v(\alpha) < 0$ for v in Ξ_∞ and $\sigma_v(\alpha) > 0$ for any real place v of F not in Ξ_∞.*

Further, $\mathfrak{q}$ can be picked to avoid any finite set of primes.

LEMMA 7.1.7. *Let F be a number field and $\ell > 5$ a rational prime unramified in F. Suppose that Ξ_f is a finite subset of Σ_{fin} all whose elements are prime to ℓ and $\Xi_\infty \subseteq \Sigma_\infty$ is a set of real infinite places. Let I be an ideal in $\mathcal{O}_F$ prime to ℓ and the primes in Ξ_f. Then there exists a prime ideal $\mathfrak{q} \subset \mathcal{O}_F$ such that $I \cdot \mathfrak{q} = (\alpha)$ is principal with α satisfying:*

(a) *α is a square of a unit at the primes in Ξ_f and at all primes $\mathfrak{l}$ above ℓ.*

(b) *$\sigma_v(\alpha) < 0$ for v in Ξ_∞ and $\sigma_v(\alpha) > 0$ for v any real place of F not in Ξ_∞.*

(c) $\mathrm{N}\mathfrak{q} \not\equiv 0, \pm 1 \pmod{\ell}$.

Further, $\mathfrak{q}$ can be picked to avoid any finite set of primes.

We first prove Lemma 7.1.6 and then explain the modifications needed to prove Lemma 7.1.7.

PROOF (OF LEMMA 7.1.6). Let $\mathfrak{m}$ be the product of all the real places of F and the primes in Ξ_f, each raised to a sufficiently large power so that the local units congruent to 1 (mod $\mathfrak{m}$) are squares. For $\alpha \in F^\times$, let $\iota(\alpha)$ denote the principal fractional ideal generated by α. Also let $F_{\mathfrak{m},1}$ denote the set of elements in $F^\times$ that are congruent to 1 ($\mathrm{mod}^\times$ $\mathfrak{m}$). If U_F denotes the units in F and $U_{F,\mathfrak{m}}$ the units congruent to 1 ($\mathrm{mod}^\times$ $\mathfrak{m}$), then there is an exact sequence:

$$1 \to \frac{U_F}{U_{F,\mathfrak{m}}} \to \frac{F^\times}{F_{\mathfrak{m},1}} \to \frac{\iota(F^\times)}{\iota(F_{\mathfrak{m},1})} \to 1.$$

Let H be the Hilbert class field of F and $H_{\mathfrak{m}}$ the ray class field of F of conductor $\mathfrak{m}$. Then $F \subset H \subset H_{\mathfrak{m}}$ and there is a canonical isomorphism

$$\mathrm{Gal}(H_{\mathfrak{m}}/H) \simeq \frac{\iota(F^\times)}{\iota(F_{\mathfrak{m},1})}.$$

Pick an element $\beta \in F^\times$ such that $\beta \equiv 1 \pmod{\mathfrak{m}}$ and such that β is negative at the real places in Ξ_∞ and positive at the real places not in Ξ_∞. Let $\sigma_{(\beta)} \in \mathrm{Gal}(H_\mathfrak{m}/H)$ be the element corresponding to $[\iota(\beta)]$ via the isomorphism above. Let σ_I denote the image of I in $\mathrm{Gal}(H_\mathfrak{m}/F)$ under the Artin map. By Chebotarev, there exists a prime ideal $\mathfrak{q}$ in $\mathcal{O}_F$ that is prime to $\mathfrak{m}$ and such that

$$\sigma_\mathfrak{q} = \sigma_I^{-1} \cdot \sigma_{(\beta)} \quad \text{in } \mathrm{Gal}(H_\mathfrak{m}/F).$$

In particular this implies that $\sigma_\mathfrak{q} = \sigma_I^{-1}$ in $\mathrm{Gal}(H/F)$, so there exists $\alpha \in F^\times$ such that $\mathfrak{q} \cdot I = (\alpha)$. Then $\sigma_{(\alpha)} = \sigma_{(\beta)}$, which is the same as saying that

$$[\iota(\alpha)] = [\iota(\beta)] \quad \text{in } \frac{\iota(F^\times)}{\iota(F_{\mathfrak{m},1})}.$$

The exact sequence above implies then that there is a unit $u \in U_F$ such that

$$[u \cdot \alpha] = [\beta] \quad \text{in } \frac{F^\times}{F_{\mathfrak{m},1}}.$$

Replacing α by $u \cdot \alpha$, we see that it has the required properties. □

PROOF (OF LEMMA 7.1.7). We modify the proof of Lemma 7.1.6.

Let $\{\mathfrak{l}_1, \ldots, \mathfrak{l}_r\}$ be the primes of F lying over ℓ. Let $\mathfrak{m}$ be the product of all the real places of F, the primes in Ξ_f (each raised to a sufficiently large power so that the local units congruent to 1 (mod $\mathfrak{m}$) are squares) and the primes $\mathfrak{l}_2, \ldots, \mathfrak{l}_r$. Fix for the moment an element $w \in (\mathfrak{o}_F/\mathfrak{l}_1)^\times$. By the approximation theorem, we may pick $\beta \in F^\times$ such that

- β is negative at the places in Ξ_∞ and positive at the real places not in Ξ_∞.
- $\beta \equiv 1 \pmod{\mathfrak{m}}$.
- $\beta \equiv w^2 \pmod{\mathfrak{l}_1}$.

Let σ_I denote the image of I in $\mathrm{Gal}(H_{\mathfrak{m}\mathfrak{l}_1}/F)$ under the Artin map. By Chebotarev, there exists a prime ideal $\mathfrak{q}$ in $\mathcal{O}_F$ that is prime to $\mathfrak{m} \cdot \mathfrak{l}_1$ and such that

$$\sigma_\mathfrak{q} = \sigma_I^{-1} \cdot \sigma_{(\beta)} \quad \text{in } \mathrm{Gal}(H_{\mathfrak{m}\mathfrak{l}_1}/F).$$

As before then, there exists $\alpha \in F^\times$ such that $\mathfrak{q} \cdot I = (\alpha)$ and a unit $u \in U_F$ such that

$$[u \cdot \alpha] = [\beta] \quad \text{in } \frac{F^\times}{F_{\mathfrak{m}\mathfrak{l}_1,1}}.$$

Replacing α by $u \cdot \alpha$, we see that α satisfies the requirements (a), (b) of the lemma. It remains to show that w can be chosen so that $\mathfrak{q}$ satisfies (c). Clearly $\mathfrak{q}$ is prime to ℓ. But

$$\mathrm{N}\mathfrak{q} \cdot \mathrm{N}I = \pm \mathrm{N}(\alpha) = \pm \mathrm{N}(\beta) \equiv \pm \mathrm{N}_{\mathbf{F}_{\mathfrak{l}_1}/\mathbf{F}_\ell}(w^2) \pmod{\ell}.$$

Since $\mathrm{N}I$ is fixed and we only need $\mathrm{N}\mathfrak{q} \not\equiv \pm 1 \pmod{\ell}$, it suffices to show that the subgroup

$$\left\{ \mathrm{N}_{\mathbf{F}_{\mathfrak{l}_1}/\mathbf{F}_\ell}(w^2) : w \in \mathbf{F}_{\mathfrak{l}_1}^\times \right\} \subset \mathbf{F}_\ell^\times$$

contains at least 3 elements. But this subgroup is just $(\mathbf{F}_\ell^\times)^2$ (since $\mathfrak{l}_1$ is unramified over ℓ) and has cardinality $\frac{\ell-1}{2} > 2$ since $\ell > 5$ by assumption. □

Now we prove Prop. 7.1.4 and then explain the modifications needed to prove Prop. 7.1.5.

PROOF (OF PROP. 7.1.4). Let $\mathfrak{f} = \mathfrak{f}_1 \cdots \mathfrak{f}_n$ and $\mathfrak{S} = 2\ell\mathfrak{d}_{B_1 \cup B_2}\mathfrak{f}$. We begin by using Lemma 7.1.6 above to pick:

- A prime ideal $\mathfrak{q}_{B_1 \smallsetminus B_2}$ (prime to $\mathfrak{S}$) such that $\mathfrak{d}_{B_1 \smallsetminus B_2} \cdot \mathfrak{q}_{B_1 \smallsetminus B_2} = (\alpha_{B_1 \smallsetminus B_2})$, with $\alpha_{B_1 \smallsetminus B_2}$ satisfying the following conditions:
  - $\alpha_{B_1 \smallsetminus B_2}$ is a square of a unit at the primes dividing $\ell\mathfrak{d}_{B_2}\mathfrak{f}$ and the primes above 2 not dividing $\mathfrak{d}_{B_1 \smallsetminus B_2}$.
  - For $v \in \Sigma_\infty$,
$$\sigma_v(\alpha_{B_1 \smallsetminus B_2}) < 0, \text{ if } v \in \Sigma_{B_1 \smallsetminus B_2}; \quad \sigma_v(\alpha_{B_1 \smallsetminus B_2}) > 0, \text{ if } v \notin \Sigma_{B_1 \smallsetminus B_2}.$$
- A prime ideal $\mathfrak{q}_{B_2 \smallsetminus B_1}$ (prime to $\mathfrak{S}$) such that $\mathfrak{d}_{B_2 \smallsetminus B_1} \cdot \mathfrak{q}_{B_2 \smallsetminus B_1} = (\alpha_{B_2 \smallsetminus B_1})$, with $\alpha_{B_2 \smallsetminus B_1}$ satisfying the following conditions:
  - $\alpha_{B_2 \smallsetminus B_1}$ is a square of a unit at the primes dividing $\ell\mathfrak{d}_{B_1}\mathfrak{f}$ and the primes above 2 not dividing $\mathfrak{d}_{B_2 \smallsetminus B_1}$.
  - For $v \in \Sigma_\infty$,
$$\sigma_v(\alpha_{B_2 \smallsetminus B_1}) < 0, \text{ if } v \in \Sigma_{B_2 \smallsetminus B_1}; \quad \sigma_v(\alpha_{B_2 \smallsetminus B_1}) > 0, \text{ if } v \notin \Sigma_{B_2 \smallsetminus B_1}.$$
- A prime ideal $\mathfrak{q}_{B_1 \cap B_2}$ (prime to $\mathfrak{S}$) such that $\mathfrak{d}_{B_1 \cap B_2} \cdot \mathfrak{q}_{B_1 \cap B_2} = (\alpha_{B_1 \cap B_2})$, with $\alpha_{B_1 \cap B_2}$ satisfying the following conditions:
  - $\alpha_{B_1 \cap B_2}$ is a square of a unit at the primes dividing $\ell\mathfrak{d}_{B_1 \smallsetminus B_2}\mathfrak{d}_{B_2 \smallsetminus B_1}\mathfrak{f}$ and the primes above 2 not dividing $\mathfrak{d}_{B_1 \cap B_2}$.
  - For $v \in \Sigma_\infty$,
$$\sigma_v(\alpha_{B_1 \cap B_2}) < 0, \text{ if } v \in \Sigma_{B_1 \cap B_2}; \quad \sigma_v(\alpha_{B_1 \cap B_2}) > 0, \text{ if } v \notin \Sigma_{B_1 \cap B_2}.$$

Let $\mathfrak{R}$ denote the ideal
$$\mathfrak{R} := \mathfrak{q}_{B_1 \smallsetminus B_2} \cdot \mathfrak{q}_{B_2 \smallsetminus B_1} \cdot \mathfrak{q}_{B_1 \cap B_2}.$$
Next, we use the approximation theorem to pick $\alpha \in F^\times$ satisfying the following properties:

(I) $\alpha \gg 0$.

(II) $-\alpha$ is a square of a unit at the primes dividing $\ell\mathfrak{R}\mathfrak{f}$.

(III) If $\mathfrak{q}$ is a prime dividing $\mathfrak{d}_{B_1 \cup B_2}$, then $-\alpha$ is a unit at $\mathfrak{q}$ but not a square. If further $\mathfrak{q}$ divides 2, then we also require that $\sqrt{-\alpha}$ generate the unique unramified extension of $F_\mathfrak{q}$.

(IV) If $\mathfrak{q}$ is a prime dividing 2 but not dividing $\mathfrak{d}_{B_1 \cup B_2}$, then $-\alpha$ is a square of a unit at $\mathfrak{q}$.

Let
$$\mathfrak{m} := 2^a\ell \cdot \mathfrak{d}_{B_1 \cup B_2}\mathfrak{f} \cdot \mathfrak{R} \cdot \prod_{v \in \Sigma_\infty} v,$$
with the power 2^a being chosen large enough so that locally at any prime above 2, the units congruent to 1 modulo 2^a are squares. By Chebotarev, there exists a prime ideal $\mathfrak{Q} \subset \mathcal{O}_F$ (prime to $\mathfrak{m}$) such that
$$\sigma_{(\alpha)} \cdot \sigma_\mathfrak{Q} = 1 \quad \text{in } \mathrm{Gal}(H_\mathfrak{m}/F).$$
This implies that
$$(\alpha) \cdot \mathfrak{Q} = (\beta),$$
for some $\beta \equiv 1 \pmod{^\times\, \mathfrak{m}}$. Now, take
$$u := -\alpha^{-1}\beta, \quad J_1 := \alpha_{B_1 \smallsetminus B_2} \cdot \alpha_{B_1 \cap B_2}, \quad J_2 := \alpha_{B_2 \smallsetminus B_1} \cdot \alpha_{B_1 \cap B_2}.$$

Since

$$(u) = \mathfrak{Q}, \quad (J_1) = \mathfrak{d}_{B_1} \cdot \mathfrak{q}_{B_1 \smallsetminus B_2} \cdot \mathfrak{q}_{B_1 \cap B_2}, \quad (J_2) = \mathfrak{d}_{B_2} \cdot \mathfrak{q}_{B_2 \smallsetminus B_1} \cdot \mathfrak{q}_{B_1 \cap B_2},$$

we see that u, J_1, J_2 lie in $\mathcal{O}_F$, which shows that (i) is satisfied. Let E/F be the quadratic extension $E = F + F\mathbf{i}$ with $\mathbf{i}^2 = u$. Since $\alpha \gg 0$ and $\beta \gg 0$, we have $u \ll 0$, which shows (iii), whence E is a CM quadratic extension of F. The conditions (III) and (IV) above imply that if $\mathfrak{q}$ is a prime above 2, then $E_\mathfrak{q}$ is the unique unramified quadratic extension of $F_\mathfrak{q}$ if $\mathfrak{q}$ divides $\mathfrak{d}_{B_1 \cup B_2}$ and otherwise is split, which shows (v). Since $(u) = \mathfrak{Q}$, it follows that E is ramified exactly at the prime $\mathfrak{Q}$, and in particular satisfies (iv). Now we check that

$$B_1 \simeq E + E\mathbf{j}_1, \quad \mathbf{j}_1^2 = J_1, \quad \mathbf{i}\mathbf{j}_1 = -\mathbf{j}_1\mathbf{i}.$$

To show this, it suffices to check that the Hilbert symbol $(u, J_1)_v$ equals -1 exactly for those v at which B_1 is ramified. At the archimedean places this is clear since $u \ll 0$ and J_1 is negative exactly at the places at which B_1 is ramified. As for the finite places, we only need to check this for v dividing $2uJ_1$, since outside of these primes B_1 is split and $(u, J_1) = 1$ since both u and J_1 are units at such places. At the primes dividing $\mathfrak{q}_{B_1 \smallsetminus B_2} \cdot \mathfrak{q}_{B_1 \cap B_2}$, the algebra B_1 is split and u is a square of a unit, so this is clear. For $\mathfrak{q} \mid \mathfrak{d}_{B_1}$, the algebra B_1 is ramified, J_1 is a uniformizer and by (III) above, we have $(u, J_1)_\mathfrak{q} = (-\alpha, J_1)_\mathfrak{q} = -1$. Next we consider the primes $\mathfrak{q}$ above 2. If $\mathfrak{q} \mid \mathfrak{d}_{B_1}$, this is done already. If $\mathfrak{q} \mid \mathfrak{d}_{B_2 \smallsetminus B_1}$, then J_1 is a square at $\mathfrak{q}$, so $(u, J_1)_\mathfrak{q} = 1$ as required. This leaves the primes $\mathfrak{q}$ above 2 which do not divide $\mathfrak{d}_{B_1 \cup B_2}$. At such primes, u is a square of a unit, so $(u, J_1)_\mathfrak{q} = 1$. The only prime left is $\mathfrak{Q}$ at which the required equality follows from the product formula! The isomorphism $B_2 \simeq E + E\mathbf{j}_2$ follows similarly, and then the isomorphism $B \simeq E + E\mathbf{j}$ follows from the equality $B = B_1 \cdot B_2$ in the Brauer group. This completes the proof of (vi). The conditions (vii) and (viii) are easily verified, which leaves (ii).

Finally, we check that (ii) is satisfied, namely that at every place v of F, at least one of u, J_1, J_2 or J is a square. At the archimedean places, this is obvious. At the primes dividing $\mathfrak{d}_{B_1 \cup B_2}$, this follows from (vii). Let $\mathfrak{q}$ be a finite prime not dividing $\mathfrak{d}_{B_1 \cup B_2}$. If such a $\mathfrak{q}$ divides 2, then all of u, J_1, J_2, J are squares at $\mathfrak{q}$. So let $\mathfrak{q}$ be prime to $2\mathfrak{d}_{B_1 \cup B_2}$. If E is split at $\mathfrak{q}$, then u is a square at $\mathfrak{q}$. If E is inert at $\mathfrak{q}$, then J_1, J_2 lie in $\mathrm{N}_{E_\mathfrak{q}/F_\mathfrak{q}}(E_\mathfrak{q}^\times)$ since B_1 and B_2 are split at $\mathfrak{q}$. If J_1 and J_2 are both not squares at such $\mathfrak{q}$, it must be the case that $J = J_1 J_2$ is a square. Finally, we deal with $\mathfrak{q} = \mathfrak{Q}$, the only ramified prime in E. At this prime, J_1 and J_2 are both units. Again, if both of them are non-squares, their product must be a square. This completes the proof. □

We will now prove Prop 7.1.5.

Proof (of Prop. 7.1.5). We will show that we can pick u, J_1, J_2 such that (ix) is satisfied in addition to (i)-(viii). The proof is almost the same as that of Prop. 7.1.4 with some minor modifications which we now describe. First, we pick as before the prime ideals $\mathfrak{q}_{B_1 \smallsetminus B_2}, \mathfrak{q}_{B_2 \smallsetminus B_1}, \mathfrak{q}_{B_1 \cap B_2}$ and the elements $\alpha_{B_1 \smallsetminus B_2}, \alpha_{B_2 \smallsetminus B_1}, \alpha_{B_1 \cap B_2}$. Using Lemma 7.1.7, we can ensure that for $\mathfrak{q} = \mathfrak{q}_{B_1 \smallsetminus B_2}$ and $\mathfrak{q}_{B_2 \smallsetminus B_1}$, we have

$$\mathrm{N}\mathfrak{q} \not\equiv 0, \pm 1 \pmod{\ell}.$$

Fix a prime $\mathfrak{l}_1$ of F above ℓ. Let $\mathfrak{N}$ be as before and then using the approximation theorem, pick $\alpha \in F^\times$ satisfying the properties (I) through (IV) of the proof of Prop. 7.1.4 and the following additional conditions:

(V) $-\alpha \equiv w^2 \pmod{\mathfrak{l}_1}$, where $w \in \mathbf{F}_{\mathfrak{l}_1}^\times$ is an element such that $\mathrm{N}_{\mathbf{F}_{\mathfrak{l}_1}/\mathbf{F}_\ell}(w^2) \neq \pm 1$.

(VI) $-\alpha \equiv 1 \pmod{\mathfrak{l}}$ for all primes $\mathfrak{l} \neq \mathfrak{l}_1$ dividing ℓ.

Note that this uses the assumption that ℓ is unramified in F and $\ell > 5$. Next, as before, we pick $\mathfrak{m}$, β and $\mathfrak{Q}$ and set:

$$u := -\alpha^{-1}\beta, \quad J_1 := \alpha_{B_1 \smallsetminus B_2} \cdot \alpha_{B_1 \cap B_2}, \quad J_2 := \alpha_{B_2 \smallsetminus B_1} \cdot \alpha_{B_1 \cap B_2}.$$

It is easy to see (using arguments similar to those of Prop. 7.1.4) that (i)-(viii) hold. We verify (ix) now.

The field E is only ramified at $\mathfrak{Q}$. Also,

$$\mathrm{N}\mathfrak{Q} = \pm \mathrm{N}(\beta) \cdot \mathrm{N}(\alpha^{-1}) \equiv \pm \mathrm{N}(\alpha^{-1}) \equiv \pm \mathrm{N}_{\mathbf{F}_{\mathfrak{l}_1}/\mathbf{F}_\ell}(w^{-2}) \not\equiv 0, \pm 1 \pmod{\ell},$$

which proves what we need for the field E. Now let us consider the field K. Since $J = J_1 J_2$, we have

$$\begin{aligned}(J) &= \mathfrak{d}_{B_1} \cdot \mathfrak{q}_{B_1 \smallsetminus B_2} \cdot \mathfrak{q}_{B_1 \cap B_2} \cdot \mathfrak{d}_{B_2} \cdot \mathfrak{q}_{B_2 \smallsetminus B_1} \cdot \mathfrak{q}_{B_1 \cap B_2} \\ &= \mathfrak{d}_{B_1 \smallsetminus B_2} \cdot \mathfrak{d}_{B_2 \smallsetminus B_1} \cdot \mathfrak{q}_{B_1 \smallsetminus B_2} \cdot \mathfrak{q}_{B_2 \smallsetminus B_1} \cdot \mathfrak{d}_{B_1 \cap B_2}^2 \cdot \mathfrak{q}_{B_1 \cap B_2}^2.\end{aligned}$$

We claim that K is ramified exactly at the primes dividing

$$\mathfrak{d}_K := \mathfrak{d}_{B_1 \smallsetminus B_2} \cdot \mathfrak{d}_{B_2 \smallsetminus B_1} \cdot \mathfrak{q}_{B_1 \smallsetminus B_2} \cdot \mathfrak{q}_{B_2 \smallsetminus B_1}.$$

This will follow if we show that K/F is unramified at the primes $\mathfrak{q}$ over 2 that do not divide $\mathfrak{d}_{B_1 \smallsetminus B_2} \cdot \mathfrak{d}_{B_2 \smallsetminus B_1}$. We claim that J is a square (and hence K is split) at such primes $\mathfrak{q}$. Indeed, if a prime $\mathfrak{q}$ above 2 does not divide $\mathfrak{d}_{B_1 \cup B_2}$, then $\alpha_{B_1 \smallsetminus B_2}$, $\alpha_{B_2 \smallsetminus B_1}$ and $\alpha_{B_1 \cap B_2}$ are all squares of units at $\mathfrak{q}$, hence J_1, J_2 and J are squares at $\mathfrak{q}$. On the other hand, if a prime $\mathfrak{q}$ above 2 divides $\mathfrak{d}_{B_1 \cap B_2}$, then $\alpha_{B_1 \smallsetminus B_2}$, $\alpha_{B_2 \smallsetminus B_1}$ are squares of units at $\mathfrak{q}$ and thus

$$J = \alpha_{B_1 \smallsetminus B_2} \cdot \alpha_{B_2 \smallsetminus B_1} \cdot \alpha_{B_1 \cap B_2}^2$$

is a square at $\mathfrak{q}$.

Thus it suffices to consider the primes $\mathfrak{q}$ dividing $\mathfrak{d}_K$. For $\mathfrak{q}$ dividing either $\mathfrak{d}_{B_1 \smallsetminus B_2}$ or $\mathfrak{d}_{B_2 \smallsetminus B_1}$, it follows from the assumptions in the statement of the proposition that $\mathrm{N}\mathfrak{q} \not\equiv 0, \pm 1 \pmod{\ell}$. For $\mathfrak{q}$ equal to either $\mathfrak{q}_{B_1 \smallsetminus B_2}$ or $\mathfrak{q}_{B_2 \smallsetminus B_1}$, the same follows from the choice of these prime ideals. □

7.2. The main conjecture

Finally, in this section we come to the main conjecture. Our starting data will be the totally real field F, the automorphic representation Π (from the introduction) of conductor $\mathfrak{N} = \mathfrak{N}_s \cdot \mathfrak{N}_{\mathrm{ps}}$ and the two quaternion algebras B_1 and B_2. We assume that Π admits a Jacquet–Langlands transfer to both B_1 and B_2. We also assume the following condition holds (see Remark 6 of the introduction):

- $\mathfrak{N}$ is prime to $2\mathfrak{D}_{F/\mathbb{Q}}$, where $\mathfrak{D}_{F/\mathbb{Q}}$ denotes the *different* of $F/\mathbb{Q}$.

7.2.1. The conjecture will in addition depend on several auxiliary choices which we now make completely explicit in the following series of steps.

(i) Let ℓ be a rational prime such that $(\ell, N(\Pi)) = 1$. Recall from the introduction that
$$N(\Pi) = 2 \cdot h_F \cdot D_F \cdot N_\Pi \cdot k_\Pi!,$$
where $N_\Pi = \mathrm{N}\mathfrak{N}$ and $k_\Pi = \max k_i$.

(ii) Pick u, J_1 and J_2 satisfying all the conditions of Prop. 7.1.4, with $\{\mathfrak{f}_1, \dots, \mathfrak{f}_n\}$ being the set of primes of F dividing $2\mathfrak{D}_{F/\mathbb{Q}}$. In other words, we pick a well adapted coordinate system for $(F, B_1, B_2, (2\ell)\mathfrak{D}_{F/\mathbb{Q}})$. As mentioned in Remark 7.1.3 this ensures that the list of possibilities in the table of §5.5 covers all possible ramification scenarios at any place v of F.

(iii) Set $E = F + F\mathbf{i}$ where $\mathbf{i}^2 = u$. An explicit model for B_i, $i = 1, 2$, is $B_i = E + E\mathbf{j}_i$ where $\mathbf{j}_i^2 = J_i$ and $\alpha\mathbf{j}_i = \mathbf{j}_i\alpha^\rho$ for $\alpha \in E$. Likewise an explicit model for $B = B_1 \cdot B_2$ is $B = E + E\mathbf{j}$ where $\mathbf{j}^2 = J := J_1 J_2$.

(iv) Let B' denote any one of the quaternion algebras B, B_1 or B_2. Let $\mathfrak{d}_{B'}$ denote the discriminant of B' and define $\mathfrak{N}_{B'}$ by
$$\mathfrak{N} = \mathfrak{d}_{B'} \cdot \mathfrak{N}_{B'}.$$
Thus $\mathfrak{d}_{B'}$ divides $\mathfrak{N}_{\mathrm{s}}$, and $\mathfrak{N}_{\mathrm{ps}}$ divides $\mathfrak{N}_{B'}$.

(v) Given the choices of u, J_1 and J_2, in §5.6 we have picked local maximal orders and *oriented* Eichler orders of level $\mathfrak{N}_{B'}$. (Here the orientation is only picked at places dividing $\mathfrak{N}_{\mathrm{ps}}$.) Let $\mathcal{O}_{B'}$ (resp. $\mathcal{O}_{B'}(\mathfrak{N}_{B'})$) denote the unique maximal order (resp. Eichler order of level $\mathfrak{N}_{B'}$) corresponding to these choices. Also denote by $o_{B'}$ the corresponding orientation on $\mathcal{O}_{B'}(\mathfrak{N}_{B'})$. This defines open compact subgroups $\mathcal{K}^{B'} = \prod_v \mathcal{K}_v^{B'}$ and $\tilde{\mathcal{K}}^{B'} = \prod_v \tilde{\mathcal{K}}_v^{B'}$ of $B'^\times(\mathbb{A}_f)$ with
$$\mathcal{K}_v^{B'} = \ker\left[o_{B',v} : (\mathcal{O}_{B'}(\mathfrak{N}_{B'}) \otimes_{\mathcal{O}_F} \mathcal{O}_{F,v})^\times \to (\mathcal{O}_{F,v}/\mathfrak{N}_{\mathrm{ps}}\mathcal{O}_{F,v})^\times\right]$$
and
$$\tilde{\mathcal{K}}_v^{B'} = (\mathcal{O}_{B'}(\mathfrak{N}_{B'}) \otimes_{\mathcal{O}_F} \mathcal{O}_{F,v})^\times.$$
For future use, we record that with our choices of u, J_1 and J_2 and local orders, we have
$$\mathcal{O}_E \otimes \mathbb{Z}_{(\ell)} \subseteq \mathcal{O}_{B'} \otimes \mathbb{Z}_{(\ell)}. \tag{7.3}$$

(vi) As in §1.4, we pick a large enough number field L and isomorphisms ϕ_B, ϕ_{B_1} and ϕ_{B_2} satisfying (1.18). We recall that $\phi_{B'}$ gives an isomorphism:
$$\mathcal{O}_{B'} \otimes \mathcal{O}_{L,(\ell)} \simeq \prod_\sigma \mathrm{M}_2(\mathcal{O}_{L,(\ell)}) \tag{7.4}$$
for $B' = B$, B_1 and B_2. As explained in §1.4, we can define automorphic vector bundles on $\mathrm{Sh}_{\mathcal{K}^B}(G_B, X_B)$, $\mathrm{Sh}_{\mathcal{K}^{B_1}}(G_{B_1}, X_{B_1})$ and $\mathrm{Sh}_{\mathcal{K}^{B_2}}(G_{B_2}, X_{B_2})$ as well as sections s_B, s_{B_1} and s_{B_2} of these bundles that are ℓ-normalized. (See Prop. 1.3.4.) For $B' = B, B_1$ or B_2 let us denote the corresponding bundle by $\mathcal{V}^{B'}_{\underline{k}_{B'},r}$ to indicate that is a bundle on $X_{B'}$.

(vii) So far, we have not had to pick a base point of $X_{B'}$ but we will now need to do so. Fix isomorphisms $E \otimes_{F,\sigma} \mathbb{R} \simeq \mathbb{C}$ for all infinite places σ of F as in

§5.6. For $B' = B, B_1, B_2$ we define $h_{B'}$ as follows: take the composite maps

$$\mathbb{C} \to \prod_{\sigma \in \Sigma_\infty} \mathbb{C} \simeq E \otimes \mathbb{R} \to B' \otimes \mathbb{R}$$

where the first map sends

$$z \mapsto (z_\sigma)_\sigma$$

with $z_\sigma = z$ if σ is split in B' and $z_\sigma = 1$ is σ is ramified in B'.

(viii) For $B' = B, B_1$ or B_2, let $F_{B'} = \mathrm{Lift}_{h_{B'}}(s_{B'})$.

(ix) To move to scalar valued forms, we need to pick an embedding of $(\mathbb{C}^\times)^d$ in $(B' \otimes_{\mathbb{Q}} \mathbb{R})^\times$ as in §1.2.3. We use the following embedding:

$$(\mathbb{C}^\times)^d \simeq \prod_\sigma (E \otimes_{F,\sigma} \mathbb{R})^\times \simeq (E \otimes_{\mathbb{Q}} \mathbb{R})^\times \hookrightarrow (B' \otimes_{\mathbb{Q}} \mathbb{R})^\times.$$

For $\sigma \in \Sigma_\infty \smallsetminus \Sigma_{B',\infty}$, the embedding $\mathbb{C}^\times \hookrightarrow (B' \otimes_{F,\sigma} \mathbb{R})^\times$ is simply $h_{B',\sigma}$. For $\sigma \in \Sigma_{B',\infty}$, we denote the embedding $\mathbb{C}^\times \hookrightarrow (B' \otimes_{F,\sigma} \mathbb{R})^\times$ by $h'_{B',\sigma}$ following the notation in §1.2.3.

(x) For each σ and each B', we pick a vector $v^{B'}_{\sigma,k_{B',\sigma}} \in V_{\sigma,k_{B',\sigma},r}$ satisfying (1.13) and that is integrally normalized with respect to the ℓ-integral structure given by

$$\begin{aligned}\mathrm{Sym}^{k_{B',\sigma}}(\mathcal{O}^2_{L,(\ell)}) \otimes \det(\mathcal{O}^2_{L,(\ell)})^{\frac{r-k_{B',\sigma}}{2}} &\subset \mathrm{Sym}^{k_{B',\sigma}}(L^2) \otimes \det(L^2)^{\frac{r-k_{B',\sigma}}{2}} \\ &\subset V_{\sigma,k_{B',\sigma},r}.\end{aligned}$$

This means that $v^{B'}_{\sigma,k_{B',\sigma}}$ lies in the $\mathcal{O}_{L,(\ell)}$-module on the left and is not divisible in it by any prime of $\mathcal{O}_{L,(\ell)}$ above (ℓ). (At this point we also need to ensure that $\mathrm{Sym}^{k_{B',\sigma}}(L^2) \otimes \det(L^2)^{\frac{r-k_{B',\sigma}}{2}}$ contains an eigenvector for the action of $(E \otimes_{F,\sigma} \mathbb{R})^\times = \mathbb{C}^\times$. Since $E^\times$ is dense in $(E \otimes_{F,\sigma} \mathbb{R})^\times$, it suffices to have an eigenvector for the $E^\times$-action; this can be arranged by extending L if necessary to a larger number field.)

(xi) Let $v^{B'}_{\underline{k}_{B'}} = \otimes_\sigma v^{B'}_{\sigma,k_{B',\sigma}}$. Now we can define $\phi_{F_{B'}} = (F_{B'}(g), v^{B'}_{\underline{k}_{B'}})$. Let $f_{B'}$ denote the corresponding element of $\pi_{B'}$:

$$f_{B'}(g) = \phi_{F_{B'}}(g) \cdot \nu_{B'}(g)^{-r/2}.$$

Then $f_{B'}$ is a new-vector as defined in §6.3.

From Defn. 1.2.8, Prop. 1.2.10 and Prop. 1.2.14, we find that

$$\begin{aligned}\langle\!\langle s_{B'}, s_{B'}\rangle\!\rangle_{\tilde{\mathcal{K}}^{B'}} &= 2^{d_{B'}} h_F [\mathcal{K}_0^{B'} : \tilde{\mathcal{K}}^{B'}] \cdot \langle F_{B'}, F_{B'} \rangle_{h_{B'}} \\ &= 2^{d_{B'}} h_F [\mathcal{K}_0^{B'} : \tilde{\mathcal{K}}^{B'}] \cdot \frac{\operatorname{rank} \mathcal{V}^{B'}_{\underline{k}_{B'},r}}{\langle v^{B'}_{\underline{k}_{B'}}, v^{B'}_{\underline{k}_{B'}} \rangle_{h_{B'}}} \cdot \langle f_{B'}, f_{B'} \rangle,\end{aligned} \tag{7.5}$$

where $d_{B'}$ is the number of infinite places of F where B' is split, and $\mathcal{K}_0^{B'}$ is the maximal open compact subgroup of $B'^\times(\mathbb{A}_f)$ defined by $\mathcal{K}_0^{B'} = \prod_v \mathcal{K}^{B'}_{0,v}$, with $\mathcal{K}^{B'}_{0,v} = (\mathcal{O}_{B'} \otimes \mathcal{O}_{F,v})^\times$.

We note the following proposition about the values $\langle v^{B'}_{\underline{k}_{B'}}, v^{B'}_{\underline{k}_{B'}} \rangle_{h_{B'}}$.

PROPOSITION 7.2.1. *For $B' = B, B_1, B_2$, the value $\langle v^{B'}_{\underline{k}_{B'}}, v^{B'}_{\underline{k}_{B'}} \rangle_{h_{B'}}$ is a unit at all primes above ℓ, i.e., is an element of $R^\times_{(\ell)}$.*

PROOF. Since $v^{B'}_{\underline{k}_{B'}} = \otimes_\sigma v^{B'}_{\sigma,k_{B',\sigma}}$, it suffices to show that $\langle v^{B'}_{\sigma,k_{B',\sigma}}, v^{B'}_{\sigma,k_{B',\sigma}}\rangle$ is in $R^\times_{(\ell)}$ for all σ. Recall that this inner product is defined using the hermitian forms H and H_h of §1.2.1 in the cases $\sigma \in \Sigma_{B',\infty}$ and $\sigma \in \Sigma_\infty \smallsetminus \Sigma_{B',\infty}$ respectively. We consider these cases separately. In either case, for $B' = B, B_1, B_2$ respectively, we write J' for J, J_1, J_2 and $\mathbf{j}'$ for $\mathbf{j}, \mathbf{j}_1, \mathbf{j}_2$ respectively.

First consider the case $\sigma \in \Sigma_{B',\infty}$. Let $\mathcal{L}_\sigma = \mathcal{O}^2_{L,(\ell)}$. The vector $v^{B'}_{\sigma,k_{B',\sigma}}$ is characterized as the unique element (up to scaling) in $\mathrm{Sym}^{k_{B',\sigma}}(\mathcal{L}_\sigma)\otimes\det(\mathcal{L}_\sigma)^{(r-k_{B',\sigma})/2}$ on which $\mathbb{C}^\times$ acts by $ze^{i\theta} \mapsto z^r e^{ik_{B',\sigma}\theta}$. Since the hermitian form on $\det(\mathcal{L}_\sigma)$ is regular and $\mathbb{C}^\times$ acts on $\det(\mathcal{L}_\sigma)$ as $ze^{i\theta} \mapsto z^2$, it suffices to consider instead just $\mathrm{Sym}^k(\mathcal{L}_\sigma)$ (for any integer k with $0 \le k < \ell$) and show that the norm of an integrally normalized vector in $\mathrm{Sym}^k(\mathcal{L}_\sigma)$ on which $\mathbb{C}^\times$ acts by $ze^{i\theta} \mapsto z^k e^{ik\theta}$ lies in $R^\times_{(\ell)}$. Now the hermitian form on $\mathrm{Sym}^k(\mathcal{L}_\sigma)$ is defined by first using the section s of equation (1.10) to embed it into $V^{\otimes k}_{\sigma,\mathbb{C}}$ and then taking the restriction of the tensor product form. Since $\ell > k$, this section maps $\mathrm{Sym}^k(\mathcal{L}_\sigma)$ into $\mathcal{L}^{\otimes k}_\sigma$ and thus the hermitian form on $\mathrm{Sym}^k(\mathcal{L}_\sigma)$ takes values in $R_{(\ell)}$. It is thus enough to show that there is a non-zero vector $\mathbf{v}_k$ in $\mathrm{Sym}^k(\mathcal{L}_\sigma)$ on which $\mathbb{C}^\times$ acts by $ze^{i\theta} \mapsto z^k e^{ik\theta}$ and whose norm lies in $R^\times_{(\ell)}$. Moreover, it suffices to consider the case $k = 1$ since for general k, we can simply take $\mathbf{v}_k$ to be the image of $\mathbf{v}_1^{\otimes k}$ in $\mathrm{Sym}^k(\mathcal{L}_\sigma)$.

Recall that we have an embedding $E \to B'$ and hence an embedding ϕ_σ defined by the diagram below:

$$\begin{tikzcd} \mathbb{C}^\times \arrow[r, "\simeq"] \arrow[rr, bend right, "h'_{B',\sigma}"'] & (E\otimes_{F,\sigma}\mathbb{R})^\times \arrow[r] \arrow[rr, bend left, "\phi_\sigma"] & (B'\otimes_{F,\sigma}\mathbb{R})^\times \arrow[r, "\simeq", "\phi_{B',\sigma}"'] & \mathbb{H}^\times, \end{tikzcd}$$

where the (inverse of) first isomorphism is the map that sends $\mathbf{i} \in E^\times$ to $i\sqrt{|\sigma u|} \in \mathbb{C}^\times$. Also, the line spanned by the vector $\mathbf{v}_1$ is distinguished simply by the action of $\phi_\sigma(\mathbf{i})$. On account of (7.3) and (7.4), we have $\phi_\sigma(\mathbf{i}) \in \mathrm{GL}_2(\mathcal{O}_{L,(\ell)})$.

Since $\sigma \in \Sigma_{B',\infty}$, we have $\sigma J' < 0$. To prove the proposition, we may extend L if needed to a larger number field and thus we may assume that $i\sqrt{|\sigma u|}, i\sqrt{|\sigma J'|} \in L$. Now consider the isomorphism

$$\phi' : B' \otimes_{F,\sigma} \mathbb{R} \to \mathbb{H}$$

defined by

$$\phi'(\mathbf{i}) = \begin{bmatrix} i\sqrt{|\sigma u|} & \\ & -i\sqrt{|\sigma u|} \end{bmatrix}, \quad \phi'(\mathbf{j}') = \begin{bmatrix} & i\sqrt{|\sigma J'|} \\ i\sqrt{|\sigma J'|} & \end{bmatrix}.$$

Then, as in the middle line of the second commutative diagram in the proof of Lemma 1.3.1, ϕ' gives an isomorphism of Azumaya algebras over the semi-local ring $\mathcal{O}_{L_0,(\ell)}$:

$$\phi' : \mathcal{O}_{B'} \otimes_{\mathcal{O}_F,\sigma} \mathcal{O}_{L_0,(\ell)} \simeq \mathrm{M}_2(\mathcal{O}_{L,(\ell)}) \cap \mathbb{H},$$

where $L_0 = L \cap \mathbb{R}$. Since $\phi_{B',\sigma}$ also gives such an isomorphism, ϕ' and $\phi_{B',\sigma}$ must be conjugate by an element

$$A \in (\mathrm{M}_2(\mathcal{O}_{L,(\ell)}) \cap \mathbb{H})^\times = \mathrm{GL}_2(\mathcal{O}_{L,(\ell)}) \cap \mathbb{H}^\times.$$

Now $\mathbf{e}_1 := \begin{bmatrix} 1 \\ 0 \end{bmatrix}$ is an eigenvector of $\phi'(\mathbf{i})$ with eigenvalue $i\sqrt{|\sigma u|}$ and so $A\mathbf{e}_1$ is an eigenvector of $\phi_\sigma(\mathbf{i})$ with the same eigenvalue. Since $A \in \mathrm{GL}_2(\mathcal{O}_{L,(\ell)}) \cap \mathbb{H}^\times$, we have $\mathbf{v}_1 := A\mathbf{e}_1 \in \mathcal{L}_\sigma$ and

$$\langle \mathbf{v}_1, \mathbf{v}_1 \rangle = \langle A\mathbf{e}_1, A\mathbf{e}_1 \rangle = \nu(A)\langle \mathbf{e}_1, \mathbf{e}_1 \rangle = \nu(A) \in \mathcal{O}_{L,(\ell)}^\times.$$

Next we consider the case $\sigma \in \Sigma_\infty \smallsetminus \Sigma_{B',\infty}$. This time, it suffices to consider the hermitian inner product on the space $(V_{\sigma,h}^{-1,0})^{\otimes k}$ with $h = \phi_{B',\sigma} \circ h_{B',\sigma}$ and $k = k_{B',\sigma}$. Again, it is enough to produce a non-zero vector $\mathbf{v}_k$ in $(V_{\sigma,h}^{-1,0})^{\otimes k}$ whose image lies in $\mathrm{Sym}^k(\mathcal{L}_\sigma)$ and whose norm is an ℓ-unit. (We are using here that $\ell > k$ as in the earlier case. We also extend L if needed to a larger number field so that $i, \sqrt{|\sigma u|}, \sqrt{|\sigma J'|} \in L$.) As in the previous case, it also suffices to consider only the case $k = 1$. Consider the composite map

$$V_{\sigma,\mathbb{R}} \to V_{\sigma,\mathbb{C}} \to V_{\sigma,h}^{-1,0}$$

which is an $\mathbb{R}$-linear isomorphism that is used to define the hermitian inner product on $V_{\sigma,h}^{-1,0}$. As before we have an embedding $E \to B'$ and hence an embedding ϕ_σ which fits into the diagram below:

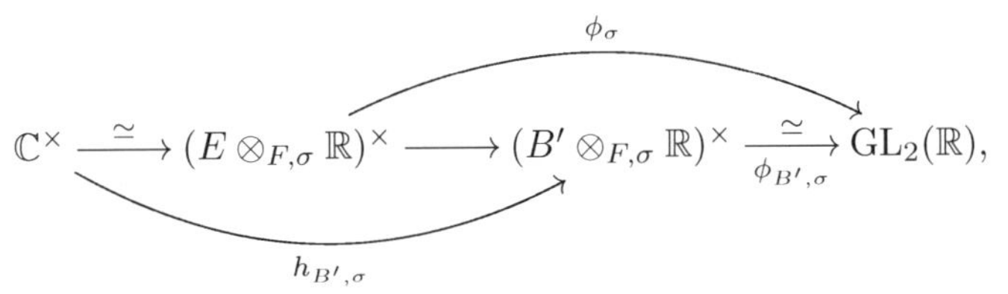

where the (inverse of) first isomorphism is the map that sends $\mathbf{i} \in E^\times$ to $i\sqrt{|\sigma u|} \in \mathbb{C}^\times$. By definition $V_{\sigma,h}^{-1,0}$ is the subspace of $V_{\sigma,\mathbb{C}}$ on which $\mathbb{C}^\times = (E \otimes_{F,\sigma} \mathbb{R})^\times$ acts by $z \mapsto z$. Now, $V_{\sigma,h}^{-1,0}$ may more simply be identified as the subspace of $V_{\sigma,\mathbb{C}} = \mathbb{C}^2$ on which $\phi_\sigma(\mathbf{i})$ acts as $i\sqrt{|\sigma u|}$. Since B' is split, we have $\sigma J' > 0$. Consider the isomorphism

$$\phi' : B' \otimes_{F,\sigma} \mathbb{R} \to \mathrm{M}_2(\mathbb{R})$$

defined by

$$\phi'(\mathbf{i}) = \begin{bmatrix} & \sqrt{|\sigma u|} \\ -\sqrt{|\sigma u|} & \end{bmatrix}, \quad \phi'(\mathbf{j}') = \begin{bmatrix} \sqrt{|\sigma J'|} & \\ & -\sqrt{|\sigma J'|} \end{bmatrix}.$$

Then ϕ' gives an isomorphism of Azumaya algebras over $\mathcal{O}_{L_0,(\ell)}$:

$$\phi' : \mathcal{O}_{B'} \otimes_{\mathcal{O}_F,\sigma} \mathcal{O}_{L_0,(\ell)} \simeq \mathrm{M}_2(\mathcal{O}_{L,(\ell)}) \cap \mathrm{M}_2(\mathbb{R}) = \mathrm{M}_2(\mathcal{O}_{L_0,(\ell)}).$$

Since $\phi_{B',\sigma}$ also gives such an isomorphism, ϕ' and $\phi_{B',\sigma}$ must be conjugate by an element $A \in \mathrm{GL}_2(\mathcal{O}_{L_0,(\ell)})$. Now $\mathbf{e}_1 := \frac{1}{2}\begin{bmatrix} 1 \\ i \end{bmatrix}$ is an eigenvector of $\phi'(\mathbf{i})$ with eigenvalue $i\sqrt{|\sigma u|}$ and so $A\mathbf{e}_1$ is an eigenvector of $\phi_\sigma(\mathbf{i}) = \phi_{B',\sigma}(\mathbf{i})$ with the same eigenvalue. Since $A \in \mathrm{GL}_2(\mathcal{O}_{L_0,(\ell)})$, we have $\mathbf{v}_1 := A\mathbf{e}_1 \in \mathcal{L}_\sigma$ and

$$\langle \mathbf{v}_1, \mathbf{v}_1 \rangle_h = \langle A\mathbf{e}_1, A\mathbf{e}_1 \rangle_h = \det(A)\langle \mathbf{e}_1, \mathbf{e}_1 \rangle_{A^{-1}\cdot h}.$$

Now

$$(A^{-1}\cdot h)(i) = A^{-1}(\phi_{B',\sigma}\circ h_{B',\sigma})(i)A = \frac{1}{\sqrt{|\sigma u|}}A^{-1}\phi_{B',\sigma}(\mathbf{i})A$$
$$= \frac{1}{\sqrt{|\sigma u|}}\phi'(\mathbf{i}) = \begin{bmatrix} & 1 \\ -1 & \end{bmatrix}.$$

Since $\mathbf{e}_1$ corresponds to the vector $e_1 := \begin{bmatrix} 1 \\ 0 \end{bmatrix} \in V_{\sigma,\mathbb{R}}$ via the $\mathbb{R}$-linear isomorphism $V_{\sigma,\mathbb{R}} \simeq V^{-1,0}_{\sigma,A^{-1}\cdot h}$, we have

$$\langle \mathbf{e}_1, \mathbf{e}_1\rangle_{A^{-1}\cdot h} = {}^t e_1 \begin{bmatrix} & -1 \\ 1 & \end{bmatrix} (A^{-1}\cdot h)(i)e_1 = {}^t e_1 e_1 = 1.$$

Thus $\langle \mathbf{v}_1, \mathbf{v}_1\rangle_h = \det(A) \in \mathcal{O}^\times_{L_0,(\ell)}$. □

7.2.2. In order to state the main conjecture, we will need to renormalize the measure and the Schwartz function in the definition of the theta lift. First we renormalize the Schwartz function. Let φ_v be the Schwartz function on $\mathcal{S}(\mathbb{X}_v)$ defined in §5.7. We set $\boldsymbol{\varphi} = \otimes_v \boldsymbol{\varphi}_v$ where

$$\boldsymbol{\varphi}_v = \sqrt{C^{\varphi}_v}\cdot\varphi_v$$

with

$$C^{\varphi}_v = \begin{cases} 1 & \text{if } \Pi_v \text{ is unramified principal series,} \\ \dfrac{q_v - 1}{q_v^{2n_v+1}} & \text{if } \Pi_v \text{ is ramified principal series with conductor } q_v^{n_v}, \\ q_v^{-2} & \text{if } \Pi_v \text{ is special,} \\ \dfrac{k_v!}{2^{k_v}\pi^{k_v}} & \text{if } v \in \Sigma_\infty \smallsetminus (\Sigma_{B,\infty}\cup\Sigma_{B_1,\infty}\cup\Sigma_{B_2,\infty}), \\ \dfrac{(k_v-2)!}{2^{k_v-2}\pi^{k_v-2}} & \text{if } v\in\Sigma_{B_1,\infty}\cap\Sigma_{B_2,\infty} \\ \dfrac{(k_v-2)!}{2^{k_v-4}\pi^{k_v-2}} & \text{if } v\in\Sigma_{B,\infty}. \end{cases}$$

As for the measure used in the theta lift (4.1) we renormalize the measure on $B^{(1)}(\mathbb{A})$ to

$$[\mathcal{K}^B_0 : \tilde{\mathcal{K}}^B]\cdot \operatorname{rank}\mathcal{V}^B_{\underline{k}_B,r}\cdot \text{standard measure on } B^{(1)}(\mathbb{A}).$$

With this choice of measure, let us define $\boldsymbol{\alpha}(B_1, B_2)$ by

$$\theta_{\boldsymbol{\varphi}}(f_B) = \boldsymbol{\alpha}(B_1,B_2)\cdot(f_{B_1}\times f_{B_2}). \tag{7.6}$$

REMARK 7.2.2. The definition of the constant C^{φ}_v is rigged so that (7.8) below holds. It would be better to have a conceptual definition of C^{φ}_v but for the moment it may simply be viewed as a local correction factor that is 1 at almost all places.

THEOREM 7.2.3. *Suppose $B \neq \mathrm{M}_2(F)$. Then*

$$|\boldsymbol{\alpha}(B_1,B_2)|^2\cdot\langle\langle s_{B_1},s_{B_1}\rangle\rangle_{\tilde{\mathcal{K}}^{B_1}}\cdot\langle\langle s_{B_2},s_{B_2}\rangle\rangle_{\tilde{\mathcal{K}}^{B_2}} = \Lambda(1,\Pi,\mathrm{ad})\cdot\langle\langle s_B,s_B\rangle\rangle_{\tilde{\mathcal{K}}^{B}} \tag{7.7}$$

in $\mathbb{C}^\times/R^\times_{(\ell)}$.

PROOF. Recall that $\Lambda(s,\Pi,\mathrm{ad}) = \prod_{v\in\Sigma_\infty} L(s,\Pi_v,\mathrm{ad})\cdot L(s,\Pi,\mathrm{ad})$ with

$$L(s,\Pi_v,\mathrm{ad}) = \Gamma_{\mathbb{R}}(s+1)\Gamma_{\mathbb{C}}(s+k_v-1),$$

where $\Gamma_{\mathbb{R}}(s) = \pi^{-\frac{s}{2}}\Gamma(\frac{s}{2})$ and $\Gamma_{\mathbb{C}}(s) = 2(2\pi)^{-s}\Gamma(s)$. Since

$$L(1,\Pi_v,\mathrm{ad}) = \frac{(k_v-1)!}{2^{k_v-1}\pi^{k_v+1}}$$

for $v\in\Sigma_\infty$, it follows from Proposition 6.3.1 that

$$\langle f,f\rangle = 2|D_F|\cdot\prod_v C_v^\Lambda\cdot\Lambda(1,\Pi,\mathrm{ad})$$

with

$$C_v^\Lambda = \begin{cases} 1 & \text{if } \Pi_v \text{ is unramified principal series,}\\ \dfrac{q_v}{q_v+1} & \text{if } \Pi_v \text{ is ramified principal series or special,}\\ \dfrac{1}{2^{k_v+2}} & \text{if } v\in\Sigma_\infty. \end{cases}$$

We define a constant $C_v^{B'}$ for $B' = B_1, B_2, B$ by

$$C_v^{B'} = \begin{cases} 1 & \text{if } \Pi_v \text{ is unramified principal series,}\\ q_v^{n_v-1}(q_v+1) & \text{if } \Pi_v \text{ is ramified principal series with conductor } q_v^{n_v},\\ q_v+1 & \text{if } \Pi_v \text{ is special and } v\notin\Sigma_{B',\mathrm{fin}},\\ 1 & \text{if } \Pi_v \text{ is special and } v\in\Sigma_{B',\mathrm{fin}},\\ 1 & \text{if } v\in\Sigma_\infty\smallsetminus\Sigma_{B',\infty},\\ k_v-1 & \text{if } v\in\Sigma_{B',\infty}, \end{cases}$$

so that we have

$$\langle\!\langle s_{B'},s_{B'}\rangle\!\rangle_{\tilde{\mathcal{K}}^{B'}} = 2^{d_{B'}}h_F\cdot\prod_v C_v^{B'}\cdot\frac{\langle f_{B'},f_{B'}\rangle}{\langle v_{\underline{k}_{B'}}^{B'},v_{\underline{k}_{B'}}^{B'}\rangle_{h_{B'}}}$$

by (7.5). Then, by the definition of $\alpha(B_1,B_2)$ and $\boldsymbol{\alpha}(B_1,B_2)$ (see (6.4) and (7.6) respectively), we have

$$|\boldsymbol{\alpha}(B_1,B_2)|^2 = \prod_v (C_v^B)^2 C_v^\varphi\cdot|\alpha(B_1,B_2)|^2.$$

By Theorem 6.3.2, we have

$$|\alpha(B_1,B_2)|^2\cdot\langle f_{B_1},f_{B_1}\rangle\cdot\langle f_{B_2},f_{B_2}\rangle = |D_F|^2\cdot\prod_v C_v\cdot\langle f_B,f_B\rangle\cdot\langle f,f\rangle$$

with the constant C_v defined there, and hence

$$\begin{aligned}&\frac{|\boldsymbol{\alpha}(B_1,B_2)|^2}{\prod_v (C_v^B)^2C_v^\varphi}\cdot\frac{\langle v_{\underline{k}_{B_1}}^{B_1},v_{\underline{k}_{B_1}}^{B_1}\rangle_{h_{B_1}}\cdot\langle\!\langle s_{B_1},s_{B_1}\rangle\!\rangle_{\tilde{\mathcal{K}}^{B_1}}}{2^{d_{B_1}}h_F\cdot\prod_v C_v^{B_1}}\cdot\frac{\langle v_{\underline{k}_{B_2}}^{B_2},v_{\underline{k}_{B_2}}^{B_2}\rangle_{h_{B_2}}\cdot\langle\!\langle s_{B_2},s_{B_2}\rangle\!\rangle_{\tilde{\mathcal{K}}^{B_2}}}{2^{d_{B_2}}h_F\cdot\prod_v C_v^{B_2}}\\ &= |D_F|^2\cdot\prod_v C_v\cdot\frac{\langle v_{\underline{k}_B}^B,v_{\underline{k}_B}^B\rangle_{h_B}\cdot\langle\!\langle s_B,s_B\rangle\!\rangle_{\tilde{\mathcal{K}}^B}}{2^{d_B}h_F\cdot\prod_v C_v^B}\cdot 2|D_F|\cdot\prod_v C_v^\Lambda\cdot\Lambda(1,\Pi,\mathrm{ad}).\end{aligned}$$

Since $2, h_F, |D_F|, \langle v^{B'}_{\underline{k}_{B'}}, v^{B'}_{\underline{k}_{B'}} \rangle_{h_{B'}} \in R^{\times}_{(\ell)}$ for $B' = B_1, B_2, B$, this implies that

$$|\boldsymbol{\alpha}(B_1, B_2)|^2 \cdot \langle\!\langle s_{B_1}, s_{B_1} \rangle\!\rangle_{\tilde{\mathcal{K}}^{B_1}} \cdot \langle\!\langle s_{B_2}, s_{B_2} \rangle\!\rangle_{\tilde{\mathcal{K}}^{B_2}} = \prod_v C_v^{B_1} C_v^{B_2} C_v^{B} C_v^{\Lambda} C_v^{\varphi} C_v \cdot \Lambda(1, \Pi, \mathrm{ad}) \cdot \langle\!\langle s_B, s_B \rangle\!\rangle_{\tilde{\mathcal{K}}^{B}}$$

in $\mathbb{C}^\times / R^{\times}_{(\ell)}$. Now the theorem follows from this and the fact that

$$C_v^{B_1} C_v^{B_2} C_v^{B} C_v^{\Lambda} C_v^{\varphi} C_v = 1 \tag{7.8}$$

for all v (see the table below).

Π_v	$B_{1,v}$	$B_{2,v}$	B_v	$C_v^{B_1}$	$C_v^{B_2}$	C_v^{B}
ur	spl	spl	spl	1	1	1
rps	spl	spl	spl	$q_v^{n_v-1}(q_v+1)$	$q_v^{n_v-1}(q_v+1)$	$q_v^{n_v-1}(q_v+1)$
st	spl	spl	spl	q_v+1	q_v+1	q_v+1
st	ram	ram	spl	1	1	q_v+1
st	ram	spl	ram	1	q_v+1	1
st	spl	ram	ram	q_v+1	1	1
ds	spl	spl	spl	1	1	1
ds	ram	ram	spl	k_v-1	k_v-1	1
ds	ram	spl	ram	k_v-1	1	k_v-1
ds	spl	ram	ram	1	k_v-1	k_v-1

Π_v	$B_{1,v}$	$B_{2,v}$	B_v	C_v^{Λ}	C_v^{φ}	C_v
ur	spl	spl	spl	1	1	1
rps	spl	spl	spl	$\frac{q_v}{q_v+1}$	$\frac{q_v-1}{q_v^{2n_v+1}}$	$\frac{1}{q_v^{n_v-3}(q_v-1)(q_v+1)^2}$
st	spl	spl	spl	$\frac{q_v}{q_v+1}$	q_v^{-2}	$\frac{q_v}{(q_v+1)^2}$
st	ram	ram	spl	$\frac{q_v}{q_v+1}$	q_v^{-2}	q_v
st	ram	spl	ram	$\frac{q_v}{q_v+1}$	q_v^{-2}	q_v
st	spl	ram	ram	$\frac{q_v}{q_v+1}$	q_v^{-2}	q_v
ds	spl	spl	spl	$\frac{1}{2^{k_v+2}}$	$\frac{k_v!}{2^{k_v}\pi^{k_v}}$	$\frac{2^{2k_v+2}\pi^{k_v}}{k_v!}$
ds	ram	ram	spl	$\frac{1}{2^{k_v+2}}$	$\frac{(k_v-2)!}{2^{k_v-2}\pi^{k_v-2}}$	$\frac{2^{2k_v}\pi^{k_v-2}}{(k_v-1)^2\cdot(k_v-2)!}$
ds	ram	spl	ram	$\frac{1}{2^{k_v+2}}$	$\frac{(k_v-2)!}{2^{k_v-4}\pi^{k_v-2}}$	$\frac{2^{2k_v-2}\pi^{k_v-2}}{(k_v-1)^2\cdot(k_v-2)!}$
ds	spl	ram	ram	$\frac{1}{2^{k_v+2}}$	$\frac{(k_v-2)!}{2^{k_v-4}\pi^{k_v-2}}$	$\frac{2^{2k_v-2}\pi^{k_v-2}}{(k_v-1)^2\cdot(k_v-2)!}$

□

7.2.3. We now motivate the main conjecture of this paper. Let us set (as in (I.1))

$$\Lambda(\Pi) := \Lambda(1, \Pi, \mathrm{ad}).$$

Thus from (7.7), we see that for $B_1 \neq B_2$, we have

$$|\boldsymbol{\alpha}(B_1, B_2)|^2 \cdot q_{B_1}(\Pi, \ell) \cdot q_{B_2}(\Pi, \ell) = \Lambda(\Pi) \cdot q_B(\Pi, \ell) \quad \text{in} \quad \mathbb{C}^\times / R^{\times}_{(\ell)}, \tag{7.9}$$

and consequently,

$$|\boldsymbol{\alpha}(B_1, B_2)|^2 \cdot q_{B_1}(\Pi) \cdot q_{B_2}(\Pi) = \Lambda(\Pi) \cdot q_B(\Pi) \quad \text{in} \quad \mathbb{C}^\times / R^{\times}_{(\ell)}. \tag{7.10}$$

If we combine this with Conjecture A(ii) of the introduction, we are lead to the following conjectural expression for $|\boldsymbol{\alpha}(B_1, B_2)|^2$:

$$|\boldsymbol{\alpha}(B_1, B_2)|^2 \stackrel{?}{=} \frac{\Lambda(\Pi) \cdot \dfrac{\Lambda(\Pi)}{\prod_{v\in\Sigma_B} c_v(\Pi)}}{\dfrac{\Lambda(\Pi)}{\prod_{v\in\Sigma_{B_1}} c_v(\Pi)} \cdot \dfrac{\Lambda(\Pi)}{\prod_{v\in\Sigma_{B_2}} c_v(\Pi)}} = \left[\prod_{v\in\Sigma_{B_1}\cap\Sigma_{B_2}} c_v(\Pi)\right]^2 \text{ in } \mathbb{C}^\times/R_{(\ell)}^\times.$$

Combining this last expression with Conjecture A(i) suggests Conjecture D of the introduction on the arithmetic nature of the constants $\boldsymbol{\alpha}(B_1, B_2)$. We restate it below for the convenience of the reader.

CONJECTURE 7.2.4. *Suppose that $B_1 \neq B_2$ and $\Sigma_{B_1} \cap \Sigma_{B_2} \cap \Sigma_\infty = \varnothing$, that is B_1 and B_2 have no infinite places of ramification in common. Then*

(i) *$\boldsymbol{\alpha}(B_1, B_2)$ lies in $\overline{\mathbb{Q}}^\times$.*
(ii) *Moreover, $\boldsymbol{\alpha}(B_1, B_2)$ belongs to $R_{(\ell)}$.*
(iii) *If in addition B_1 and B_2 have no finite places of ramification in common, then $\boldsymbol{\alpha}(B_1, B_2)$ lies in $R_{(\ell)}^\times$.*

Note that this conjecture makes absolutely no reference to the constants $c_v(\Pi)$. However, we shall show now that the truth of this conjecture (for all ℓ prime to $N(\Pi)$) implies the truth of Conj. A.

THEOREM 7.2.5. *Suppose that Conj. 7.2.4 is true for all ℓ prime to $N(\Pi)$. Then Conj. A is true.*

REMARK 7.2.6. The proof below will show that the validity of Conj. 7.2.4 for a single ℓ implies a version of Conj. A with $R^\times$ replaced by $R_{(\ell)}^\times$.

REMARK 7.2.7. The constant $\boldsymbol{\alpha}(B_1, B_2)$ depends implicitly on the many choices made above including that of a well adapted coordinate system for $(F, B_1, B_2, (2\ell)\mathfrak{D}_F)$, the base points $h_{B'}$ and the vectors $v^{B'}_{\underline{k}_{B'}}$ for $B' = B, B_1, B_2$. The relation (7.10) shows however that $|\boldsymbol{\alpha}(B_1, B_2)|^2$, viewed as an element of $\mathbb{C}^\times/R_{(\ell)}^\times$ does not depend on any choices. This in turn implies (from the proof below) that in the statement of Theorem 7.2.5 above, for a given ℓ, it suffices to check the validity of Conj. 7.2.4 for a single set of choices.

PROOF. Recall from Remark 3 of the introduction that

$$q_{\mathrm{M}_2(F)}(\Pi) = \Lambda(\Pi) \quad \text{in } \mathbb{C}^\times/R^\times. \tag{7.11}$$

Note that if $|\Sigma_\Pi| = 0$ or 1, then Π does not transfer to any non-split quaternion algebra, so the conjecture follows from (7.11).

If $|\Sigma_\Pi| = 2$, say $\Sigma_\Pi = \{v, w\}$, then there is a unique non-split quaternion algebra B with $\Sigma_B \subseteq \Sigma_\Pi$, given by $\Sigma_B = \Sigma_\Pi$. In this case, we need to pick two elements $c_v(\Pi)$ and $c_w(\Pi)$ in $\mathbb{C}^\times/R^\times$ such that the relation

$$q_B(\Pi) = \frac{\Lambda(\Pi)}{c_v(\Pi) \cdot c_w(\Pi)}$$

is satisfied (in addition to (7.11)), and there are obviously many ways to do this. Note that in this case, the invariants $c_v(\Pi)$ and $c_w(\Pi)$ are not uniquely determined by the single relation above, so in order to get canonical invariants one would need to rigidify the choices by imposing other constraints on them. We do not pursue this here.

Thus we may assume that $|\Sigma_\Pi| \geq 3$. We need to first define the constants $c_v(\Pi)$ in this case. First, for any subset $\Sigma \subseteq \Sigma_\Pi$ of *even* cardinality let us define $c_\Sigma(\Pi) \in \mathbb{C}^\times/R^\times$ by

$$c_\Sigma(\Pi) := \frac{\Lambda(\Pi)}{q_{B_\Sigma}(\Pi)},$$

where B_Σ denotes the unique quaternion algebra ramified exactly at Σ. Note that from (7.11), we have

$$c_\varnothing(\Pi) = 1 \quad \text{in } \mathbb{C}^\times/R^\times. \tag{7.12}$$

Now let v be any element in Σ_Π. We will define $c_v(\Pi)$ as follows. Pick any two other elements $u, w \in \Sigma_\Pi$ and define $c_v(\Pi)$ to be the unique element in $\mathbb{C}^\times/R^\times$ such that

$$c_v(\Pi)^2 = \frac{c_{\{v,u\}}(\Pi) \cdot c_{\{v,w\}}(\Pi)}{c_{\{u,w\}}(\Pi)}. \tag{7.13}$$

We will show that the truth of Conj. 7.2.4 for a single ℓ implies that $c_v(\Pi)$ is well defined in $\mathbb{C}^\times/R_{(\ell)}^\times$, that it lies in $R_{(\ell)}$ if v is a finite place and that the relation

$$q_B(\Pi) = \frac{\Lambda(\Pi)}{\prod_{v\in\Sigma_B} c_v(\Pi)} \quad \text{in} \quad \mathbb{C}^\times/R_{(\ell)}^\times \tag{7.14}$$

is satisfied. It follows from this that the truth of Conj. 7.2.4 for *all* ℓ prime to $N(\Pi)$ implies that the $c_v(\Pi)$ is well defined in $\mathbb{C}^\times/R^\times$, that it lies in R if v is a finite place and that the relation

$$q_B(\Pi) = \frac{\Lambda(\Pi)}{\prod_{v\in\Sigma_B} c_v(\Pi)} \quad \text{in} \quad \mathbb{C}^\times/R^\times$$

is satisfied, which would complete the proof of the theorem.

Thus let ℓ be any prime not dividing $N(\Pi)$ and let us assume the truth of Conj. 7.2.4 for this fixed ℓ. If Σ_1 and Σ_2 are two distinct subsets of Σ_Π of even cardinality and if B_1 and B_2 are the corresponding quaternion algebras, the relation (7.10) gives

$$|\boldsymbol{\alpha}(B_1, B_2)|^2 \cdot c_\Sigma(\Pi) = c_{\Sigma_1}(\Pi) \cdot c_{\Sigma_2}(\Pi) \quad \text{in} \quad \mathbb{C}^\times/R_{(\ell)}^\times,$$

where $\Sigma := (\Sigma_1 \smallsetminus \Sigma_2) \sqcup (\Sigma_2 \smallsetminus \Sigma_1)$. If moreover Σ_1 and Σ_2 are disjoint, then Conj. 7.2.4 implies that $\boldsymbol{\alpha}(B_1, B_2)$ lies in $R_{(\ell)}^\times$. Thus we get the key multiplicative relation:

$$c_{\Sigma_1}(\Pi) \cdot c_{\Sigma_2}(\Pi) = c_{\Sigma_1\sqcup\Sigma_2}(\Pi) \quad \text{in} \quad \mathbb{C}^\times/R_{(\ell)}^\times, \quad \text{if} \quad \Sigma_1 \cap \Sigma_2 = \varnothing, \tag{7.15}$$

including the case $\Sigma_1 = \Sigma_2 = \varnothing$ on account of (7.12). We can use this to check that $c_v(\Pi)$ defined via (7.13) is independent of the choice of u and w, viewed as an element in $\mathbb{C}^\times/R_{(\ell)}^\times$. If $|\Sigma_\Pi| = 3$, then it is clear that there is a unique way to define $c_v(\Pi)$, so we may assume that $|\Sigma_\Pi| \geq 4$. Since the definition is symmetric in u and w, it suffices to show that it remains invariant under changing u to some other u' distinct from u and w. However, this follows from the equality

$$c_{\{v,u\}}(\Pi) \cdot c_{\{u',w\}}(\Pi) = c_{\{v,u,u',w\}}(\Pi) = c_{\{v,u'\}}(\Pi) \cdot c_{\{u,w\}}(\Pi) \quad \text{in} \quad \mathbb{C}^\times/R_{(\ell)}^\times,$$

which is implied by (7.15).

Next we check that if v is a finite place, then $c_v(\Pi)$ lies in $R_{(\ell)}$. If B_1, B_2 and B are the quaternion algebras with $\Sigma_{B_1} = \{v, u\}$, $\Sigma_{B_2} = \{v, w\}$ and $\Sigma_B = \{u, w\}$, then $B = B_1 \cdot B_2$ and

$$c_v(\Pi)^2 = \frac{\dfrac{\Lambda(\Pi)}{q_{B_1}(\Pi)} \cdot \dfrac{\Lambda(\Pi)}{q_{B_2}(\Pi)}}{\dfrac{\Lambda(\Pi)}{q_B(\Pi)}} = \frac{\Lambda(\Pi) \cdot q_B(\Pi)}{q_{B_1}(\Pi) \cdot q_{B_2}(\Pi)} = |\boldsymbol{\alpha}(B_1, B_2)|^2 \quad \text{in } \mathbb{C}^\times / R_{(\ell)}^\times.$$

Since B_1 and B_2 have no infinite places of ramification in common, it follows from (i) and (ii) of Conj. 7.2.4 that $c_v(\Pi)$ lies in $R_{(\ell)}$.

Finally, let us check that $c_u(\Pi) \cdot c_v(\Pi) = c_{\{u,v\}}(\Pi)$ in $\mathbb{C}^\times / R_{(\ell)}^\times$ if u, v are distinct elements in Σ_B. Indeed, picking any w distinct from u and v, we have

$$\begin{aligned} c_u(\Pi)^2 \cdot c_v(\Pi)^2 &= \frac{c_{\{u,v\}}(\Pi) \cdot c_{\{u,w\}}(\Pi)}{c_{\{v,w\}}(\Pi)} \cdot \frac{c_{\{v,u\}}(\Pi) \cdot c_{\{v,w\}}(\Pi)}{c_{\{u,w\}}(\Pi)} \\ &= c_{\{u,v\}}(\Pi)^2 \quad \text{in} \quad \mathbb{C}^\times / R_{(\ell)}^\times, \end{aligned}$$

as claimed. From this, (7.12) and (7.15) it follows immediately that for any subset $\Sigma \subseteq \Sigma_\Pi$ of even cardinality, we have

$$c_\Sigma(\Pi) = \prod_{v \in \Sigma} c_v(\Pi) \quad \text{in} \quad \mathbb{C}^\times / R_{(\ell)}^\times,$$

from which (7.14) follows immediately. $\square$

APPENDIX A

Abelian Varieties, Polarizations and Hermitian Forms

In this section we discuss polarizations and the action of complex conjugation on Hodge structures attached to abelian varieties. This material is completely standard, so the purpose of this section is simply to carefully fix our conventions and motivate some of our constructions in Chapter 1.

A.1. Abelian varieties and Hodge structures

If A is a complex abelian variety, there is a natural Hodge structure on $\Lambda = H_1(A, \mathbb{Z})$. If $V = \Lambda \otimes \mathbb{Q}$, we have

$$V_{\mathbb{C}} = H_1(A, \mathbb{C}) = V^{-1,0} \oplus V^{0,-1}$$

where $V^{-1,0} = \mathrm{Lie}(A)$ and $V^{0,-1} = F^0(V) = \overline{V^{-1,0}}$ is identified with $H^1(A, \mathcal{O}_A)^\vee$. In fact, the exact sequence

$$0 \to V^{0,-1} \to V_{\mathbb{C}} \to V^{-1,0} \to 0$$

is dual to

$$0 \to H^0(A, \Omega_A^1) \to H^1(A, \mathbb{C}) \to H^1(A, \mathcal{O}_A) \to 0$$

which describes the Hodge filtration on $H^1(A, \mathbb{C})$. As a complex torus, A is recovered as

$$A = V^{0,-1} \backslash V_{\mathbb{C}} / \Lambda \simeq V^{-1,0}/\Lambda.$$

Let $h : \mathbb{S} \to \mathrm{GL}(V_{\mathbb{R}})$ be the homomorphism of the Deligne torus into $\mathrm{GL}(V_{\mathbb{R}})$ corresponding to the Hodge structure on $H_1(A, \mathbb{Z})$. Let $C = h(i)$. Recall that according to our conventions, the operator $C \otimes 1$ on $V_{\mathbb{R}} \otimes_{\mathbb{R}} \mathbb{C}$ acts on $V^{-1,0}$ as i and on $V^{0,-1}$ as $-i$. We write F for $F^0V = V^{0,-1}$ so that $\bar{F} = V^{-1,0} = \mathrm{Lie}(A)$. Then the composite maps

$$\Lambda \otimes \mathbb{R} \to \Lambda \otimes \mathbb{C} \to F, \quad \Lambda \otimes \mathbb{R} \to \Lambda \otimes \mathbb{C} \to \bar{F} \tag{A.1}$$

are $\mathbb{R}$-linear isomorphisms.

A.2. Polarizations and hermitian forms

Let Ψ be a skew-symmetric form

$$\Psi : \Lambda \times \Lambda \to \mathbb{Z}(1)$$

whose $\mathbb{R}$-linear extension $\Psi_{\mathbb{R}} : V_{\mathbb{R}} \times V_{\mathbb{R}} \to \mathbb{R}(1)$ satisfies

$$\Psi_{\mathbb{R}}(Cv, Cw) = \Psi_{\mathbb{R}}(v, w).$$

Define

$$B : \Lambda \times \Lambda \to \mathbb{Z}, \quad B(v, w) = \frac{1}{2\pi i}\Psi(v, w).$$

REMARK A.2.1. Note that the discussion up to this point was in fact independent of a choice of i. However, in the definition of B above and in the sequel, we need to fix such a choice. For any element $x+yi \in \mathbb{C}$ let us also set

$$\mathrm{Im}(x+yi) = yi, \quad \mathrm{im}(x+yi) = y.$$

Let $B_{\mathbb{R}}$ and $B_{\mathbb{C}}$ denote the $\mathbb{R}$-linear and $\mathbb{C}$-linear extensions of B to $V_{\mathbb{R}}$ and $V_{\mathbb{C}}$ respectively. Let B_C denote the hermitian form on $V_{\mathbb{C}}$ given by

$$B_C(v,w) := B_{\mathbb{C}}(v, C\bar{w}).$$

Finally, we let B_F and $B_{\bar{F}}$ denote the bilinear forms on F and $\bar{F}$ obtained from $B_{\mathbb{R}}$ via the isomorphisms (A.1) above.

PROPOSITION A.2.2. *The forms $B_{\mathbb{C}}$ and B_C have the following properties:*
(i) *The subspaces F and $\bar{F}$ of $V_{\mathbb{C}}$ are isotropic for $B_{\mathbb{C}}$.*
(ii) *The form B_C pairs $F \times \bar{F}$ to zero.*
(iii) $2 \cdot \mathrm{im}(B_C)|_F = B_F$ *and* $2 \cdot \mathrm{im}(B_C)|_{\bar{F}} = -B_{\bar{F}}$.

PROOF. For $v, w \in F$, we have

$$B_{\mathbb{C}}(v,w) = B_{\mathbb{C}}(h(i)v, h(i)w) = B_{\mathbb{C}}(-iv, -iw) = -B_{\mathbb{C}}(v,w),$$

so F is isotropic for $B_{\mathbb{C}}$. The argument for $\bar{F}$ is similar. Part (ii) follows immediately from part (i). For part (iii), suppose $v, w \in F$. Then

$$\begin{aligned} 2\,\mathrm{Im}(B_C)(v,w) &= B_C(v,w) - \overline{B_C(v,w)} = B_{\mathbb{C}}(v, C\bar{w}) - B_{\mathbb{C}}(\bar{v}, Cw) \\ &= B_{\mathbb{C}}(v, i\bar{w}) - B_{\mathbb{C}}(\bar{v}, -iw) = i\left(B_{\mathbb{C}}(v,\bar{w}) + B_{\mathbb{C}}(\bar{v}, w)\right). \end{aligned}$$

On the other hand, under the isomorphism $V_{\mathbb{R}} \simeq F$, the element $v \in F$ corresponds to $v + \bar{v} \in V_{\mathbb{R}}$. Thus

$$B_F(v,w) = B_{\mathbb{R}}(v+\bar{v}, w+\bar{w}) = B_{\mathbb{C}}(v+\bar{v}, w+\bar{w}) = B_{\mathbb{C}}(v,\bar{w}) + B_{\mathbb{C}}(\bar{v}, w)$$

from part (i). This shows that $2 \cdot \mathrm{im}(B_C)|_F = B_F$. The proof for $\bar{F}$ is similar. ☐

PROPOSITION A.2.3. *The following are equivalent:*
(i) *The bilinear form $(v,w) \mapsto B_{\mathbb{R}}(v, Cw)$ on $V_{\mathbb{R}}$ is positive definite.*
(ii) *The hermitian form B_C on $V_{\mathbb{C}}$ is positive definite and induces by restriction positive definite hermitian forms on both F and $\bar{F}$.*

PROOF. Let $v, w \in V_{\mathbb{C}}$. Suppose $v = v_1 + v_2$ and $w = w_1 + w_2$ with $v_1, w_1 \in F$ and $v_2, w_2 \in \bar{F}$. Then

$$\begin{aligned} B_C(v,w) &= B_C(v_1+v_2, w_1+w_2) = B_C(v_1,w_1) + B_C(v_2,w_2) \\ &= B_{\mathbb{C}}(v_1, C\bar{w}_1) + B_{\mathbb{C}}(v_2, C\bar{w}_2) \end{aligned}$$

so in particular,

$$B_C(v,v) = B_{\mathbb{C}}(v_1, C\bar{v}_1) + B_{\mathbb{C}}(v_2, C\bar{v}_2).$$

On the other hand,

$$\begin{aligned} B_{\mathbb{R}}(v_1+\bar{v}_1, C(v_1+\bar{v}_1)) &= B_{\mathbb{C}}(v_1, C\bar{v}_1) + B_{\mathbb{C}}(\bar{v}_1, Cv_1) \\ &= B_{\mathbb{C}}(v_1, C\bar{v}_1) - B_{\mathbb{C}}(C^2\bar{v}_1, Cv_1) \\ &= B_{\mathbb{C}}(v_1, C\bar{v}_1) - B_{\mathbb{C}}(C\bar{v}_1, v_1) \\ &= 2B_{\mathbb{C}}(v_1, C\bar{v}_1). \end{aligned}$$

Likewise,

$$B_{\mathbb{R}}(v_2 + \bar{v}_2, C(v_2 + \bar{v}_2)) = 2B_{\mathbb{C}}(v_2, C\bar{v}_2).$$

The implication (i) $\iff$ (ii) is clear from this. □

DEFINITION A.2.4. We will say that Ψ or B is a polarization if either of the equivalent conditions of the proposition above are satisfied.

REMARK A.2.5. In the classical theory of complex abelian varieties, one considers hermitian forms H on F or $\bar{F}$ whose imaginary part im H equals a given skew-symmetric form. A polarization corresponds to the choice of a skew-symmetric form such that H is either positive or negative definite. This can lead to some confusion: note for example that the skew-symmetric form B_F is the imaginary part of the positive definite form $2 \cdot B_C|_F$, while the skew-symmetric form $B_{\bar{F}}$ is the imaginary part of the negative definite form $-2 \cdot B_C|_{\bar{F}}$. We will always use the form B_C which is positive definite on both F and $\bar{F}$.

APPENDIX B

Metaplectic Covers of Symplectic Similitude Groups

In this section, we recall some basic properties of the metaplectic covers of symplectic similitude groups.

B.1. Setup

Let F be a local field of characteristic zero. Fix a nontrivial additive character ψ of F.

Let $\mathbb{V}$ be a $2n$-dimensional symplectic space over F. Let $\mathrm{GSp}(\mathbb{V})$ and $\mathrm{Sp}(\mathbb{V}) := \ker \nu$ be the symplectic similitude group and the symplectic group of $\mathbb{V}$ respectively, where $\nu : \mathrm{GSp}(\mathbb{V}) \to F^\times$ is the similitude character.

Fix a complete polarization $\mathbb{V} = \mathbb{X} \oplus \mathbb{Y}$. Choose a basis $\mathbf{e}_1, \dots, \mathbf{e}_n, \mathbf{e}_1^*, \dots, \mathbf{e}_n^*$ of $\mathbb{V}$ such that

$$\mathbb{X} = F\mathbf{e}_1 + \cdots + F\mathbf{e}_n, \quad \mathbb{Y} = F\mathbf{e}_1^* + \cdots + F\mathbf{e}_n^*, \quad \langle\langle \mathbf{e}_i, \mathbf{e}_j^* \rangle\rangle = \delta_{ij}.$$

Using this basis, we may write

$$\mathrm{GSp}(\mathbb{V}) = \left\{ g \in \mathrm{GL}_{2n}(F) \;\middle|\; g \begin{pmatrix} & \mathbf{1}_n \\ -\mathbf{1}_n & \end{pmatrix} {}^t g = \nu(g) \cdot \begin{pmatrix} & \mathbf{1}_n \\ -\mathbf{1}_n & \end{pmatrix} \right\}.$$

For $\nu \in F^\times$, we define $d(\nu) = d_{\mathbb{Y}}(\nu) \in \mathrm{GSp}(\mathbb{V})$ by

$$d(\nu) := \begin{pmatrix} \mathbf{1}_n & \\ & \nu \cdot \mathbf{1}_n \end{pmatrix}.$$

Let $P = P_{\mathbb{Y}}$ be the maximal parabolic subgroup of $\mathrm{Sp}(\mathbb{V})$ stabilizing $\mathbb{Y}$:

$$P = \left\{ \begin{pmatrix} a & * \\ & {}^t a^{-1} \end{pmatrix} \;\middle|\; a \in \mathrm{GL}_n(F) \right\}.$$

We have a Bruhat decomposition

$$\mathrm{Sp}(\mathbb{V}) = \coprod_{j=0}^{n} P\tau_j P,$$

where

$$\tau_j := \begin{pmatrix} \mathbf{1}_{n-j} & & & \\ & & & -\mathbf{1}_j \\ & & \mathbf{1}_{n-j} & \\ & \mathbf{1}_j & & \end{pmatrix}.$$

For $h \in \mathrm{Sp}(\mathbb{V})$, put $j(h) := j$ if $h \in P\tau_j P$. We define a map

$$x : \mathrm{Sp}(\mathbb{V}) \longrightarrow F^\times/(F^\times)^2$$

by

$$x(p_1 \tau_j p_2) := \det(a_1 a_2) \bmod (F^\times)^2,$$

where

$$p_i = \begin{pmatrix} a_i & * \\ & {}^t a_i^{-1} \end{pmatrix} \in P.$$

In particular, we have $x(p_1 h p_2) = x(p_1)x(h)x(p_2)$ for $p_1, p_2 \in P$ and $h \in \mathrm{Sp}(\mathbb{V})$.

Let $z_{\mathbb{Y}} = z_{\mathbb{Y}}^{\mathrm{Sp}}$ be the 2-cocycle given by

$$z_{\mathbb{Y}}(h_1, h_2) := \gamma_F(\frac{1}{2}\psi \circ q(\mathbb{Y}, \mathbb{Y}h_2^{-1}, \mathbb{Y}h_1))$$

for $h_1, h_2 \in \mathrm{Sp}(\mathbb{V})$.

LEMMA B.1.1. *We have*

- $z_{\mathbb{Y}}(h, h^{-1}) = 1$ *for* $h \in \mathrm{Sp}(\mathbb{V})$,
- $z_{\mathbb{Y}}(p_1 h_1 p, p^{-1} h_2 p_2) = z_{\mathbb{Y}}(h_1, h_2)$ *for* $p, p_i \in P$ *and* $h_i \in \mathrm{Sp}(\mathbb{V})$,
- $z_{\mathbb{Y}}(\tau_i, \tau_j) = 1$,
- $z_{\mathbb{Y}}(\tau_n, \mathbf{n}(\beta)\tau_n) = \gamma_F(\frac{1}{2}\psi \circ q_\beta)$ *for* $\mathbf{n}(\beta) = \begin{pmatrix} \mathbf{1}_n & \beta \\ & \mathbf{1}_n \end{pmatrix}$ *with* $\beta \in \mathrm{Hom}(\mathbb{X}, \mathbb{Y})$ *if* q_β *is non-degenerate, where* q_β *is a symmetric bilinear form on* $\mathbb{X}$ *defined by* $q_\beta(x, y) = \langle\langle x, y\beta \rangle\rangle$.

PROOF. See [**63**, Theorem 4.1, Corollary 4.2]. □

Suppose that $\mathbb{V} = \mathbb{V}_1 \oplus \mathbb{V}_2$, where each $\mathbb{V}_i$ is a non-degenerate symplectic subspace over F. If $\mathbb{V}_i = \mathbb{X}_i \oplus \mathbb{Y}_i$ is a complete polarization and

$$\mathbb{X} = \mathbb{X}_1 \oplus \mathbb{X}_2, \quad \mathbb{Y} = \mathbb{Y}_1 \oplus \mathbb{Y}_2,$$

then we have

$$z_{\mathbb{Y}_1}(h_1, h_1') \cdot z_{\mathbb{Y}_2}(h_2, h_2') = z_{\mathbb{Y}}(h_1 h_2, h_1' h_2')$$

for $h_i, h_i' \in \mathrm{Sp}(\mathbb{V}_i)$ (see Theorem 4.1 of [**63**]).

B.2. Action of outer automorphisms on the 2-cocycle

For $\nu \in F^\times$, let $\alpha_\nu = \alpha_{\mathbb{Y},\nu}$ be the outer automorphism of $\mathrm{Sp}(\mathbb{V})$ given by

$$\alpha_\nu(h) = d(\nu) \cdot h \cdot d(\nu)^{-1}$$

for $h \in \mathrm{Sp}(\mathbb{V})$. This induces an action of $F^\times$ on $\mathrm{Sp}(\mathbb{V})$ and thus we have an isomorphism

$$\begin{aligned} \mathrm{Sp}(\mathbb{V}) \rtimes F^\times &\longrightarrow \mathrm{GSp}(\mathbb{V}). \\ (h, \nu) &\longmapsto (h, \nu)_{\mathbb{Y}} := h \cdot d(\nu) \end{aligned}$$

Note that

$$(h, \nu)_{\mathbb{Y}} \cdot (h', \nu')_{\mathbb{Y}} = (h \cdot \alpha_\nu(h'), \nu \cdot \nu')_{\mathbb{Y}}.$$

There exists a unique automorphism $\tilde{\alpha}_\nu$ of $\mathrm{Mp}(\mathbb{V})$ such that $\tilde{\alpha}_\nu|_{\mathbb{C}^1} = \mathrm{id}_{\mathbb{C}^1}$ and the diagram

$$\begin{array}{ccc} \mathrm{Mp}(\mathbb{V}) & \xrightarrow{\tilde{\alpha}_\nu} & \mathrm{Mp}(\mathbb{V}) \\ \downarrow & & \downarrow \\ \mathrm{Sp}(\mathbb{V}) & \xrightarrow{\alpha_\nu} & \mathrm{Sp}(\mathbb{V}) \end{array}$$

commutes. This implies that there exists a unique function

$$v_{\mathbb{Y}} : \mathrm{Sp}(\mathbb{V}) \times F^\times \longrightarrow \mathbb{C}^1$$

such that

$$\tilde{\alpha}_\nu(h,z) = (\alpha_\nu(h), z \cdot v_{\mathbb{Y}}(h,\nu))$$

for $(h,z) \in \mathrm{Mp}(\mathbb{V})_{\mathbb{Y}}$. Since $\tilde{\alpha}_\nu$ is an automorphism, we have

$$z_{\mathbb{Y}}(\alpha_\nu(h), \alpha_\nu(h')) = z_{\mathbb{Y}}(h,h') \cdot v_{\mathbb{Y}}(hh',\nu) \cdot v_{\mathbb{Y}}(h,\nu)^{-1} \cdot v_{\mathbb{Y}}(h',\nu)^{-1}$$

for $h, h' \in \mathrm{Sp}(\mathbb{V})$ and $\nu \in F^\times$.

LEMMA B.2.1. *We have*

$$v_{\mathbb{Y}}(h,\nu) = (x(h),\nu)_F \cdot \gamma_F(\nu, \frac{1}{2}\psi)^{-j(h)}$$

for $h \in \mathrm{Sp}(\mathbb{V})$ *and* $\nu \in F^\times$.

PROOF. See [**2**, Proposition 1.2.A]. For convenience, we recall the proof of [**2**, Proposition 1.2.A]. *Warning: our convention differs from that in* [**2**].

Note that $z_{\mathbb{Y}}(p,h) = z_{\mathbb{Y}}(h,p) = 1$ for $p \in P$ and $h \in \mathrm{Sp}(\mathbb{V})$. This implies that

$$v_{\mathbb{Y}}(php',\nu) = v_{\mathbb{Y}}(p,\nu) \cdot v_{\mathbb{Y}}(h,\nu) \cdot v_{\mathbb{Y}}(p',\nu)$$

for $p, p' \in P$ and $h \in \mathrm{Sp}(\mathbb{V})$. Moreover, there exist a character ξ_ν of $F^\times$ and an element $\gamma_\nu \in \mathbb{C}^1$ such that

$$v_{\mathbb{Y}}(p,\nu) = \xi_\nu(x(p)), \quad v_{\mathbb{Y}}(\tau_j,\nu) = \gamma_\nu^j.$$

To determine ξ_ν and γ_ν, we may assume that $\dim \mathbb{V} = 2$ as explained in the proof of [**2**, Proposition 1.2.A]. Put

$$\underline{\mathbf{n}}(x) := \begin{pmatrix} 1 & 0 \\ x & 1 \end{pmatrix}.$$

If $x \neq 0$, then we have

$$\underline{\mathbf{n}}(x) = \begin{pmatrix} 1 & x^{-1} \\ 0 & 1 \end{pmatrix} \begin{pmatrix} 0 & -1 \\ 1 & 0 \end{pmatrix} \begin{pmatrix} x & 1 \\ 0 & x^{-1} \end{pmatrix},$$

so that

$$v_{\mathbb{Y}}(\underline{\mathbf{n}}(x),\nu) = \xi_\nu(x) \cdot \gamma_\nu.$$

Let $x, y \in F$ such that $x \neq 0$, $y \neq 0$, $x + y \neq 0$. Since $\alpha_\nu(\underline{\mathbf{n}}(x)) = \underline{\mathbf{n}}(\nu x)$, we have

$$\frac{z_{\mathbb{Y}}(\underline{\mathbf{n}}(\nu x), \underline{\mathbf{n}}(\nu y))}{z_{\mathbb{Y}}(\underline{\mathbf{n}}(x), \underline{\mathbf{n}}(y))} = \frac{v_{\mathbb{Y}}(\underline{\mathbf{n}}(x+y),\nu)}{v_{\mathbb{Y}}(\underline{\mathbf{n}}(x),\nu) \cdot v_{\mathbb{Y}}(\underline{\mathbf{n}}(y),\nu)} = \xi_\nu\left(\frac{x+y}{xy}\right) \cdot \gamma_\nu^{-1}.$$

By [**63**, Corollary 4.3], we have

$$z_{\mathbb{Y}}(\underline{\mathbf{n}}(x), \underline{\mathbf{n}}(y)) = \gamma_F(\frac{1}{2}xy(x+y) \cdot \psi)$$

and hence

$$\begin{aligned}
\frac{z_{\mathbb{Y}}(\underline{\mathbf{n}}(\nu x), \underline{\mathbf{n}}(\nu y))}{z_{\mathbb{Y}}(\underline{\mathbf{n}}(x), \underline{\mathbf{n}}(y))} &= \frac{\gamma_F(\frac{1}{2}\nu^3 xy(x+y) \cdot \psi)}{\gamma_F(\frac{1}{2}xy(x+y) \cdot \psi)} \\
&= \frac{\gamma_F(\nu^3 xy(x+y), \frac{1}{2}\psi)}{\gamma_F(xy(x+y), \frac{1}{2}\psi)} \\
&= \gamma_F(\nu^3, \frac{1}{2}\psi) \cdot (xy(x+y), \nu^3)_F \\
&= \gamma_F(\nu, \frac{1}{2}\psi) \cdot (xy(x+y), \nu)_F.
\end{aligned}$$

Thus we obtain

$$\gamma_F(\nu, \frac{1}{2}\psi) \cdot \left(\frac{x+y}{xy}, \nu\right)_F = \xi_\nu\left(\frac{x+y}{xy}\right) \cdot \gamma_\nu^{-1}.$$

Taking $x = y = 2$, we have

$$\gamma_\nu = \gamma_F(\nu, \frac{1}{2}\psi)^{-1}$$

and hence

$$\xi_\nu(a) = (a, \nu)_F$$

for all $a \in F^\times$. □

B.3. Metaplectic groups

For each $\nu \in F^\times$, we have an automorphism $\tilde{\alpha}_\nu$ of $\mathrm{Mp}(\mathbb{V})$. This induces an action of $F^\times$ on $\mathrm{Mp}(\mathbb{V})$ and thus we have a topological group

$$\mathrm{Mp}(\mathbb{V}) \rtimes F^\times.$$

We define a bijection

$$\begin{aligned} \mathrm{Mp}(\mathbb{V})_{\mathbb{Y}} \rtimes F^\times &\longrightarrow \mathrm{GMp}(\mathbb{V})_{\mathbb{Y}} := \mathrm{GSp}(\mathbb{V}) \times \mathbb{C}^1 \\ ((h,z),\nu) &\longmapsto ((h,\nu)_{\mathbb{Y}}, z) \end{aligned}$$

as sets. Via this bijection, we regard $\mathrm{GMp}(\mathbb{V})_{\mathbb{Y}}$ as a group. Note that the diagram

$$\begin{array}{ccc} \mathrm{Mp}(\mathbb{V})_{\mathbb{Y}} \rtimes F^\times & \longrightarrow & \mathrm{GMp}(\mathbb{V})_{\mathbb{Y}} \\ \downarrow & & \downarrow \\ \mathrm{Sp}(\mathbb{V}) \rtimes F^\times & \longrightarrow & \mathrm{GSp}(\mathbb{V}) \end{array}$$

commutes. Let $z_{\mathbb{Y}}^{\mathrm{GSp}}$ be the 2-cocycle associated to $\mathrm{GMp}(\mathbb{V})_{\mathbb{Y}}$. By definition, one can see that

$$z_{\mathbb{Y}}^{\mathrm{GSp}}(g, g') = z_{\mathbb{Y}}^{\mathrm{Sp}}(h, \alpha_\nu(h')) \cdot v_{\mathbb{Y}}(h', \nu)$$

for $g = (h,\nu)_{\mathbb{Y}}, g' = (h',\nu')_{\mathbb{Y}} \in \mathrm{GSp}(\mathbb{V})$. In particular, the restriction of $z_{\mathbb{Y}}^{\mathrm{GSp}}$ to $\mathrm{Sp}(\mathbb{V}) \times \mathrm{Sp}(\mathbb{V})$ is equal to $z_{\mathbb{Y}}^{\mathrm{Sp}}$. Thus we omit the superscripts GSp and Sp from the notation.

We shall see that the isomorphism class of $\mathrm{GMp}(\mathbb{V})_{\mathbb{Y}}$ does not depend on the choice of the complete polarization. If there is no confusion, we write $\mathrm{GMp}(\mathbb{V}) = \mathrm{GMp}(\mathbb{V})_{\mathbb{Y}}$.

B.4. Change of polarizations

Let $\mathbb{V} = \mathbb{X}' + \mathbb{Y}'$ be another complete polarization. Fix an element $h_0 \in \mathrm{Sp}(\mathbb{V})$ such that $\mathbb{X}' = \mathbb{X}h_0$ and $\mathbb{Y}' = \mathbb{Y}h_0$. Let α_0 be the inner automorphism of $\mathrm{GSp}(\mathbb{V})$ given by

$$\alpha_0(g) = h_0 \cdot g \cdot h_0^{-1}$$

for $g \in \mathrm{GSp}(\mathbb{V})$. Note that $\alpha_0|_{\mathrm{Sp}(\mathbb{V})}$ is an inner automorphism of $\mathrm{Sp}(\mathbb{V})$. We have

$$d_{\mathbb{Y}'}(\nu) = h_0^{-1} \cdot d_{\mathbb{Y}}(\nu) \cdot h_0, \quad \alpha_{\mathbb{Y}',\nu} = \alpha_0^{-1} \circ \alpha_{\mathbb{Y},\nu} \circ \alpha_0.$$

By [**40**, Lemma 4.2], we have

$$z_{\mathbb{Y}'}(h, h') = z_{\mathbb{Y}}(\alpha_0(h), \alpha_0(h'))$$

for $h, h' \in \mathrm{Sp}(\mathbb{V})$, and an isomorphism

$$\begin{aligned} \mathrm{Mp}(\mathbb{V})_{\mathbb{Y}} &\longrightarrow \mathrm{Mp}(\mathbb{V})_{\mathbb{Y}'}, \\ (h, z) &\longmapsto (h, z \cdot \mu(h)) \end{aligned}$$

where

$$\mu(h) = z_{\mathbb{Y}}(h_0, hh_0^{-1}) \cdot z_{\mathbb{Y}}(h, h_0^{-1})$$

for $h \in \mathrm{Sp}(\mathbb{V})$.

LEMMA B.4.1. *We have*

$$v_{\mathbb{Y}'}(h, \nu) = v_{\mathbb{Y}}(\alpha_0(h), \nu)$$

for $h \in \mathrm{Sp}(\mathbb{V})$ *and* $\nu \in F^\times$.

PROOF. We have

$$\begin{aligned} &z_{\mathbb{Y}'}(\alpha_{\mathbb{Y}',\nu}(h), \alpha_{\mathbb{Y}',\nu}(h')) \\ &= z_{\mathbb{Y}'}((\alpha_0^{-1} \circ \alpha_{\mathbb{Y},\nu} \circ \alpha_0)(h), (\alpha_0^{-1} \circ \alpha_{\mathbb{Y},\nu} \circ \alpha_0)(h')) \\ &= z_{\mathbb{Y}}((\alpha_{\mathbb{Y},\nu} \circ \alpha_0)(h), (\alpha_{\mathbb{Y},\nu} \circ \alpha_0)(h')) \\ &= z_{\mathbb{Y}}(\alpha_0(h), \alpha_0(h')) \cdot v_{\mathbb{Y}}(\alpha_0(h) \cdot \alpha_0(h'), \nu) \cdot v_{\mathbb{Y}}(\alpha_0(h), \nu)^{-1} \cdot v_{\mathbb{Y}}(\alpha_0(h'), \nu)^{-1} \\ &= z_{\mathbb{Y}'}(h, h') \cdot v_{\mathbb{Y}}(\alpha_0(hh'), \nu) \cdot v_{\mathbb{Y}}(\alpha_0(h), \nu)^{-1} \cdot v_{\mathbb{Y}}(\alpha_0(h'), \nu)^{-1}. \end{aligned}$$

Thus the assertion follows from the characterization of $v_{\mathbb{Y}'}$. □

LEMMA B.4.2. *We have*

$$z_{\mathbb{Y}'}(g, g') = z_{\mathbb{Y}}(\alpha_0(g), \alpha_0(g'))$$

for $g, g' \in \mathrm{GSp}(\mathbb{V})$.

PROOF. Let $g = (h, \nu)_{\mathbb{Y}'}, g' = (h', \nu')_{\mathbb{Y}'} \in \mathrm{GSp}(\mathbb{V})$. Then we have

$$\begin{aligned} z_{\mathbb{Y}'}(g, g') &= z_{\mathbb{Y}'}(h, \alpha_{\mathbb{Y}',\nu}(h')) \cdot v_{\mathbb{Y}'}(h', \nu) \\ &= z_{\mathbb{Y}}(\alpha_0(h), (\alpha_0 \circ \alpha_{\mathbb{Y}',\nu})(h')) \cdot v_{\mathbb{Y}}(\alpha_0(h'), \nu) \\ &= z_{\mathbb{Y}}(\alpha_0(h), (\alpha_{\mathbb{Y},\nu} \circ \alpha_0)(h')) \cdot v_{\mathbb{Y}}(\alpha_0(h'), \nu) \\ &= z_{\mathbb{Y}}((\alpha_0(h), \nu)_{\mathbb{Y}}, (\alpha_0(h'), \nu')_{\mathbb{Y}}). \end{aligned}$$

Since

$$\alpha_0(g) = h_0 \cdot h \cdot d_{\mathbb{Y}'}(\nu) \cdot h_0^{-1} = h_0 \cdot h \cdot h_0^{-1} \cdot d_{\mathbb{Y}}(\nu) = (\alpha_0(h), \nu)_{\mathbb{Y}},$$

the assertion follows. □

Put

$$\mu(g) = z_{\mathbb{Y}}(g, h_0^{-1}) \cdot z_{\mathbb{Y}}(h_0, gh_0^{-1}) = z_{\mathbb{Y}'}(h_0^{-1}gh_0, h_0^{-1}) \cdot z_{\mathbb{Y}'}(h_0^{-1}, g)^{-1}$$

for $g \in \mathrm{GSp}(\mathbb{V})$. Note that μ depends on the choice of h_0. By a direct calculation, one can see that

$$z_{\mathbb{Y}'}(g, g') = z_{\mathbb{Y}}(g, g') \cdot \mu(gg') \cdot \mu(g)^{-1} \cdot \mu(g')^{-1}$$

for $g, g' \in \mathrm{GSp}(\mathbb{V})$. Thus we obtain an isomorphism

$$\begin{aligned} \mathrm{GMp}(\mathbb{V})_{\mathbb{Y}} &\longrightarrow \mathrm{GMp}(\mathbb{V})_{\mathbb{Y}'}. \\ (g, z) &\longmapsto (g, z \cdot \mu(g)) \end{aligned}$$

APPENDIX C

Splittings: The Case $\dim_B V = 2$ and $\dim_B W = 1$

In this section, we construct a splitting of the metaplectic cover over $\mathrm{GU}(V)^0 \times \mathrm{GU}(W)$, where V is the 2-dimensional skew-hermitian B-space given in §2.2 and W is the 1-dimensional hermitian B-space given in §3.3.1, and show that it satisfies natural properties. In particular, we prove the product formula (Proposition C.4.4), which plays a crucial role in the definition of the global theta lift.

C.1. Setup

Let F be a number field. Recall that

$$\begin{array}{llll} E = F + F\mathbf{i}, & B = E + E\mathbf{j}, & B_1 = E + E\mathbf{j}_1, & B_2 = E + E\mathbf{j}_2, \\ u := \mathbf{i}^2, & J := \mathbf{j}^2, & J_1 := \mathbf{j}_1^2, & J_2 := \mathbf{j}_2^2, \end{array}$$

where

$$J = J_1 J_2.$$

Recall that

$$V = B_1 \otimes_E B_2 \quad \text{and} \quad W = B$$

are a right skew-hermitian B-space and a left hermitian B-space respectively, and

$$\mathbb{V} = V \otimes_B W$$

is an F-space with a symplectic form

$$\langle\!\langle \cdot, \cdot \rangle\!\rangle = \frac{1}{2} \operatorname{tr}_{B/F}(\langle \cdot, \cdot \rangle \otimes \langle \cdot, \cdot \rangle^*).$$

Recall that $\mathbb{V} = \mathbb{X} + \mathbb{Y}$ is a complete polarization, where

$$\mathbb{X} = F\mathbf{e}_1 + F\mathbf{e}_2 + F\mathbf{e}_3 + F\mathbf{e}_4, \quad \mathbb{Y} = F\mathbf{e}_1^* + F\mathbf{e}_2^* + F\mathbf{e}_3^* + F\mathbf{e}_4^*.$$

The actions of B, B_1, B_2 on $\mathbb{V}$ are given as follows:

- B-action

$$\begin{array}{llll} \mathbf{e}_1\mathbf{i} = u\mathbf{e}_1^* & \mathbf{e}_2\mathbf{i} = -uJ_1\mathbf{e}_2^* & \mathbf{e}_3\mathbf{i} = -uJ_2\mathbf{e}_3^* & \mathbf{e}_4\mathbf{i} = uJ\mathbf{e}_4^* \\ \mathbf{e}_1^*\mathbf{i} = \mathbf{e}_1 & \mathbf{e}_2^*\mathbf{i} = -\dfrac{1}{J_1}\mathbf{e}_2 & \mathbf{e}_3^*\mathbf{i} = -\dfrac{1}{J_2}\mathbf{e}_3 & \mathbf{e}_4^*\mathbf{i} = \dfrac{1}{J}\mathbf{e}_4 \\ \mathbf{e}_1\mathbf{j} = \mathbf{e}_4 & \mathbf{e}_2\mathbf{j} = J_1\mathbf{e}_3 & \mathbf{e}_3\mathbf{j} = J_2\mathbf{e}_2 & \mathbf{e}_4\mathbf{j} = J\mathbf{e}_1 \\ \mathbf{e}_1^*\mathbf{j} = -J\mathbf{e}_4^* & \mathbf{e}_2^*\mathbf{j} = -J_2\mathbf{e}_3^* & \mathbf{e}_3^*\mathbf{j} = -J_1\mathbf{e}_2^* & \mathbf{e}_4^*\mathbf{j} = -\mathbf{e}_1^* \\ \mathbf{e}_1\mathbf{ij} = -uJ\mathbf{e}_4^* & \mathbf{e}_2\mathbf{ij} = uJ\mathbf{e}_3^* & \mathbf{e}_3\mathbf{ij} = uJ\mathbf{e}_2^* & \mathbf{e}_4\mathbf{ij} = -uJ\mathbf{e}_1^* \\ \mathbf{e}_1^*\mathbf{ij} = \mathbf{e}_4 & \mathbf{e}_2^*\mathbf{ij} = -\mathbf{e}_3 & \mathbf{e}_3^*\mathbf{ij} = -\mathbf{e}_2 & \mathbf{e}_4^*\mathbf{ij} = \mathbf{e}_1 \end{array}$$

- B_1-action

$$\begin{array}{llll}
\mathbf{ie}_1 = u\mathbf{e}_1^* & \mathbf{ie}_2 = uJ_1\mathbf{e}_2^* & \mathbf{ie}_3 = -uJ_2\mathbf{e}_3^* & \mathbf{ie}_4 = -uJ\mathbf{e}_4^* \\
\mathbf{ie}_1^* = \mathbf{e}_1 & \mathbf{ie}_2^* = \dfrac{1}{J_1}\mathbf{e}_2 & \mathbf{ie}_3^* = -\dfrac{1}{J_2}\mathbf{e}_3 & \mathbf{ie}_4^* = -\dfrac{1}{J}\mathbf{e}_4 \\
\mathbf{j}_1\mathbf{e}_1 = \mathbf{e}_2 & \mathbf{j}_1\mathbf{e}_2 = J_1\mathbf{e}_1 & \mathbf{j}_1\mathbf{e}_3 = \mathbf{e}_4 & \mathbf{j}_1\mathbf{e}_4 = J_1\mathbf{e}_3 \\
\mathbf{j}_1\mathbf{e}_1^* = -J_1\mathbf{e}_2^* & \mathbf{j}_1\mathbf{e}_2^* = -\mathbf{e}_1^* & \mathbf{j}_1\mathbf{e}_3^* = -J_1\mathbf{e}_4^* & \mathbf{j}_1\mathbf{e}_4^* = -\mathbf{e}_3^* \\
\mathbf{ij}_1\mathbf{e}_1 = uJ_1\mathbf{e}_2^* & \mathbf{ij}_1\mathbf{e}_2 = uJ_1\mathbf{e}_1^* & \mathbf{ij}_1\mathbf{e}_3 = -uJ\mathbf{e}_4^* & \mathbf{ij}_1\mathbf{e}_4 = -uJ\mathbf{e}_3^* \\
\mathbf{ij}_1\mathbf{e}_1^* = -\mathbf{e}_2 & \mathbf{ij}_1\mathbf{e}_2^* = -\mathbf{e}_1 & \mathbf{ij}_1\mathbf{e}_3^* = \dfrac{1}{J_2}\mathbf{e}_4 & \mathbf{ij}_1\mathbf{e}_4^* = \dfrac{1}{J_2}\mathbf{e}_3
\end{array}$$

- B_2-action

$$\begin{array}{llll}
\mathbf{ie}_1 = u\mathbf{e}_1^* & \mathbf{ie}_2 = -uJ_1\mathbf{e}_2^* & \mathbf{ie}_3 = uJ_2\mathbf{e}_3^* & \mathbf{ie}_4 = -uJ\mathbf{e}_4^* \\
\mathbf{ie}_1^* = \mathbf{e}_1 & \mathbf{ie}_2^* = -\dfrac{1}{J_1}\mathbf{e}_2 & \mathbf{ie}_3^* = \dfrac{1}{J_2}\mathbf{e}_3 & \mathbf{ie}_4^* = -\dfrac{1}{J}\mathbf{e}_4 \\
\mathbf{j}_2\mathbf{e}_1 = \mathbf{e}_3 & \mathbf{j}_2\mathbf{e}_2 = \mathbf{e}_4 & \mathbf{j}_2\mathbf{e}_3 = J_2\mathbf{e}_1 & \mathbf{j}_2\mathbf{e}_4 = J_2\mathbf{e}_2 \\
\mathbf{j}_2\mathbf{e}_1^* = -J_2\mathbf{e}_3^* & \mathbf{j}_2\mathbf{e}_2^* = -J_2\mathbf{e}_4^* & \mathbf{j}_2\mathbf{e}_3^* = -\mathbf{e}_1^* & \mathbf{j}_2\mathbf{e}_4^* = -\mathbf{e}_2^* \\
\mathbf{ij}_2\mathbf{e}_1 = uJ_2\mathbf{e}_3^* & \mathbf{ij}_2\mathbf{e}_2 = -uJ\mathbf{e}_4^* & \mathbf{ij}_2\mathbf{e}_3 = uJ_2\mathbf{e}_1^* & \mathbf{ij}_2\mathbf{e}_4 = -uJ\mathbf{e}_2^* \\
\mathbf{ij}_2\mathbf{e}_1^* = -\mathbf{e}_3 & \mathbf{ij}_2\mathbf{e}_2^* = \dfrac{1}{J_1}\mathbf{e}_4 & \mathbf{ij}_2\mathbf{e}_3^* = -\mathbf{e}_1 & \mathbf{ij}_2\mathbf{e}_4^* = \dfrac{1}{J_1}\mathbf{e}_2
\end{array}$$

Let $\boldsymbol{\alpha}_i \in B_i^\times$ and $\boldsymbol{\alpha} \in B^\times$. We write

$$\begin{aligned}
\boldsymbol{\alpha}_1 &= a_1 + b_1\mathbf{i} + c_1\mathbf{j}_1 + d_1\mathbf{ij}_1, \\
\boldsymbol{\alpha}_2 &= a_2 + b_2\mathbf{i} + c_2\mathbf{j}_2 + d_2\mathbf{ij}_2, \\
\boldsymbol{\alpha} &= a + b\mathbf{i} + c\mathbf{j} + d\mathbf{ij},
\end{aligned}$$

where $a_i, a, b_i, b, c_i, c, d_i, d \in F$. Then we have

$$\begin{bmatrix} \boldsymbol{\alpha}_1\mathbf{e}_1 \\ \boldsymbol{\alpha}_1\mathbf{e}_2 \\ \boldsymbol{\alpha}_1\mathbf{e}_3 \\ \boldsymbol{\alpha}_1\mathbf{e}_4 \\ \boldsymbol{\alpha}_1\mathbf{e}_1^* \\ \boldsymbol{\alpha}_1\mathbf{e}_2^* \\ \boldsymbol{\alpha}_1\mathbf{e}_3^* \\ \boldsymbol{\alpha}_1\mathbf{e}_4^* \end{bmatrix} = \mathbf{g}_1 \begin{bmatrix} \mathbf{e}_1 \\ \mathbf{e}_2 \\ \mathbf{e}_3 \\ \mathbf{e}_4 \\ \mathbf{e}_1^* \\ \mathbf{e}_2^* \\ \mathbf{e}_3^* \\ \mathbf{e}_4^* \end{bmatrix}, \qquad \begin{bmatrix} \boldsymbol{\alpha}_2\mathbf{e}_1 \\ \boldsymbol{\alpha}_2\mathbf{e}_2 \\ \boldsymbol{\alpha}_2\mathbf{e}_3 \\ \boldsymbol{\alpha}_2\mathbf{e}_4 \\ \boldsymbol{\alpha}_2\mathbf{e}_1^* \\ \boldsymbol{\alpha}_2\mathbf{e}_2^* \\ \boldsymbol{\alpha}_2\mathbf{e}_3^* \\ \boldsymbol{\alpha}_2\mathbf{e}_4^* \end{bmatrix} = \mathbf{g}_2 \begin{bmatrix} \mathbf{e}_1 \\ \mathbf{e}_2 \\ \mathbf{e}_3 \\ \mathbf{e}_4 \\ \mathbf{e}_1^* \\ \mathbf{e}_2^* \\ \mathbf{e}_3^* \\ \mathbf{e}_4^* \end{bmatrix}, \qquad \begin{bmatrix} \mathbf{e}_1\boldsymbol{\alpha} \\ \mathbf{e}_2\boldsymbol{\alpha} \\ \mathbf{e}_3\boldsymbol{\alpha} \\ \mathbf{e}_4\boldsymbol{\alpha} \\ \mathbf{e}_1^*\boldsymbol{\alpha} \\ \mathbf{e}_2^*\boldsymbol{\alpha} \\ \mathbf{e}_3^*\boldsymbol{\alpha} \\ \mathbf{e}_4^*\boldsymbol{\alpha} \end{bmatrix} = \mathbf{g} \begin{bmatrix} \mathbf{e}_1 \\ \mathbf{e}_2 \\ \mathbf{e}_3 \\ \mathbf{e}_4 \\ \mathbf{e}_1^* \\ \mathbf{e}_2^* \\ \mathbf{e}_3^* \\ \mathbf{e}_4^* \end{bmatrix},$$

where

$$\mathbf{g}_1 = \begin{pmatrix} a_1 & c_1 & & & b_1 u & d_1 u J_1 & & \\ c_1 J_1 & a_1 & & & d_1 u J_1 & b_1 u J_1 & & \\ & & a_1 & c_1 & & & -b_1 u J_2 & -d_1 u J \\ & & c_1 J_1 & a_1 & & & -d_1 u J & -b_1 u J \\ b_1 & -d_1 & & & a_1 & -c_1 J_1 & & \\ -d_1 & \frac{b_1}{J_1} & & & -c_1 & a_1 & & \\ & & -\frac{b_1}{J_2} & \frac{d_1}{J_2} & & & a_1 & -c_1 J_1 \\ & & \frac{d_1}{J_2} & -\frac{b_1}{J} & & & -c_1 & a_1 \end{pmatrix},$$

$$\mathbf{g}_2 = \begin{pmatrix} a_2 & & c_2 & & b_2 u & & d_2 u J_2 & \\ & a_2 & & c_2 & & -b_2 u J_1 & & -d_2 u J \\ c_2 J_2 & & a_2 & & d_2 u J_2 & & b_2 u J_2 & \\ & c_2 J_2 & & a_2 & & -d_2 u J & & -b_2 u J \\ b_2 & & -d_2 & & a_2 & & -c_2 J_2 & \\ & -\frac{b_2}{J_1} & & \frac{d_2}{J_1} & & a_2 & & -c_2 J_2 \\ -d_2 & & \frac{b_2}{J_2} & & -c_2 & & a_2 & \\ & \frac{d_2}{J_1} & & -\frac{b_2}{J} & & -c_2 & & a_2 \end{pmatrix},$$

$$\mathbf{g} = \begin{pmatrix} a & & & c & bu & & & -duJ \\ & a & cJ_1 & & & -buJ_1 & duJ & \\ & cJ_2 & a & & & duJ & -buJ_2 & \\ cJ & & & a & -duJ & & & buJ \\ b & & & d & a & & & -cJ \\ & -\frac{b}{J_1} & -d & & & a & -cJ_2 & \\ & -d & -\frac{b}{J_2} & & & -cJ_1 & a & \\ d & & & \frac{b}{J} & -c & & & a \end{pmatrix}.$$

For $\mathbf{a} \in \mathrm{GL}_4(F)$ and $\mathbf{b} \in \mathrm{Sym}_4(F)$, put

$$\mathbf{m}(\mathbf{a}) := \begin{pmatrix} \mathbf{a} & \\ & {}^t\mathbf{a}^{-1} \end{pmatrix}, \quad \mathbf{n}(\mathbf{b}) := \begin{pmatrix} \mathbf{1}_4 & \mathbf{b} \\ & \mathbf{1}_4 \end{pmatrix}.$$

Fix a place v of F. In §§C.2, C.3, we shall suppress the subscript v from the notation. Thus $F = F_v$ will be a local field of characteristic zero.

C.2. The case $u \in (F^\times)^2$ or $J \in (F^\times)^2$

First we explicate Morita theory. Fix an isomorphism

$$\mathfrak{i} : B \longrightarrow \mathrm{M}_2(F)$$

of F-algebras such that

$$\mathfrak{i}(\boldsymbol{\alpha}^*) = \mathfrak{i}(\boldsymbol{\alpha})^*$$

for $\boldsymbol{\alpha} \in B$. Put

$$e := \mathfrak{i}^{-1}\begin{pmatrix} 1 & 0 \\ 0 & 0 \end{pmatrix}, \quad e' := \mathfrak{i}^{-1}\begin{pmatrix} 0 & 1 \\ 0 & 0 \end{pmatrix}, \quad e'' := \mathfrak{i}^{-1}\begin{pmatrix} 0 & 0 \\ 1 & 0 \end{pmatrix}.$$

Then we have

$$e^* = \mathfrak{i}^{-1}\begin{pmatrix} 0 & 0 \\ 0 & 1 \end{pmatrix}$$

and

$$\begin{array}{llll} e^2 = e, & ee' = e', & ee'' = 0, & ee^* = 0, \\ e'e = 0, & (e')^2 = 0, & e'e'' = e, & e'e^* = e', \\ e''e = e'', & e''e' = e^*, & (e'')^2 = 0, & e''e^* = 0, \\ e^*e = 0, & e^*e' = 0, & e^*e'' = e'', & (e^*)^2 = e^*. \end{array}$$

Thus we obtain

$$B = Fe + Fe' + Fe'' + Fe^*, \quad eB = Fe + Fe', \quad Be = Fe + Fe'', \quad eBe = Fe$$

and

$$\begin{bmatrix} e \cdot \boldsymbol{\alpha} \\ e' \cdot \boldsymbol{\alpha} \end{bmatrix} = \mathfrak{i}(\boldsymbol{\alpha}) \cdot \begin{bmatrix} e \\ e' \end{bmatrix}$$

for $\boldsymbol{\alpha} \in B$.

Now we consider an F-space $W^\dagger := eW$. Since $eBe^* = Fe'$ and $(e')^* = -e'$, we have

$$\langle x, y \rangle \in Fe', \quad \langle y, x \rangle = -\langle x, y \rangle$$

for $x, y \in W^\dagger$. Hence we can define a symplectic form

$$\langle \cdot, \cdot \rangle^\dagger : W^\dagger \times W^\dagger \longrightarrow F$$

by

$$\langle x, y \rangle^* = \langle x, y \rangle^\dagger \cdot e'$$

for $x, y \in W^\dagger$. Moreover, we see that $\langle \cdot, \cdot \rangle^\dagger$ is non-degenerate.

We have $W^\dagger = Fe + Fe'$ and

$$\langle e, e \rangle^\dagger = \langle e', e' \rangle^\dagger = 0, \quad \langle e, e' \rangle^\dagger = 1.$$

Thus we may identify $W^\dagger$ with the space of row vectors F^2 so that

$$\langle x, y \rangle^\dagger = x_1 y_2 - x_2 y_1$$

for $x = (x_1, x_2)$, $y = (y_1, y_2) \in W^\dagger$.

LEMMA C.2.1. *The restriction to $W^\dagger$ induces an isomorphism*

$$\mathrm{GU}(W) \xrightarrow{\cong} \mathrm{GSp}(W^\dagger).$$

PROOF. One can see that the restriction to $W^\dagger$ induces a homomorphism $\mathrm{GU}(W) \to \mathrm{GSp}(W^\dagger)$. Since

$$B \cdot W^\dagger = B \cdot eW = BeB \cdot W = B \cdot W = W,$$

this homomorphism is injective. Let $h \in \mathrm{GSp}(W^\dagger)$. Since $W = W^\dagger \oplus e''W$, we can define a map $\tilde{h} : W \to W$ by

$$\tilde{h}(x) = \begin{cases} h(x) & \text{if } x \in W^\dagger, \\ e''h(e'x) & \text{if } x \in e''W. \end{cases}$$

Then one can check that $\tilde{h} \in \mathrm{GU}(W)$. This yields the lemma. □

Thus we may identify $\mathrm{GU}(W)$ with $\mathrm{GSp}(W^\dagger)$. Similarly, we consider an F-space $V^\dagger := Ve$ with a non-degenerate symmetric bilinear form

$$\langle \cdot, \cdot \rangle^\dagger : V^\dagger \times V^\dagger \longrightarrow F$$

defined by

$$\frac{1}{2} \cdot \langle x, y \rangle = \langle x, y \rangle^\dagger \cdot e''$$

for $x, y \in V^\dagger$. As in Lemma C.2.1, the restriction to $V^\dagger$ induces an isomorphism

$$\mathrm{GU}(V) \xrightarrow{\cong} \mathrm{GO}(V^\dagger).$$

Thus we may identify $\mathrm{GU}(V)$ with $\mathrm{GO}(V^\dagger)$.

One can see that the natural map

$$V^\dagger \otimes_F W^\dagger \longrightarrow V \otimes_B W$$

is an isomorphism. Thus we may identify $\mathbb{V}$ with $V^\dagger \otimes_F W^\dagger$.

LEMMA C.2.2. *We have*

$$\langle\!\langle \cdot, \cdot \rangle\!\rangle = \langle \cdot, \cdot \rangle^\dagger \otimes \langle \cdot, \cdot \rangle^\dagger.$$

PROOF. Let $a = \langle x, x' \rangle^\dagger$ and $b = \langle y, y' \rangle^\dagger$ for $x, x' \in V^\dagger$ and $y, y' \in W^\dagger$. Then we have

$$\begin{aligned} \langle\!\langle x \otimes y, x' \otimes y' \rangle\!\rangle &= \frac{1}{2} \cdot \mathrm{tr}_{B/F}(\langle x, x' \rangle \cdot \langle y, y' \rangle^*) \\ &= \mathrm{tr}_{B/F}(ae'' \cdot be') \\ &= ab \cdot \mathrm{tr}_{B/F}(e^*) \\ &= ab. \end{aligned}$$

□

Thus we obtain a commutative diagram

$$\begin{array}{ccc} \mathrm{GU}(V) \times \mathrm{GU}(W) & \longrightarrow & \mathrm{GSp}(\mathbb{V}) \\ \downarrow & & \| \\ \mathrm{GO}(V^\dagger) \times \mathrm{GSp}(W^\dagger) & \longrightarrow & \mathrm{GSp}(\mathbb{V}) \end{array}.$$

Let $W^\dagger = X + Y$ be a complete polarization given by

$$X = Fe, \quad Y = Fe'.$$

Put

$$\mathbb{X}' = V^\dagger \otimes_F X, \quad \mathbb{Y}' = V^\dagger \otimes_F Y.$$

Then we have a complete polarization $\mathbb{V} = \mathbb{X}' + \mathbb{Y}'$. Put

$$s'(h) := \gamma^{j(h)}$$

for $h \in \mathrm{GSp}(W^\dagger)$, where

$$\gamma = \begin{cases} 1 & \text{if } B_1 \text{ and } B_2 \text{ are split,} \\ -1 & \text{if } B_1 \text{ and } B_2 \text{ are ramified,} \end{cases}$$

and

$$j(h) = \begin{cases} 0 & \text{if } \mathfrak{i}(h) = \left(\begin{smallmatrix} * & * \\ 0 & * \end{smallmatrix}\right), \\ 1 & \text{otherwise.} \end{cases}$$

LEMMA C.2.3. *We have*

$$z_{\mathbb{Y}'}(h,h') = s'(hh') \cdot s'(h)^{-1} \cdot s'(h')^{-1}$$

for $h, h' \in \mathrm{GSp}(W^\dagger)$.

PROOF. The lemma follows from [**40**, Theorem 3.1, case 1_+] and [**66**, Proposition 2.1]. We shall give a proof for the sake of completeness.

Recall that $\dim_F V^\dagger = 4$ and $\det V^\dagger = 1$. By [**40**, Theorem 3.1, case 1_+], we have

$$z_{\mathbb{Y}'}(h,h') = s'(hh') \cdot s'(h)^{-1} \cdot s'(h')^{-1} \tag{C.1}$$

for $h, h' \in \mathrm{Sp}(W^\dagger)$.

Let $g, g' \in \mathrm{GSp}(W^\dagger)$. For $\nu \in F^\times$, put

$$d(\nu) = \begin{pmatrix} 1 & \\ & \nu \end{pmatrix} \in \mathrm{GSp}(W^\dagger).$$

We write

$$g = h \cdot d(\nu), \quad g' = h' \cdot d(\nu')$$

with $h, h' \in \mathrm{Sp}(W^\dagger)$ and $\nu, \nu' \in F^\times$. Then we have

$$z_{\mathbb{Y}'}(g,g') = z_{\mathbb{Y}'}(h,h'') \cdot v_{\mathbb{Y}'}(h',\nu),$$

where

$$h'' = d(\nu) \cdot h' \cdot d(\nu)^{-1}.$$

By (C.1), we have

$$z_{\mathbb{Y}'}(h,h'') = s'(hh'') \cdot s'(h)^{-1} \cdot s'(h'')^{-1}.$$

We have $s'(h) = s'(g)$, and since $j(h'') = j(h')$, we have $s'(h'') = s'(h') = s'(g')$. Moreover, since $gg' = hh'' \cdot d(\nu\nu')$, we have $s'(hh'') = s'(gg')$. Thus we obtain

$$z_{\mathbb{Y}'}(h,h'') = s'(gg') \cdot s'(g)^{-1} \cdot s'(g')^{-1}.$$

By Lemma B.2.1, we have

$$v_{\mathbb{Y}'}(h',\nu) = (x_{\mathbb{Y}'}(h'),\nu)_F \cdot \gamma_F(\nu, \frac{1}{2}\psi)^{-j_{\mathbb{Y}'}(h')},$$

where $x_{\mathbb{Y}'}$ and $j_{\mathbb{Y}'}$ are as in §B.1 with respect to the complete polarization $\mathbb{V} = \mathbb{X}' + \mathbb{Y}'$. Since $\dim_F V^\dagger = 4$ and $\det V^\dagger = 1$, one can see that $x_{\mathbb{Y}'}(h') \equiv 1 \bmod (F^\times)^2$ and $j_{\mathbb{Y}'}(h') = 4 \cdot j(h')$. Hence we have

$$v_{\mathbb{Y}'}(h',\nu) = 1.$$

This completes the proof. □

LEMMA C.2.4. *We have*

$$z_{\mathbb{Y}'}(g,g') = 1$$

for $g, g' \in \mathrm{GO}(V^\dagger)^0$.

PROOF. For $g, g' \in \mathrm{GO}(V^\dagger)^0$, we have

$$z_{\mathbb{Y}'}(g,g') = z_{\mathbb{Y}'}(h,h'') \cdot v_{\mathbb{Y}'}(h',\nu),$$

where

$$h = g \cdot d_{\mathbb{Y}'}(\nu)^{-1}, \qquad h' = g' \cdot d_{\mathbb{Y}'}(\nu')^{-1}, \qquad h'' = d_{\mathbb{Y}'}(\nu) \cdot h' \cdot d_{\mathbb{Y}'}(\nu)^{-1},$$
$$\nu = \nu(g), \qquad \nu' = \nu(g').$$

We have $h, h' \in P_{\mathbb{Y}'}$ and $z_{\mathbb{Y}'}(h, h'') = 1$. Since $g' \in \mathrm{GO}(V^\dagger)^0$, we have

$$x_{\mathbb{Y}'}(h') \equiv \det g' \equiv 1 \bmod (F^\times)^2,$$

so that $v_{\mathbb{Y}'}(h', \nu) = 1$ by Lemma B.2.1. This completes the proof. □

LEMMA C.2.5. *We have*

$$z_{\mathbb{Y}'}(g, h) = z_{\mathbb{Y}'}(h, g) = 1$$

for $g \in \mathrm{GO}(V^\dagger)^0$ *and* $h \in \mathrm{GSp}(W^\dagger)$.

PROOF. See [**2**, Proposition 2.2.A]. We shall give a proof for the sake of completeness.

For $g \in \mathrm{GO}(V^\dagger)^0$ and $h \in \mathrm{GSp}(W^\dagger)$, we have

$$z_{\mathbb{Y}'}(g, h) = z_{\mathbb{Y}'}(g', h'') \cdot v_{\mathbb{Y}'}(h', \nu), \quad z_{\mathbb{Y}'}(h, g) = z_{\mathbb{Y}'}(h', g'') \cdot v_{\mathbb{Y}'}(g', \nu'),$$

where

$$\begin{aligned} g' &= g \cdot d_{\mathbb{Y}'}(\nu)^{-1}, & g'' &= d_{\mathbb{Y}'}(\nu') \cdot g' \cdot d_{\mathbb{Y}'}(\nu')^{-1}, & \nu &= \nu(g), \\ h' &= h \cdot d(\nu')^{-1}, & h'' &= d(\nu) \cdot h' \cdot d(\nu)^{-1}, & \nu' &= \nu(h). \end{aligned}$$

Since $g', g'' \in P_{\mathbb{Y}'}$, we have $z_{\mathbb{Y}'}(g', h'') = z_{\mathbb{Y}'}(h', g'') = 1$. As in the proof of Lemma C.2.3, we have $v_{\mathbb{Y}'}(h', \nu) = 1$. As in the proof of Lemma C.2.4, we have $v_{\mathbb{Y}'}(g', \nu') = 1$. This completes the proof. □

We define a map $s' : \mathrm{GO}(V^\dagger)^0 \times \mathrm{GSp}(W^\dagger) \to \mathbb{C}^1$ by

$$s'(\mathbf{g}) = \gamma^{j(h)}$$

for $\mathbf{g} = (g, h) \in \mathrm{GO}(V^\dagger)^0 \times \mathrm{GSp}(W^\dagger)$. By Lemmas C.2.3, C.2.4, C.2.5, we see that

$$z_{\mathbb{Y}'}(\mathbf{g}, \mathbf{g}') = s'(\mathbf{g}\mathbf{g}') \cdot s'(\mathbf{g})^{-1} \cdot s'(\mathbf{g}')^{-1}$$

for $\mathbf{g}, \mathbf{g}' \in \mathrm{GO}(V^\dagger)^0 \times \mathrm{GSp}(W^\dagger)$.

Recall that we have two complete polarizations $\mathbb{V} = \mathbb{X} + \mathbb{Y} = \mathbb{X}' + \mathbb{Y}'$, where

$$\begin{aligned} \mathbb{X} &= F\mathbf{e}_1 + F\mathbf{e}_2 + F\mathbf{e}_3 + F\mathbf{e}_4, & \mathbb{Y} &= F\mathbf{e}_1^* + F\mathbf{e}_2^* + F\mathbf{e}_3^* + F\mathbf{e}_4^*, \\ \mathbb{X}' &= F\mathbf{e}_1 e + F\mathbf{e}_1 e'' + F\mathbf{e}_2 e + F\mathbf{e}_2 e'', & \mathbb{Y}' &= F\mathbf{e}_1 e' + F\mathbf{e}_1 e^* + F\mathbf{e}_2 e' + F\mathbf{e}_2 e^*. \end{aligned}$$

Fix $\mathbf{h}_0 \in \mathrm{Sp}(\mathbb{V})$ such that $\mathbb{X}' = \mathbb{X}\mathbf{h}_0$ and $\mathbb{Y}' = \mathbb{Y}\mathbf{h}_0$, and put

$$s(\mathbf{g}) := s'(\mathbf{g}) \cdot \mu(\mathbf{g}),$$

where

$$\mu(\mathbf{g}) := z_{\mathbb{Y}}(\mathbf{h}_0 \mathbf{g} \mathbf{h}_0^{-1}, \mathbf{h}_0) \cdot z_{\mathbb{Y}}(\mathbf{h}_0, \mathbf{g})^{-1}$$

for $\mathbf{g} \in \mathrm{GU}(V)^0 \times \mathrm{GU}(W)$. Then we have

$$z_{\mathbb{Y}}(\mathbf{g}, \mathbf{g}') = s(\mathbf{g}\mathbf{g}') \cdot s(\mathbf{g})^{-1} \cdot s(\mathbf{g}')^{-1}$$

for $\mathbf{g}, \mathbf{g}' \in \mathrm{GU}(V)^0 \times \mathrm{GU}(W)$.

C.2.1. The case $u \in (F^\times)^2$. Choose $t \in F^\times$ such that $u = t^2$. We take an isomorphism $\mathrm{i} : B \to \mathrm{M}_2(F)$ determined by

$$\mathrm{i}(1) = \begin{pmatrix} 1 & \\ & 1 \end{pmatrix}, \quad \mathrm{i}(\mathbf{i}) = \begin{pmatrix} t & \\ & -t \end{pmatrix}, \quad \mathrm{i}(\mathbf{j}) = \begin{pmatrix} & 1 \\ J & \end{pmatrix}, \quad \mathrm{i}(\mathbf{ij}) = \begin{pmatrix} & t \\ -tJ & \end{pmatrix}.$$

Then we have

$$e = \frac{1}{2} + \frac{1}{2t}\mathbf{i}, \quad e' = \frac{1}{2}\mathbf{j} + \frac{1}{2t}\mathbf{ij}, \quad e'' = \frac{1}{2J}\mathbf{j} - \frac{1}{2tJ}\mathbf{ij}, \quad e^* = \frac{1}{2} - \frac{1}{2t}\mathbf{i}.$$

Put

$$\mathbf{h}_0 = \begin{pmatrix} -\frac{1}{2t} & & & & -\frac{1}{2} & & & \\ & \frac{1}{2tJ_1} & & & & -\frac{1}{2} & & \\ & & \frac{1}{2tJ_2} & & & & -\frac{1}{2} & \\ & & & -\frac{1}{2tJ} & & & & -\frac{1}{2} \\ 1 & & & & -t & & & \\ & 1 & & & & tJ_1 & & \\ & & 1 & & & & tJ_2 & \\ & & & 1 & & & & -tJ \end{pmatrix} \in \mathrm{Sp}(\mathbb{V}).$$

Then we have

$$\begin{bmatrix} -\frac{1}{t}\mathbf{e}_1 e \\ \frac{1}{tJ_1}\mathbf{e}_2 e \\ \frac{1}{t}\mathbf{e}_2 e'' \\ -\frac{1}{t}\mathbf{e}_1 e'' \\ 2\mathbf{e}_1 e^* \\ 2\mathbf{e}_2 e^* \\ \frac{2}{J_1}\mathbf{e}_2 e' \\ 2\mathbf{e}_1 e' \end{bmatrix} = \mathbf{h}_0 \begin{bmatrix} \mathbf{e}_1 \\ \mathbf{e}_2 \\ \mathbf{e}_3 \\ \mathbf{e}_4 \\ \mathbf{e}_1^* \\ \mathbf{e}_2^* \\ \mathbf{e}_3^* \\ \mathbf{e}_4^* \end{bmatrix},$$

and hence $\mathbb{X}' = \mathbb{X}\mathbf{h}_0$ and $\mathbb{Y}' = \mathbb{Y}\mathbf{h}_0$.

LEMMA C.2.6. *Let $\mathbf{g}_i := \boldsymbol{\alpha}_i^{-1} \in \mathrm{GU}(V)^0$ with $\boldsymbol{\alpha}_i = a_i + b_i\mathbf{i} + c_i\mathbf{j}_i + d_i\mathbf{ij}_i \in B_i^\times$. Then we have*

$$\mu(\mathbf{g}_i) = \begin{cases} 1 & \text{if } b_i = d_i = 0, \\ \gamma_F(J_j, \frac{1}{2}\psi) \cdot ((a_i b_i + c_i d_i J_i)\nu_i J_i, J_j)_F & \text{if } (b_i, d_i) \neq (0,0),\ b_i^2 - d_i^2 J_i = 0, \\ (-(b_i^2 - d_i^2 J_i)\nu_i J_i, J_j)_F & \text{if } (b_i, d_i) \neq (0,0),\ b_i^2 - d_i^2 J_i \neq 0, \end{cases}$$

where $\nu_i = \nu(\boldsymbol{\alpha}_i)$ and $\{i, j\} = \{1, 2\}$.

PROOF. We only consider the case $i = 1$; the other case is similar. Put $\mathbf{d} := d_{\mathbb{Y}}(\nu_1) \in \mathrm{GSp}(\mathbb{V})$. We have

$$z_{\mathbb{Y}}(\mathbf{h}_0 \mathbf{g}_1 \mathbf{h}_0^{-1}, \mathbf{h}_0) = z_{\mathbb{Y}}(\mathbf{h}_0 \mathbf{g}_1 \mathbf{h}_0^{-1} \cdot \mathbf{d}^{-1}, \mathbf{d} \cdot \mathbf{h}_0 \cdot \mathbf{d}^{-1}) \cdot v_{\mathbb{Y}}(\mathbf{h}_0, \nu_1).$$

Since $\mathbb{Y}'\mathbf{g}_1 = \mathbb{Y}'$, we have $\mathbf{h}_0 \mathbf{g}_1 \mathbf{h}_0^{-1} \cdot \mathbf{d}^{-1} \in P_{\mathbb{Y}}$ and hence

$$z_{\mathbb{Y}}(\mathbf{h}_0 \mathbf{g}_1 \mathbf{h}_0^{-1} \cdot \mathbf{d}^{-1}, \mathbf{d} \cdot \mathbf{h}_0 \cdot \mathbf{d}^{-1}) = 1.$$

We have $\mathbf{h}_0 = \mathbf{n}(\mathbf{b}_1) \cdot \tau_4 \cdot \mathbf{n}(\mathbf{b}_2)$, where

$$\mathbf{b}_1 = \frac{1}{2tJ} \cdot \begin{pmatrix} -J & & & \\ & J_2 & & \\ & & J_1 & \\ & & & -1 \end{pmatrix},$$

$$\mathbf{b}_2 = t \cdot \begin{pmatrix} -1 & & & \\ & J_1 & & \\ & & J_2 & \\ & & & -J \end{pmatrix},$$

so that $x_{\mathbb{Y}}(\mathbf{h}_0) \equiv 1 \bmod (F^\times)^2$ and $j_{\mathbb{Y}}(\mathbf{h}_0) = 4$. Hence we have $v_{\mathbb{Y}}(\mathbf{h}_0, \nu_1) = 1$. Thus we obtain

$$z_{\mathbb{Y}}(\mathbf{h}_0 \mathbf{g}_1 \mathbf{h}_0^{-1}, \mathbf{h}_0) = 1.$$

Now we compute $z_{\mathbb{Y}}(\mathbf{h}_0, \mathbf{g}_1)$. We have

$$z_{\mathbb{Y}}(\mathbf{h}_0, \mathbf{g}_1) = z_{\mathbb{Y}}(\mathbf{h}_0, \mathbf{g}_1 \cdot \mathbf{d}^{-1}).$$

First assume that $b_1 = d_1 = 0$. Then we have $\mathbf{g}_1 \cdot \mathbf{d}^{-1} \in P_{\mathbb{Y}}$ and hence

$$z_{\mathbb{Y}}(\mathbf{h}_0, \mathbf{g}_1 \cdot \mathbf{d}^{-1}) = 1.$$

Next assume that $(b_1, d_1) \neq (0,0)$ and $b_1^2 - d_1^2 J_1 = 0$. Then we have $b_1 \neq 0$, $d_1 \neq 0$, and $\nu_1 = a_1^2 - c_1^2 J_1 \neq 0$. Since

$$\begin{aligned}(a_1 d_1 - b_1 c_1) \cdot (a_1 b_1 + c_1 d_1 J_1) &= a_1^2 b_1 d_1 + a_1 c_1 d_1^2 J_1 - a_1 b_1^2 c_1 - b_1 c_1^2 d_1 J_1 \\ &= a_1^2 b_1 d_1 - b_1 c_1^2 d_1 J_1 \\ &= \nu_1 b_1 d_1 \\ &\neq 0,\end{aligned}$$

we have $a_1 d_1 - b_1 c_1 \neq 0$ and $a_1 b_1 + c_1 d_1 J_1 \neq 0$. We have $\mathbf{g}_1 \cdot \mathbf{d}^{-1} \in \mathbf{m}(\mathbf{a}_1) \cdot \mathbf{n}(\mathbf{b}_3) \cdot \tau_2 \cdot P_{\mathbb{Y}}$, where

$$\mathbf{a}_1 = \begin{pmatrix} b_1 & & & \\ d_1 J_1 & & 1 & \\ & b_1 & & \\ & d_1 J_1 & & 1 \end{pmatrix},$$

$$\mathbf{b}_3 = \frac{a_1 d_1 - b_1 c_1}{b_1 d_1} \cdot \begin{pmatrix} 0 & & & \\ & 0 & & \\ & & J_1 & \\ & & & -J \end{pmatrix}.$$

Hence we have

$$\begin{aligned}z_{\mathbb{Y}}(\mathbf{h}_0, \mathbf{g}_1 \cdot \mathbf{d}^{-1}) &= z_{\mathbb{Y}}(\tau_4 \cdot \mathbf{n}(\mathbf{b}_2), \mathbf{m}(\mathbf{a}_1) \cdot \mathbf{n}(\mathbf{b}_3) \cdot \tau_2) \\ &= z_{\mathbb{Y}}(\tau_4 \cdot \mathbf{m}(\mathbf{a}_1), \mathbf{m}(\mathbf{a}_1)^{-1} \cdot \mathbf{n}(\mathbf{b}_2) \cdot \mathbf{m}(\mathbf{a}_1) \cdot \mathbf{n}(\mathbf{b}_3) \cdot \tau_2) \\ &= z_{\mathbb{Y}}(\tau_4, \mathbf{n}(\mathbf{b}_4 + \mathbf{b}_3) \cdot \tau_2),\end{aligned}$$

where

$$\mathbf{b}_4 = \mathbf{a}_1^{-1} \cdot \mathbf{b}_2 \cdot {}^t\mathbf{a}_1^{-1} = \frac{t}{b_1^2} \cdot \begin{pmatrix} -1 & & d_1 J_1 & \\ & J_2 & & -d_1 J \\ d_1 J_1 & & & \\ & -d_1 J & & \end{pmatrix}.$$

Since $\tau_2^{-1} \cdot \mathbf{n}(\mathbf{b}_4) \cdot \tau_2 \in P_{\mathbb{Y}}$, we have

$$z_{\mathbb{Y}}(\mathbf{h}_0, \mathbf{g}_1 \cdot \mathbf{d}^{-1}) = z_{\mathbb{Y}}(\tau_4, \mathbf{n}(\mathbf{b}_3) \cdot \tau_2) = \gamma_F(\frac{1}{2}\psi \circ q_1),$$

where q_1 is a non-degenerate symmetric bilinear form associated to

$$\frac{a_1 d_1 - b_1 c_1}{b_1 d_1} \cdot \begin{pmatrix} J_1 & \\ & -J \end{pmatrix}.$$

Since $\det q_1 \equiv -J_2 \bmod (F^\times)^2$ and $h_F(q_1) = (\frac{a_1 d_1 - b_1 c_1}{b_1 d_1} \cdot J_1, J_2)_F$, we have

$$\begin{aligned}
\gamma_F(\frac{1}{2}\psi \circ q_1) &= \gamma_F(\frac{1}{2}\psi)^2 \cdot \gamma_F(-J_2, \frac{1}{2}\psi) \cdot (\frac{a_1 d_1 - b_1 c_1}{b_1 d_1} \cdot J_1, J_2)_F \\
&= \gamma_F(J_2, \frac{1}{2}\psi)^{-1} \cdot (\frac{a_1 d_1 - b_1 c_1}{b_1 d_1} \cdot J_1, J_2)_F \\
&= \gamma_F(J_2, \frac{1}{2}\psi)^{-1} \cdot (\frac{\nu_1}{a_1 b_1 + c_1 d_1 J_1} \cdot J_1, J_2)_F \\
&= \gamma_F(J_2, \frac{1}{2}\psi)^{-1} \cdot ((a_1 b_1 + c_1 d_1 J_1)\nu_1 J_1, J_2)_F.
\end{aligned}$$

Finally assume that $(b_1, d_1) \neq (0, 0)$ and $b_1^2 - d_1^2 J_1 \neq 0$. We have $\mathbf{g}_1 \cdot \mathbf{d}^{-1} \in \mathbf{n}(\mathbf{b}_5) \cdot \tau_4 \cdot P_{\mathbb{Y}}$, where

$$\mathbf{b}_5 = \begin{pmatrix} \mathbf{q} & \\ & -J_2 \cdot \mathbf{q} \end{pmatrix}$$

with

$$\mathbf{q} = \frac{1}{b_1^2 - d_1^2 J_1} \cdot \begin{pmatrix} a_1 b_1 + c_1 d_1 J_1 & (a_1 d_1 + b_1 c_1) J_1 \\ (a_1 d_1 + b_1 c_1) J_1 & (a_1 b_1 + c_1 d_1 J_1) J_1 \end{pmatrix}.$$

Hence we have

$$\begin{aligned}
z_{\mathbb{Y}}(\mathbf{h}_0, \mathbf{g}_1 \cdot \mathbf{d}^{-1}) &= z_{\mathbb{Y}}(\tau_4 \cdot \mathbf{n}(\mathbf{b}_2), \mathbf{n}(\mathbf{b}_5) \cdot \tau_4) \\
&= z_{\mathbb{Y}}(\tau_4, \mathbf{n}(\mathbf{b}_2) \cdot \mathbf{n}(\mathbf{b}_5) \cdot \tau_4) \\
&= \gamma_F(\frac{1}{2}\psi \circ q_2),
\end{aligned}$$

where q_2 is a non-degenerate symmetric bilinear form associated to $\mathbf{b}_2 + \mathbf{b}_5$. We have

$$\mathbf{b}_2 + \mathbf{b}_5 = \begin{pmatrix} \mathbf{q}' & \\ & -J_2 \cdot \mathbf{q}' \end{pmatrix},$$

where

$$\mathbf{q}' = t \cdot \begin{pmatrix} -1 & \\ & J_1 \end{pmatrix} + \frac{1}{b_1^2 - d_1^2 J_1} \cdot \begin{pmatrix} a_1 b_1 + c_1 d_1 J_1 & (a_1 d_1 + b_1 c_1) J_1 \\ (a_1 d_1 + b_1 c_1) J_1 & (a_1 b_1 + c_1 d_1 J_1) J_1 \end{pmatrix}.$$

Since

$$\det \mathbf{q}' = \frac{\nu_1 J_1}{b_1^2 - d_1^2 J_1} \neq 0,$$

we have $\det q_2 \equiv 1 \bmod (F^\times)^2$ and

$$\begin{aligned}
h_F(q_2) &= (\det \mathbf{q}', J_2)_F \cdot (-1, -J_2)_F \\
&= (-\frac{\nu_1 J_1}{b_1^2 - d_1^2 J_1}, J_2)_F \cdot (-1, -1)_F.
\end{aligned}$$

Hence we have

$$\begin{aligned}\gamma_F(\frac{1}{2}\psi \circ q_2) &= \gamma_F(\frac{1}{2}\psi)^4 \cdot (-\frac{\nu_1 J_1}{b_1^2 - d_1^2 J_1}, J_2)_F \cdot (-1,-1)_F \\ &= (-(b_1^2 - d_1^2 J_1)\nu_1 J_1, J_2)_F.\end{aligned}$$

This completes the proof. □

LEMMA C.2.7. *Let* $\mathbf{g} := \boldsymbol{\alpha} \in \mathrm{GU}(W)$ *with* $\boldsymbol{\alpha} = a + b\mathbf{i} + c\mathbf{j} + d\mathbf{ij} \in B^\times$. *Then we have*

$$\mu(\mathbf{g}) = \begin{cases} (\nu, J_1)_F & \text{if } b = d = 0, \\ \gamma_F(J_1, \frac{1}{2}\psi) \cdot (ab - cdJ, J_1)_F & \text{if } (b,d) \neq (0,0),\ b^2 - d^2 J = 0, \\ (-(b^2 - d^2 J)J, J_1)_F & \text{if } (b,d) \neq (0,0),\ b^2 - d^2 J \neq 0, \end{cases}$$

where $\nu = \nu(\boldsymbol{\alpha})$.

PROOF. Put $\mathbf{d} := d_{\mathbb{Y}}(\nu) \in \mathrm{GSp}(\mathbb{V})$. We have

$$z_{\mathbb{Y}}(\mathbf{h}_0 \mathbf{g} \mathbf{h}_0^{-1}, \mathbf{h}_0) = z_{\mathbb{Y}}(\mathbf{h}_0 \mathbf{g} \mathbf{h}_0^{-1} \cdot \mathbf{d}^{-1}, \mathbf{d} \cdot \mathbf{h}_0 \cdot \mathbf{d}^{-1}) \cdot v_{\mathbb{Y}}(\mathbf{h}_0, \nu).$$

As in the proof of Lemma C.2.6, we have $v_{\mathbb{Y}}(\mathbf{h}_0, \nu) = 1$. We have

$$\mathbf{h}_0 \mathbf{g} \mathbf{h}_0^{-1} = \begin{pmatrix} (a+bt) \cdot \mathbf{1}_4 & \frac{c+dt}{2t} \cdot \mathbf{y} \\ 2(c-dt)tJ \cdot \mathbf{y}^{-1} & (a-bt) \cdot \mathbf{1}_4 \end{pmatrix},$$

where

$$\mathbf{y} = \begin{pmatrix} & & & -1 \\ & & 1 & \\ & 1 & & \\ -1 & & & \end{pmatrix}.$$

If $c - dt = 0$, then we have $\mathbf{h}_0 \mathbf{g} \mathbf{h}_0^{-1} \cdot \mathbf{d}^{-1} \in P_{\mathbb{Y}}$ and hence $z_{\mathbb{Y}}(\mathbf{h}_0 \mathbf{g} \mathbf{h}_0^{-1} \cdot \mathbf{d}^{-1}, \mathbf{d} \cdot \mathbf{h}_0 \cdot \mathbf{d}^{-1}) = 1$. If $c - dt \neq 0$, then we have $\mathbf{h}_0 \mathbf{g} \mathbf{h}_0^{-1} \cdot \mathbf{d}^{-1} \in P_{\mathbb{Y}} \cdot \tau_4 \cdot \mathbf{n}(\mathbf{b}_6)$, where

$$\mathbf{b}_6 = \frac{a - bt}{2\nu t J (c - dt)} \cdot \mathbf{y}.$$

We have $\mathbf{d} \cdot \mathbf{h}_0 \cdot \mathbf{d}^{-1} \in \mathbf{n}(\nu^{-1} \cdot \mathbf{b}_1) \cdot \tau_4 \cdot P_{\mathbb{Y}}$, where $\mathbf{b}_1$ is as in the proof of Lemma C.2.6. Hence we have

$$\begin{aligned} z_{\mathbb{Y}}(\mathbf{h}_0 \mathbf{g} \mathbf{h}_0^{-1} \cdot \mathbf{d}^{-1}, \mathbf{d} \cdot \mathbf{h}_0 \cdot \mathbf{d}^{-1}) &= z_{\mathbb{Y}}(\tau_4 \cdot \mathbf{n}(\mathbf{b}_6), \mathbf{n}(\nu^{-1} \cdot \mathbf{b}_1) \cdot \tau_4) \\ &= z_{\mathbb{Y}}(\tau_4, \mathbf{n}(\mathbf{b}_7) \cdot \tau_4), \end{aligned}$$

where $\mathbf{b}_7 = \nu^{-1} \cdot \mathbf{b}_1 + \mathbf{b}_6$. Put $r = \frac{a-bt}{c-dt}$. We have

$$\mathbf{b}_7 = \frac{1}{2\nu t J} \cdot \begin{pmatrix} -J & & & -r \\ & J_2 & r & \\ & r & J_1 & \\ -r & & & -1 \end{pmatrix} = \mathbf{a}_2 \cdot \mathbf{b}_8 \cdot {}^t\mathbf{a}_2,$$

where

$$\mathbf{a}_2 = \begin{pmatrix} 1 & & & \\ & & \frac{r}{J_1} & 1 \\ & & 1 & \\ \frac{r}{J} & 1 & & \end{pmatrix},$$

$$\mathbf{b}_8 = \frac{1}{2\nu t J} \cdot \begin{pmatrix} -J & & & \\ & \frac{r^2}{J} - 1 & & \\ & & J_1 & \\ & & & J_2 - \frac{r^2}{J_1} \end{pmatrix},$$

and hence

$$\begin{aligned} z_{\mathbb{Y}}(\tau_4, \mathbf{n}(\mathbf{b}_7) \cdot \tau_4) &= z_{\mathbb{Y}}(\tau_4, \mathbf{m}(\mathbf{a}_2) \cdot \mathbf{n}(\mathbf{b}_8) \cdot \mathbf{m}(\mathbf{a}_2^{-1}) \cdot \tau_4) \\ &= z_{\mathbb{Y}}(\tau_4, \mathbf{n}(\mathbf{b}_8) \cdot \tau_4). \end{aligned}$$

If $J - r^2 = 0$, then we have $z_{\mathbb{Y}}(\tau_4, \mathbf{n}(\mathbf{b}_8) \cdot \tau_4) = \gamma_F(\frac{1}{2}\psi \circ q_3)$, where q_3 is a non-degenerate symmetric bilinear form associated to

$$\frac{1}{2\nu t J} \cdot \begin{pmatrix} -J & \\ & J_1 \end{pmatrix}.$$

We have $\det q_3 \equiv -J_2 \bmod (F^\times)^2$ and

$$h_F(q_3) = (-\frac{1}{2\nu t}, \frac{1}{2\nu t J_2})_F = (-2\nu t, J_2)_F.$$

Hence we have

$$\gamma_F(\frac{1}{2}\psi \circ q_3) = \gamma_F(\frac{1}{2}\psi)^2 \cdot \gamma_F(-J_2, \frac{1}{2}\psi) \cdot (-2\nu t, J_2)_F = \gamma_F(J_2, \frac{1}{2}\psi) \cdot (2\nu t, J_2)_F.$$

Note that $\gamma_F(J_1, \frac{1}{2}\psi) = \gamma_F(J_2, \frac{1}{2}\psi)$ and $(2\nu t, J_1)_F = (2\nu t, J_2)_F$ since $J = r^2 \in (F^\times)^2$. If $J - r^2 \neq 0$, then we have $z_{\mathbb{Y}}(\tau_4, \mathbf{n}(\mathbf{b}_8) \cdot \tau_4) = \gamma_F(\frac{1}{2}\psi \circ q_4)$, where q_4 is a non-degenerate symmetric bilinear form associated to $\mathbf{b}_8$. We have $\det q_4 \equiv 1 \bmod (F^\times)^2$ and

$$\begin{aligned} h_F(q_4) &= (\det q_4, \frac{1}{2\nu t J})_F \cdot (-J, \frac{r^2}{J} - 1)_F \\ &\quad \times (-J(\frac{r^2}{J} - 1), J_1(J_2 - \frac{r^2}{J_1}))_F \cdot (J_1, J_2 - \frac{r^2}{J_1})_F \\ &= (-J, J - r^2)_F \cdot (-J, -\frac{1}{J})_F \\ &\quad \times (J - r^2, J - r^2)_F \cdot (J_1, J - r^2)_F \cdot (J_1, \frac{1}{J_1})_F \\ &= (-J, J - r^2)_F \cdot (-J, -1)_F \\ &\quad \times (J - r^2, -1)_F \cdot (J_1, J - r^2)_F \cdot (J_1, -1)_F \\ &= (JJ_1, J - r^2)_F \cdot (-JJ_1, -1)_F \\ &= (J_2, J - r^2)_F \cdot (J_2, -1)_F \cdot (-1, -1)_F \\ &= (J_2, r^2 - J)_F \cdot (-1, -1)_F. \end{aligned}$$

Note that $(J_1, r^2 - J)_F = (J_2, r^2 - J)_F$ since $(J, r^2 - J)_F = 1$. Hence we have

$$\begin{aligned}\gamma_F(\frac{1}{2}\psi \circ q_4) &= \gamma_F(\frac{1}{2}\psi)^4 \cdot (J_2, r^2 - J)_F \cdot (-1, -1)_F \\ &= (J_2, r^2 - J)_F.\end{aligned}$$

Thus we obtain

$$\begin{aligned}& z_{\mathbb{Y}}(\mathbf{h}_0 \mathbf{g} \mathbf{h}_0^{-1}, \mathbf{h}_0) \\ &= \begin{cases} 1 & \text{if } c - dt = 0, \\ \gamma_F(J_1, \frac{1}{2}\psi) \cdot (2\nu t, J_1)_F & \text{if } c - dt \neq 0,\ (a - bt)^2 - (c - dt)^2 J = 0, \\ ((a - bt)^2 - (c - dt)^2 J, J_1)_F & \text{if } c - dt \neq 0,\ (a - bt)^2 - (c - dt)^2 J \neq 0. \end{cases}\end{aligned}$$

Now we compute $z_{\mathbb{Y}}(\mathbf{h}_0, \mathbf{g})$. We have

$$z_{\mathbb{Y}}(\mathbf{h}_0, \mathbf{g}) = z_{\mathbb{Y}}(\mathbf{h}_0, \mathbf{g} \cdot \mathbf{d}^{-1}).$$

First assume that $b = d = 0$. Then we have $\mathbf{g} \cdot \mathbf{d}^{-1} \in P_{\mathbb{Y}}$ and hence

$$z_{\mathbb{Y}}(\mathbf{h}_0, \mathbf{g} \cdot \mathbf{d}^{-1}) = 1.$$

Next assume that $(b, d) \neq (0, 0)$ and $b^2 - d^2 J = 0$. Then we have $b \neq 0$ and $d \neq 0$. We have $\mathbf{g} \cdot \mathbf{d}^{-1} \in \mathbf{m}(\mathbf{a}_3) \cdot \mathbf{n}(\mathbf{b}_9) \cdot \tau_2 \cdot P_{\mathbb{Y}}$, where

$$\mathbf{a}_3 = \begin{pmatrix} b & & & \\ & b & & \\ & -dJ_2 & 1 & \\ -dJ & & & 1 \end{pmatrix},$$

$$\mathbf{b}_9 = \frac{ad + bc}{bd} \cdot \begin{pmatrix} 0 & & & \\ & 0 & & \\ & & -J_2 & \\ & & & J \end{pmatrix}.$$

Hence we have

$$\begin{aligned} z_{\mathbb{Y}}(\mathbf{h}_0, \mathbf{g} \cdot \mathbf{d}^{-1}) &= z_{\mathbb{Y}}(\tau_4 \cdot \mathbf{n}(\mathbf{b}_2), \mathbf{m}(\mathbf{a}_3) \cdot \mathbf{n}(\mathbf{b}_9) \cdot \tau_2) \\ &= z_{\mathbb{Y}}(\tau_4 \cdot \mathbf{m}(\mathbf{a}_3), \mathbf{m}(\mathbf{a}_3)^{-1} \cdot \mathbf{n}(\mathbf{b}_2) \cdot \mathbf{m}(\mathbf{a}_3) \cdot \mathbf{n}(\mathbf{b}_9) \cdot \tau_2) \\ &= z_{\mathbb{Y}}(\tau_4, \mathbf{n}(\mathbf{b}_{10} + \mathbf{b}_9) \cdot \tau_2), \end{aligned}$$

where $\mathbf{b}_2$ is as in the proof of Lemma C.2.6 and

$$\mathbf{b}_{10} = \mathbf{a}_3^{-1} \cdot \mathbf{b}_2 \cdot {}^t\mathbf{a}_3^{-1} = \frac{t}{b^2} \cdot \begin{pmatrix} -1 & & & -dJ \\ & J_1 & dJ & \\ & dJ & 2b^2 J_2 & \\ -dJ & & & -2b^2 J \end{pmatrix}.$$

Put

$$r' = 2t - \frac{ad + bc}{bd}.$$

We write $\mathbf{b}_9 + \mathbf{b}_{10} = \mathbf{b}_{11} + \mathbf{b}_{12}$, where

$$\mathbf{b}_{11} = \frac{t}{b^2} \cdot \begin{pmatrix} -1 & & & -dJ \\ & J_1 & dJ & \\ & dJ & & \\ -dJ & & & \end{pmatrix},$$

$$\mathbf{b}_{12} = r' \cdot \begin{pmatrix} 0 & & & \\ & 0 & & \\ & & J_2 & \\ & & & -J \end{pmatrix}.$$

Since $\tau_2^{-1} \cdot \mathbf{n}(\mathbf{b}_{11}) \cdot \tau_2 \in P_{\mathbb{Y}}$, we have

$$z_{\mathbb{Y}}(\mathbf{h}_0, \mathbf{g} \cdot \mathbf{d}^{-1}) = z_{\mathbb{Y}}(\tau_4, \mathbf{n}(\mathbf{b}_{12}) \cdot \tau_2).$$

If $r' = 0$, then we have $z_{\mathbb{Y}}(\tau_4, \mathbf{n}(\mathbf{b}_{12}) \cdot \tau_2) = z_{\mathbb{Y}}(\tau_4, \tau_2) = 1$. If $r' \neq 0$, then we have $z_{\mathbb{Y}}(\tau_4, \mathbf{n}(\mathbf{b}_{12}) \cdot \tau_2) = \gamma_F(\frac{1}{2}\psi \circ q_5)$, where q_5 is a non-degenerate symmetric bilinear form associated to

$$r' \cdot \begin{pmatrix} J_2 & \\ & -J \end{pmatrix}.$$

Since $\det q_5 \equiv -J_1 \bmod (F^\times)^2$ and $h_F(q_5) = (r' \cdot J_2, J_1)_F$, we have

$$\begin{aligned} \gamma_F(\frac{1}{2}\psi \circ q_5) &= \gamma_F(\frac{1}{2}\psi)^2 \cdot \gamma_F(-J_1, \frac{1}{2}\psi) \cdot (r' \cdot J_2, J_1)_F \\ &= \gamma_F(J_1, \frac{1}{2}\psi)^{-1} \cdot (r' \cdot J_2, J_1)_F. \end{aligned}$$

Finally assume that $(b, d) \neq (0, 0)$ and $b^2 - d^2 J \neq 0$. We have $\mathbf{g} \cdot \mathbf{d}^{-1} \in \mathbf{n}(\mathbf{b}_{13}) \cdot \tau_4 \cdot P_{\mathbb{Y}}$, where

$$\mathbf{b}_{13} = \frac{1}{b^2 - d^2 J} \cdot \begin{pmatrix} ab - cdJ & & & -(ad-bc)J \\ & -(ab - cdJ)J_1 & (ad-bc)J & \\ & (ad-bc)J & -(ab-cdJ)J_2 & \\ -(ad-bc)J & & & (ab-cdJ)J \end{pmatrix}.$$

Hence we have

$$\begin{aligned} z_{\mathbb{Y}}(\mathbf{h}_0, \mathbf{g} \cdot \mathbf{d}^{-1}) &= z_{\mathbb{Y}}(\tau_4 \cdot \mathbf{n}(\mathbf{b}_2), \mathbf{n}(\mathbf{b}_{13}) \cdot \tau_4) \\ &= z_{\mathbb{Y}}(\tau_4, \mathbf{n}(\mathbf{b}_2 + \mathbf{b}_{13}) \cdot \tau_4). \end{aligned}$$

Put

$$\begin{aligned} l &= ab - cdJ - (b^2 - d^2 J)t, \\ l' &= (ad - bc)J, \\ r'' &= \frac{l^2 J - l'^2}{(b^2 - d^2 J)^2} = \frac{((a - bt)^2 - (c - dt)^2 J)J}{b^2 - d^2 J}. \end{aligned}$$

We have

$$\mathbf{b}_2 + \mathbf{b}_{13} = \frac{1}{b^2 - d^2 J} \cdot \begin{pmatrix} l & & & -l' \\ & -lJ_1 & l' & \\ & l' & -lJ_2 & \\ -l' & & & lJ \end{pmatrix},$$

and if $l \neq 0$, then we have $\mathbf{b}_2 + \mathbf{b}_{13} = \mathbf{a}_4 \cdot \mathbf{b}_{14} \cdot {}^t\mathbf{a}_4$, where

$$\mathbf{a}_4 = \frac{1}{l} \cdot \begin{pmatrix} l & & & \\ & & & l \\ & & -\frac{1}{J_1} & -\frac{l'}{J_1} \\ -l' & 1 & & \end{pmatrix},$$

$$\mathbf{b}_{14} = \frac{1}{b^2 - d^2J} \cdot \begin{pmatrix} l & & & \\ & (l^2J - l'^2)l & & \\ & & -(l^2J - l'^2)lJ_1 & \\ & & & -lJ_1 \end{pmatrix}.$$

If $l = l' = 0$, then we have $z_{\mathbb{Y}}(\tau_4, \mathbf{n}(\mathbf{b}_2 + \mathbf{b}_{13}) \cdot \tau_4) = z_{\mathbb{Y}}(\tau_4, \tau_4) = 1$. If $(l, l') \neq (0, 0)$ and $r'' = 0$, then we have $l \neq 0$ and $l' \neq 0$, so that $z_{\mathbb{Y}}(\tau_4, \mathbf{n}(\mathbf{b}_2 + \mathbf{b}_{13}) \cdot \tau_4) = \gamma_F(\frac{1}{2}\psi \circ q_6)$, where q_6 is a non-degenerate symmetric bilinear form associated to

$$\frac{l}{b^2 - d^2J} \cdot \begin{pmatrix} 1 & \\ & -J_1 \end{pmatrix}.$$

We have $\det q_6 \equiv -J_1 \bmod (F^\times)^2$ and

$$h_F(q_6) = (\frac{l}{b^2 - d^2J}, J_1)_F = (\frac{ab - cdJ}{b^2 - d^2J} - t, J_1)_F.$$

Hence we have

$$\begin{aligned} \gamma_F(\frac{1}{2}\psi \circ q_6) &= \gamma_F(\frac{1}{2}\psi)^2 \cdot \gamma_F(-J_1, \frac{1}{2}\psi) \cdot (\frac{ab - cdJ}{b^2 - d^2J} - t, J_1)_F \\ &= \gamma_F(J_1, \frac{1}{2}\psi)^{-1} \cdot (\frac{ab - cdJ}{b^2 - d^2J} - t, J_1)_F. \end{aligned}$$

Note that $\gamma_F(J_1, \frac{1}{2}\psi) = \gamma_F(J_2, \frac{1}{2}\psi)$ and $(\frac{ab-cdJ}{b^2-d^2J} - t, J_1)_F = (\frac{ab-cdJ}{b^2-d^2J} - t, J_2)_F$ since $r'' = 0$ and hence $J \in (F^\times)^2$. If $r'' \neq 0$, then we have $z_{\mathbb{Y}}(\tau_4, \mathbf{n}(\mathbf{b}_2 + \mathbf{b}_{13}) \cdot \tau_4) = \gamma_F(\frac{1}{2}\psi \circ q_7)$, where q_7 is a non-degenerate symmetric bilinear form associated to $\mathbf{b}_2 + \mathbf{b}_{13}$. We have $\det q_7 \equiv 1 \bmod (F^\times)^2$. Also, we have

$$\begin{aligned} h_F(q_7) = &(\frac{l}{b^2 - d^2J}, \frac{(l^2J - l'^2)l}{b^2 - d^2J})_F \\ &\times (\frac{(l^2J - l'^2)l^2}{(b^2 - d^2J)^2}, \frac{(l^2J - l'^2)l^2J_1^2}{(b^2 - d^2J)^2})_F \cdot (-\frac{(l^2J - l'^2)lJ_1}{b^2 - d^2J}, -\frac{lJ_1}{b^2 - d^2J})_F \\ = &(\frac{l}{b^2 - d^2J}, -(l^2J - l'^2))_F \\ &\times (l^2J - l'^2, l^2J - l'^2)_F \cdot (-(l^2J - l'^2), -\frac{lJ_1}{b^2 - d^2J})_F \\ = &(-J_1, -(l^2J - l'^2))_F \cdot (-1, l^2J - l'^2)_F \\ = &(J_1, -(l^2J - l'^2))_F \cdot (-1, -1)_F \\ = &(J_1, -r'') \cdot (-1, -1)_F \end{aligned}$$

if $l \neq 0$, and

$$h_F(q_7) = (-1, -1)_F = (J_1, -r'') \cdot (-1, -1)_F$$

if $l = 0$. Hence we have

$$\gamma_F(\frac{1}{2}\psi \circ q_7) = \gamma_F(\frac{1}{2}\psi)^4 \cdot (J_1, -r'') \cdot (-1, -1)_F = (J_1, -r'').$$

Note that $(J_1, -r'') = (J_2, -r'')$ since $(J, -r'') = (J, l'^2 - l^2 J) = 1$. Thus we obtain

$$z_{\mathbb{Y}}(\mathbf{h}_0, \mathbf{g})$$
$$= \begin{cases} 1 & \text{if } b = d = 0, \\ 1 & \text{if } (b,d) \neq (0,0),\ b^2 - d^2 J = 0, \\ & \quad ad + bc - 2bdt = 0, \\ \gamma_F(J_1, \tfrac{1}{2}\psi)^{-1} \cdot ((2t - \tfrac{ad+bc}{bd}) \cdot J_2, J_1)_F & \text{if } (b,d) \neq (0,0),\ b^2 - d^2 J = 0, \\ & \quad ad + bc - 2bdt \neq 0, \\ 1 & \text{if } (b,d) \neq (0,0),\ b^2 - d^2 J \neq 0, \\ & \quad ab - cdJ - (b^2 - d^2 J)t = 0, \\ & \quad ad - bc = 0, \\ \gamma_F(J_1, \tfrac{1}{2}\psi)^{-1} \cdot (\tfrac{ab - cdJ}{b^2 - d^2 J} - t, J_1)_F. & \text{if } (b,d) \neq (0,0),\ b^2 - d^2 J \neq 0, \\ & \quad (ab - cdJ - (b^2 - d^2 J)t, ad - bc) \\ & \qquad \neq (0,0), \\ & \quad (a - bt)^2 - (c - dt)^2 J = 0, \\ (-\tfrac{((a-bt)^2 - (c-dt)^2 J)J}{b^2 - d^2 J}, J_1)_F & \text{if } (b,d) \neq (0,0),\ b^2 - d^2 J \neq 0, \\ & \quad (a - bt)^2 - (c - dt)^2 J \neq 0. \end{cases}$$

Now we compute $\mu(\mathbf{g}) = z_{\mathbb{Y}}(\mathbf{h}_0 \mathbf{g} \mathbf{h}_0^{-1}, \mathbf{h}_0) \cdot z_{\mathbb{Y}}(\mathbf{h}_0, \mathbf{g})^{-1}$. Recall that $u = t^2$ and $\nu = a^2 - b^2 u - c^2 J + d^2 u J \neq 0$. We have

$$(a - bt)^2 - (c - dt)^2 J = a^2 + b^2 u - c^2 J - d^2 u J - 2t(ab - cdJ).$$

First assume that $b = d = 0$. Then we have $c - dt = c$ and

$$(a - bt)^2 - (c - dt)^2 J = a^2 - c^2 J = \nu \neq 0.$$

Hence we have

$$\mu(\mathbf{g}) = 1 \cdot 1 = (\nu, J_1)_F$$

if $c = 0$, and

$$\mu(\mathbf{g}) = ((a - bt)^2 - (c - dt)^2 J, J_1)_F \cdot 1 = (\nu, J_1)_F$$

if $c \neq 0$. Next assume that $(b,d) \neq (0,0)$ and $b^2 - d^2 J = 0$. Then we have $b \neq 0$, $d \neq 0$, $\nu = a^2 - c^2 J \neq 0$, and $J \in (F^\times)^2$. Since $(a - bt)^2 - (c - dt)^2 J = a^2 - c^2 J - 2t(ab - cdJ)$ and

$$\begin{aligned} &((a - bt)^2 - (c - dt)^2 J) \cdot bd - (ad + bc - 2bdt) \cdot (ab - cdJ) \\ &= (a^2 - c^2 J - 2abt + 2cdtJ) \cdot bd \\ &\quad - (a^2 bd + ab^2 c - 2ab^2 dt - acd^2 J - bc^2 dJ + 2bcd^2 tJ) \\ &= -ab^2 c + acd^2 J \\ &= 0, \end{aligned}$$

we have

$$(a - bt)^2 - (c - dt)^2 J = 0 \Longleftrightarrow ad + bc - 2bdt = 0.$$

If $c - dt = 0$, then we have $\nu = a^2 - b^2u = (a+bt)(a-bt) \neq 0$, so that $(a-bt)^2 - (c-dt)^2J \neq 0$. Hence we have

$$\begin{aligned}\mu(\mathbf{g}) &= 1 \cdot \gamma_F(J_1, \frac{1}{2}\psi) \cdot ((2t - \frac{ad+bc}{bd}) \cdot J_2, J_1)_F \\ &= \gamma_F(J_1, \frac{1}{2}\psi) \cdot ((\frac{2c}{d} - \frac{ad+bc}{bd}) \cdot J_2, J_1)_F \\ &= \gamma_F(J_1, \frac{1}{2}\psi) \cdot (\frac{ad-bc}{bd}, J_1)_F \\ &= \gamma_F(J_1, \frac{1}{2}\psi) \cdot (\frac{abd-b^2c}{d}, J_1)_F \\ &= \gamma_F(J_1, \frac{1}{2}\psi) \cdot (ab - cdJ, J_1)_F.\end{aligned}$$

If $c - dt \neq 0$ and $(a-bt)^2 - (c-dt)^2J = 0$, then we have

$$\nu - 2t(ab - cdJ) = (a-bt)^2 - (c-dt)^2J = 0$$

and hence

$$\mu(\mathbf{g}) = \gamma_F(J_1, \frac{1}{2}\psi) \cdot (2\nu t, J_1)_F \cdot 1 = \gamma_F(J_1, \frac{1}{2}\psi) \cdot (ab - cdJ, J_1)_F.$$

If $c - dt \neq 0$ and $(a-bt)^2 - (c-dt)^2J \neq 0$, then we have

$$(a-bt)^2 - (c-dt)^2J = (\frac{ad+bd}{bd} - 2t) \cdot (ab - cdJ)$$

and hence

$$\begin{aligned}\mu(\mathbf{g}) &= ((a-bt)^2 - (c-dt)^2J, J_1)_F \cdot \gamma_F(J_1, \frac{1}{2}\psi) \cdot ((2t - \frac{ad+bc}{bd}) \cdot J_2, J_1)_F \\ &= \gamma_F(J_1, \frac{1}{2}\psi) \cdot (-(ab - cdJ) \cdot J_2, J_1)_F \\ &= \gamma_F(J_1, \frac{1}{2}\psi) \cdot (ab - cdJ, J_1)_F.\end{aligned}$$

Finally assume that $(b, d) \neq (0, 0)$ and $b^2 - d^2J \neq 0$. Recall that

$$(ab - cdJ - (b^2 - d^2J)t)^2 - (ad - bc)^2J = ((a-bt)^2 - (c-dt)^2J) \cdot (b^2 - d^2J).$$

If $c - dt = 0$, then we have $\nu = a^2 - b^2u = (a+bt)(a-bt) \neq 0$ and

$$(a-bt)^2 - (c-dt)^2J = (a-bt)^2 \neq 0.$$

Hence we have

$$\begin{aligned}\mu(\mathbf{g}) &= 1 \cdot (-\frac{((a-bt)^2 - (c-dt)^2J)J}{b^2 - d^2J}, J_1)_F \\ &= (-\frac{(a-bt)^2J}{b^2 - d^2J}, J_1)_F \\ &= (-(b^2 - d^2J)J, J_1)_F.\end{aligned}$$

If $c - dt \neq 0$ and $(a-bt)^2 - (c-dt)^2J = 0$, then we have

$$\nu + 2(b^2 - d^2J)u = a^2 + b^2u - c^2J - d^2uJ = 2(ab - cdJ)t,$$

so that

$$ab - cdJ - (b^2 - d^2J)t \neq 0.$$

Hence we have

$$\begin{aligned}
\mu(\mathbf{g}) &= \gamma_F(J_1, \frac{1}{2}\psi) \cdot (2\nu t, J_1)_F \cdot \gamma_F(J_1, \frac{1}{2}\psi) \cdot (\frac{ab - cdJ}{b^2 - d^2J} - t, J_1)_F \\
&= (-1, J_1)_F \cdot (\frac{2(ab - cdJ)t}{b^2 - d^2J} \cdot \nu - 2\nu u, J_1)_F \\
&= (-1, J_1)_F \cdot (\frac{\nu + 2(b^2 - d^2J)u}{b^2 - d^2J} \cdot \nu - 2\nu u, J_1)_F \\
&= (-1, J_1)_F \cdot (\frac{\nu^2}{b^2 - d^2J}, J_1)_F \\
&= (-(b^2 - d^2J), J_1)_F \\
&= (-(b^2 - d^2J)J, J_1)_F.
\end{aligned}$$

If $c - dt \neq 0$ and $(a - bt)^2 - (c - dt)^2J \neq 0$, then we have

$$\begin{aligned}
\mu(\mathbf{g}) &= ((a - bt)^2 - (c - dt)^2J, J_1)_F \cdot (-\frac{((a - bt)^2 - (c - dt)^2J)J}{b^2 - d^2J}, J_1)_F \\
&= (-\frac{J}{b^2 - d^2J}, J_1)_F \\
&= (-(b^2 - d^2J)J, J_1)_F.
\end{aligned}$$

This completes the proof. $\square$

C.2.2. The case $J \in (F^\times)^2$. Choose $t \in F^\times$ such that $J = t^2$. We take an isomorphism $\mathrm{i} : B \to \mathrm{M}_2(F)$ determined by

$$\mathrm{i}(1) = \begin{pmatrix} 1 & \\ & 1 \end{pmatrix}, \quad \mathrm{i}(\mathbf{i}) = \begin{pmatrix} & 1 \\ u & \end{pmatrix}, \quad \mathrm{i}(\mathbf{j}) = \begin{pmatrix} t & \\ & -t \end{pmatrix}, \quad \mathrm{i}(\mathbf{ij}) = \begin{pmatrix} & -t \\ tu & \end{pmatrix}.$$

Then we have

$$e = \frac{1}{2} + \frac{1}{2t}\mathbf{j}, \quad e' = \frac{1}{2}\mathbf{i} - \frac{1}{2t}\mathbf{ij}, \quad e'' = \frac{1}{2u}\mathbf{i} + \frac{1}{2tu}\mathbf{ij}, \quad e^* = \frac{1}{2} - \frac{1}{2t}\mathbf{j}.$$

Put

$$\mathbf{h}_0 = \begin{pmatrix}
\frac{1}{2} & & & \frac{t}{2J} & & & & \\
& \frac{1}{2} & \frac{t}{2J_2} & & & & & \\
& & & & -\frac{1}{2} & & & \frac{t}{2} \\
& & & & & -\frac{1}{2} & \frac{t}{2J_1} & \\
& & & & 1 & & & t \\
& & & & & 1 & \frac{t}{J_1} & \\
1 & & & -\frac{t}{J} & & & & \\
& 1 & -\frac{t}{J_2} & & & & &
\end{pmatrix} \in \mathrm{Sp}(\mathbb{V}).$$

Then we have

$$\begin{bmatrix} \mathbf{e}_1 e \\ \mathbf{e}_2 e \\ -\mathbf{e}_1 e'' \\ \frac{1}{J_1}\mathbf{e}_2 e'' \\ \frac{2}{u}\mathbf{e}_1 e' \\ -\frac{2}{uJ_1}\mathbf{e}_2 e' \\ 2\mathbf{e}_1 e^* \\ 2\mathbf{e}_2 e^* \end{bmatrix} = \mathbf{h}_0 \begin{bmatrix} \mathbf{e}_1 \\ \mathbf{e}_2 \\ \mathbf{e}_3 \\ \mathbf{e}_4 \\ \mathbf{e}_1^* \\ \mathbf{e}_2^* \\ \mathbf{e}_3^* \\ \mathbf{e}_4^* \end{bmatrix},$$

and hence $\mathbb{X}' = \mathbb{X}\mathbf{h}_0$ and $\mathbb{Y}' = \mathbb{Y}\mathbf{h}_0$.

LEMMA C.2.8. *Let* $\mathbf{g}_i := \boldsymbol{\alpha}_i^{-1} \in \mathrm{GU}(V)^0$ *with* $\boldsymbol{\alpha}_i = a_i + b_i\mathbf{i} + c_i\mathbf{j}_i + d_i\mathbf{ij}_i \in B_i^\times$. *Then we have*

$$\mu(\mathbf{g}_i) = \begin{cases} 1 & \text{if } b_i = d_i = 0, \\ \gamma_F(J_j, \frac{1}{2}\psi) \cdot ((a_i b_i + c_i d_i J_i)\nu_i J_i, J_j)_F & \text{if } (b_i, d_i) \neq (0,0),\ b_i^2 - d_i^2 J_i = 0, \\ (-(b_i^2 - d_i^2 J_i)\nu_i J_i, J_j)_F & \text{if } (b_i, d_i) \neq (0,0),\ b_i^2 - d_i^2 J_i \neq 0, \end{cases}$$

where $\nu_i = \nu(\boldsymbol{\alpha}_i)$ *and* $\{i, j\} = \{1, 2\}$.

PROOF. We only consider the case $i = 1$; the other case is similar. Note that $J_1 \equiv J_2 \bmod (F^\times)^2$ since $J \in (F^\times)^2$. Put $\mathbf{d} := d_{\mathbb{Y}}(\nu_1) \in \mathrm{GSp}(\mathbb{V})$. We have

$$z_{\mathbb{Y}}(\mathbf{h}_0\mathbf{g}_1\mathbf{h}_0^{-1}, \mathbf{h}_0) = z_{\mathbb{Y}}(\mathbf{h}_0\mathbf{g}_1\mathbf{h}_0^{-1} \cdot \mathbf{d}^{-1}, \mathbf{d} \cdot \mathbf{h}_0 \cdot \mathbf{d}^{-1}) \cdot v_{\mathbb{Y}}(\mathbf{h}_0, \nu_1).$$

Since $\mathbb{Y}'\mathbf{g}_1 = \mathbb{Y}'$, we have $\mathbf{h}_0\mathbf{g}_1\mathbf{h}_0^{-1} \cdot \mathbf{d}^{-1} \in P_{\mathbb{Y}}$ and hence

$$z_{\mathbb{Y}}(\mathbf{h}_0\mathbf{g}_1\mathbf{h}_0^{-1} \cdot \mathbf{d}^{-1}, \mathbf{d} \cdot \mathbf{h}_0 \cdot \mathbf{d}^{-1}) = 1.$$

We have $\mathbf{h}_0 = \mathbf{m}(\mathbf{a}_5) \cdot \mathbf{n}(\mathbf{b}_{15}) \cdot \tau_2 \cdot \mathbf{m}(\mathbf{a}_6)$, where

$$\mathbf{a}_5 = \begin{pmatrix} 1 & & & \\ & 1 & & \\ & & & -t \\ & & -\frac{t}{J_1} & \end{pmatrix},$$

$$\mathbf{b}_{15} = \frac{1}{2t} \cdot \begin{pmatrix} & & & 1 \\ & & J_1 & \\ & J_1 & & \\ 1 & & & \end{pmatrix},$$

$$\mathbf{a}_6 = \begin{pmatrix} 1 & & & \\ & 1 & & \\ & -\frac{t}{J_1} & 1 & \\ -t & & & 1 \end{pmatrix},$$

so that $x_{\mathbb{Y}}(\mathbf{h}_0) \equiv -J_1 \bmod (F^\times)^2$ and $j_{\mathbb{Y}}(\mathbf{h}_0) = 2$. Hence we have

$$v_{\mathbb{Y}}(\mathbf{h}_0, \nu_1) = (-J_1, \nu_1)_F \cdot \gamma_F(\nu_1, \frac{1}{2}\psi)^{-2} = (J_1, \nu_1)_F.$$

Thus we obtain

$$z_{\mathbb{Y}}(\mathbf{h}_0\mathbf{g}_1\mathbf{h}_0^{-1}, \mathbf{h}_0) = (J_1, \nu_1)_F = (J_2, \nu_1)_F.$$

Moreover, if $b_1 = d_1 = 0$, then we have

$$(J_1, \nu_1)_F = (J_1, a_1^2 - c_1^2 J_1)_F = 1.$$

Now we compute $z_{\mathbb{Y}}(\mathbf{h}_0, \mathbf{g}_1)$. We have

$$z_{\mathbb{Y}}(\mathbf{h}_0, \mathbf{g}_1) = z_{\mathbb{Y}}(\mathbf{h}_0, \mathbf{g}_1 \cdot \mathbf{d}^{-1}).$$

First assume that $b_1 = d_1 = 0$. Then we have $\mathbf{g}_1 \cdot \mathbf{d}^{-1} \in P_{\mathbb{Y}}$ and hence

$$z_{\mathbb{Y}}(\mathbf{h}_0, \mathbf{g}_1 \cdot \mathbf{d}^{-1}) = 1.$$

Next assume that $(b_1, d_1) \neq (0,0)$ and $b_1^2 - d_1^2 J_1 = 0$. Then we have $b_1 \neq 0$ and $d_1 \neq 0$. As in the proof of Lemma C.2.6, we have $a_1 b_1 + c_1 d_1 J_1 \neq 0$. We have

$\mathbf{g}_1 \cdot \mathbf{d}^{-1} \in \mathbf{m}(\mathbf{a}_1) \cdot \mathbf{n}(\mathbf{b}_3) \cdot \tau_2 \cdot P_{\mathbb{Y}}$, where $\mathbf{a}_1$ and $\mathbf{b}_3$ are as in the proof of Lemma C.2.6. Hence we have

$$\begin{aligned} z_{\mathbb{Y}}(\mathbf{h}_0, \mathbf{g}_1 \cdot \mathbf{d}^{-1}) &= z_{\mathbb{Y}}(\tau_2 \cdot \mathbf{m}(\mathbf{a}_6), \mathbf{m}(\mathbf{a}_1) \cdot \mathbf{n}(\mathbf{b}_3) \cdot \tau_2) \\ &= z_{\mathbb{Y}}(\tau_2, \mathbf{m}(\mathbf{a}_6) \cdot \mathbf{m}(\mathbf{a}_1) \cdot \mathbf{n}(\mathbf{b}_3) \cdot \tau_2). \end{aligned}$$

We have

$$\mathbf{m}(\mathbf{a}_6) \cdot \mathbf{m}(\mathbf{a}_1) \cdot \mathbf{n}(\mathbf{b}_3) \cdot \tau_2 \in \mathbf{m}(\mathbf{a}_7) \cdot \mathbf{n}(\mathbf{b}_{16}) \cdot \tau' \cdot P_{\mathbb{Y}},$$

where

$$\mathbf{a}_7 = \begin{pmatrix} b_1 & & & \\ d_1 J_1 & 1 & & \\ & & b_1 & \\ & & d_1 J_1 & 1 \end{pmatrix},$$

$$\mathbf{b}_{16} = \frac{(a_1 d_1 - b_1 c_1) J_1}{b_1} \cdot \begin{pmatrix} 0 & & & \\ & \frac{1}{d_1} & & \frac{t}{b_1} \\ & & 0 & \\ & \frac{t}{b_1} & & 0 \end{pmatrix},$$

$$\tau' = \begin{pmatrix} 1 & & & & & & & \\ & 0 & & & & -1 & & \\ & & 1 & & & & & \\ & & & 0 & & & & -1 \\ & & & & 1 & & & \\ & 1 & & & & 0 & & \\ & & & & & & 1 & \\ & & & 1 & & & & 0 \end{pmatrix}.$$

Hence we have

$$\begin{aligned} z_{\mathbb{Y}}(\tau_2, \mathbf{m}(\mathbf{a}_6) \cdot \mathbf{m}(\mathbf{a}_1) \cdot \mathbf{n}(\mathbf{b}_3) \cdot \tau_2) &= z_{\mathbb{Y}}(\tau_2, \mathbf{m}(\mathbf{a}_7) \cdot \mathbf{n}(\mathbf{b}_{16}) \cdot \tau') \\ &= z_{\mathbb{Y}}(\tau_2 \cdot \mathbf{m}(\mathbf{a}_7) \cdot \mathbf{n}(\mathbf{b}_{16}), \tau'). \end{aligned}$$

Since $\tau_2 \cdot \mathbf{m}(\mathbf{a}_7) \cdot \mathbf{n}(\mathbf{b}_{16}) \cdot \tau_2^{-1} \in P_{\mathbb{Y}}$, we have

$$z_{\mathbb{Y}}(\tau_2 \cdot \mathbf{m}(\mathbf{a}_7) \cdot \mathbf{n}(\mathbf{b}_{16}), \tau') = z_{\mathbb{Y}}(\tau_2, \tau') = 1.$$

On the other hand, since $J \in (F^\times)^2$ and $J_1 \in (F^\times)^2$, we have $\gamma_F(J_2, \frac{1}{2}\psi) = 1$ and

$$((a_1 b_1 + c_1 d_1 J_1) J_1, J_2)_F = 1.$$

Finally assume that $(b_1, d_1) \neq (0, 0)$ and $b_1^2 - d_1^2 J_1 \neq 0$. We have $\mathbf{g}_1 \cdot \mathbf{d}^{-1} \in \mathbf{n}(\mathbf{b}_5) \cdot \tau_4 \cdot P_{\mathbb{Y}}$, where $\mathbf{b}_5$ is as in the proof of Lemma C.2.6. Hence we have

$$\begin{aligned} z_{\mathbb{Y}}(\mathbf{h}_0, \mathbf{g}_1 \cdot \mathbf{d}^{-1}) &= z_{\mathbb{Y}}(\tau_2 \cdot \mathbf{m}(\mathbf{a}_6), \mathbf{n}(\mathbf{b}_5) \cdot \tau_4) \\ &= z_{\mathbb{Y}}(\tau_2 \cdot \mathbf{m}(\mathbf{a}_6), \mathbf{n}(\mathbf{b}_5) \cdot \mathbf{m}(\mathbf{a}_6)^{-1} \cdot \tau_4) \\ &= z_{\mathbb{Y}}(\tau_2 \cdot \mathbf{m}(\mathbf{a}_6) \cdot \mathbf{n}(\mathbf{b}_5) \cdot \mathbf{m}(\mathbf{a}_6)^{-1}, \tau_4). \end{aligned}$$

Since $\tau_2 \cdot \mathbf{m}(\mathbf{a}_6) \cdot \mathbf{n}(\mathbf{b}_5) \cdot \mathbf{m}(\mathbf{a}_6)^{-1} \cdot \tau_2^{-1} \in P_{\mathbb{Y}}$, we have

$$z_{\mathbb{Y}}(\mathbf{h}_0, \mathbf{g}_1 \cdot \mathbf{d}^{-1}) = z_{\mathbb{Y}}(\tau_2, \tau_4) = 1.$$

On the other hand, we have

$$(-(b_1^2 - d_1^2 J_1) J_1, J_2)_F = (d_1^2 J_1^2 - b_1^2 J_1, J_1)_F = 1.$$

This competes the proof. □

LEMMA C.2.9. *Let* $\mathbf{g} := \boldsymbol{\alpha} \in \mathrm{GU}(W)$ *with* $\boldsymbol{\alpha} = a + b\mathbf{i} + c\mathbf{j} + d\mathbf{ij} \in B^\times$. *Then we have*

$$\mu(\mathbf{g}) = \begin{cases} (\nu, J_1)_F & \text{if } b = d = 0, \\ \gamma_F(J_1, \tfrac{1}{2}\psi) \cdot (ab - cdJ, J_1)_F & \text{if } (b,d) \neq (0,0),\ b^2 - d^2 J = 0, \\ (-(b^2 - d^2 J)J, J_1)_F & \text{if } (b,d) \neq (0,0),\ b^2 - d^2 J \neq 0, \end{cases}$$
$$\times \begin{cases} 1 & \text{if } b + dt = 0, \\ (u, J_1)_F & \text{if } b + dt \neq 0, \end{cases}$$

where $\nu = \nu(\boldsymbol{\alpha})$.

PROOF. Put $\mathbf{d} := d_{\mathbb{Y}}(\nu) \in \mathrm{GSp}(\mathbb{V})$. We have

$$z_{\mathbb{Y}}(\mathbf{h}_0\mathbf{g}\mathbf{h}_0^{-1}, \mathbf{h}_0) = z_{\mathbb{Y}}(\mathbf{h}_0\mathbf{g}\mathbf{h}_0^{-1} \cdot \mathbf{d}^{-1}, \mathbf{d} \cdot \mathbf{h}_0 \cdot \mathbf{d}^{-1}) \cdot v_{\mathbb{Y}}(\mathbf{h}_0, \nu).$$

As in the proof of Lemma C.2.8, we have $v_{\mathbb{Y}}(\mathbf{h}_0, \nu) = (\nu, J_1)_F$. We have

$$\mathbf{h}_0\mathbf{g}\mathbf{h}_0^{-1} = \begin{pmatrix} (a + ct) \cdot \mathbf{1}_4 & \frac{(b-dt)u}{2} \cdot \mathbf{y} \\ 2(b + dt) \cdot \mathbf{y}^{-1} & (a - ct) \cdot \mathbf{1}_4 \end{pmatrix},$$

where

$$\mathbf{y} = \begin{pmatrix} 1 & & & \\ & -J_1 & & \\ & & -\frac{1}{u} & \\ & & & \frac{1}{uJ_1} \end{pmatrix}.$$

If $b + dt = 0$, then we have $\mathbf{h}_0\mathbf{g}\mathbf{h}_0^{-1} \cdot \mathbf{d}^{-1} \in P_{\mathbb{Y}}$ and hence $z_{\mathbb{Y}}(\mathbf{h}_0\mathbf{g}\mathbf{h}_0^{-1} \cdot \mathbf{d}^{-1}, \mathbf{d} \cdot \mathbf{h}_0 \cdot \mathbf{d}^{-1}) = 1$. If $b + dt \neq 0$, then we have $\mathbf{h}_0\mathbf{g}\mathbf{h}_0^{-1} \cdot \mathbf{d}^{-1} \in P_{\mathbb{Y}} \cdot \tau_4 \cdot \mathbf{n}(\mathbf{b}_{17})$, where

$$\mathbf{b}_{17} = \frac{a - ct}{2\nu(b + dt)} \cdot \mathbf{y}.$$

We have $\mathbf{d} \cdot \mathbf{h}_0 \cdot \mathbf{d}^{-1} \in \mathbf{m}(\mathbf{a}_5) \cdot \mathbf{n}(\nu^{-1} \cdot \mathbf{b}_{15}) \cdot \tau_2 \cdot P_{\mathbb{Y}}$, where $\mathbf{a}_5$ and $\mathbf{b}_{15}$ are as in the proof of Lemma C.2.8. Hence we have

$$\begin{aligned} & z_{\mathbb{Y}}(\mathbf{h}_0\mathbf{g}\mathbf{h}_0^{-1} \cdot \mathbf{d}^{-1}, \mathbf{d} \cdot \mathbf{h}_0 \cdot \mathbf{d}^{-1}) \\ &= z_{\mathbb{Y}}(\tau_4 \cdot \mathbf{n}(\mathbf{b}_{17}), \mathbf{m}(\mathbf{a}_5) \cdot \mathbf{n}(\nu^{-1} \cdot \mathbf{b}_{15}) \cdot \tau_2) \\ &= z_{\mathbb{Y}}(\tau_4 \cdot \mathbf{m}(\mathbf{a}_5), \mathbf{m}(\mathbf{a}_5)^{-1} \cdot \mathbf{n}(\mathbf{b}_{17}) \cdot \mathbf{m}(\mathbf{a}_5) \cdot \mathbf{n}(\nu^{-1} \cdot \mathbf{b}_{15}) \cdot \tau_2) \\ &= z_{\mathbb{Y}}(\tau_4, \mathbf{n}(\mathbf{b}_{18}) \cdot \tau_2), \end{aligned}$$

where $\mathbf{b}_{18} = \nu^{-1} \cdot \mathbf{b}_{15} + \mathbf{a}_5^{-1} \cdot \mathbf{b}_{17} \cdot {}^t\mathbf{a}_5^{-1}$. Put $r = \frac{a-ct}{b+dt}$. We have

$$\mathbf{b}_{18} = \frac{1}{2\nu t} \cdot \begin{pmatrix} rt & & & 1 \\ & -rtJ_1 & J_1 & \\ & J_1 & \frac{rJ_1}{tu} & \\ 1 & & & -\frac{r}{tu} \end{pmatrix}.$$

We write $\mathbf{b}_{18} = \mathbf{b}_{19} + \mathbf{b}_{20}$, where

$$\mathbf{b}_{19} = \frac{1}{2\nu t} \cdot \begin{pmatrix} rt & & & 1 \\ & -rtJ_1 & J_1 & \\ & J_1 & & \\ 1 & & & \end{pmatrix},$$

$$\mathbf{b}_{20} = \frac{r}{2\nu uJ} \cdot \begin{pmatrix} 0 & & & \\ & 0 & & \\ & & J_1 & \\ & & & -1 \end{pmatrix}.$$

Since $\tau_2^{-1} \cdot \mathbf{n}(\mathbf{b}_{19}) \cdot \tau_2 \in P_{\mathbb{Y}}$, we have

$$z_{\mathbb{Y}}(\tau_4, \mathbf{n}(\mathbf{b}_{18}) \cdot \tau_2) = z_{\mathbb{Y}}(\tau_4, \mathbf{n}(\mathbf{b}_{20}) \cdot \tau_2).$$

If $r = 0$, then we have $z_{\mathbb{Y}}(\tau_4, \mathbf{n}(\mathbf{b}_{20}) \cdot \tau_2) = z_{\mathbb{Y}}(\tau_4, \tau_2) = 1$. If $r \neq 0$, then we have $z_{\mathbb{Y}}(\tau_4, \mathbf{n}(\mathbf{b}_{20}) \cdot \tau_2) = \gamma_F(\frac{1}{2}\psi \circ q_8)$, where q_8 is a non-degenerate symmetric bilinear form associated to

$$\frac{r}{2\nu uJ} \cdot \begin{pmatrix} J_1 & \\ & -1 \end{pmatrix}.$$

We have $\det q_8 \equiv -J_1 \bmod (F^\times)^2$ and

$$h_F(q_8) = (\frac{rJ_1}{2\nu uJ}, -\frac{r}{2\nu uJ})_F = (J_1, -2\nu ru)_F.$$

Hence we have

$$\begin{aligned} \gamma_F(\frac{1}{2}\psi \circ q_8) &= \gamma_F(\frac{1}{2}\psi)^2 \cdot \gamma_F(-J_1, \frac{1}{2}\psi) \cdot (J_1, -2\nu ru)_F \\ &= \gamma_F(J_1, \frac{1}{2}\psi) \cdot (J_1, 2\nu ru)_F. \end{aligned}$$

Thus we obtain

$$z_{\mathbb{Y}}(\mathbf{h}_0 \mathbf{g} \mathbf{h}_0^{-1}, \mathbf{h}_0) = \begin{cases} (\nu, J_1)_F & \text{if } b + dt = 0, \\ (\nu, J_1)_F & \text{if } b + dt \neq 0,\ a - ct = 0, \\ \gamma_F(J_1, \frac{1}{2}\psi) \cdot (2u \cdot \frac{a-ct}{b+dt}, J_1)_F & \text{if } b + dt \neq 0,\ a - ct \neq 0. \end{cases}$$

Now we compute $z_{\mathbb{Y}}(\mathbf{h}_0, \mathbf{g})$. We have

$$z_{\mathbb{Y}}(\mathbf{h}_0, \mathbf{g}) = z_{\mathbb{Y}}(\mathbf{h}_0, \mathbf{g} \cdot \mathbf{d}^{-1}).$$

First assume that $b = d = 0$. Then we have $\mathbf{g} \cdot \mathbf{d}^{-1} \in P_{\mathbb{Y}}$ and hence

$$z_{\mathbb{Y}}(\mathbf{h}_0, \mathbf{g} \cdot \mathbf{d}^{-1}) = 1.$$

Next assume that $(b, d) \neq (0, 0)$ and $b^2 - d^2 J = 0$. Then we have $b \neq 0$, $d \neq 0$, and $\nu = a^2 - c^2 J \neq 0$. Since

$$\begin{aligned} (ad + bc) \cdot (ab - cdJ) &= a^2 bd - acd^2 J + ab^2 c - bc^2 dJ \\ &= a^2 bd - bc^2 dJ \\ &= \nu bd \\ &\neq 0, \end{aligned}$$

we have $ad+bc \neq 0$ and $ab - cdJ \neq 0$. We have $\mathbf{g}\cdot\mathbf{d}^{-1} \in \mathbf{m}(\mathbf{a}_3)\cdot\mathbf{n}(\mathbf{b}_9)\cdot\tau_2\cdot P_{\mathbb{Y}}$, where $\mathbf{a}_3$ and $\mathbf{b}_9$ are as in the proof of Lemma C.2.7. Hence we have

$$\begin{aligned} z_{\mathbb{Y}}(\mathbf{h}_0, \mathbf{g}\cdot\mathbf{d}^{-1}) &= z_{\mathbb{Y}}(\tau_2\cdot\mathbf{m}(\mathbf{a}_6), \mathbf{m}(\mathbf{a}_3)\cdot\mathbf{n}(\mathbf{b}_9)\cdot\tau_2) \\ &= z_{\mathbb{Y}}(\tau_2\cdot\mathbf{m}(\mathbf{a}_8), \mathbf{n}(\mathbf{b}_9)\cdot\tau_2), \end{aligned}$$

where $\mathbf{a}_6$ is as in the proof of Lemma C.2.8 and

$$\mathbf{a}_8 = \mathbf{a}_6\cdot\mathbf{a}_3 = \begin{pmatrix} b & & & \\ & & b & \\ & & \frac{(b+dt)t}{J_1} & 1 \\ -(b+dt)t & & & 1 \end{pmatrix}.$$

If $b+dt=0$, then we have $\tau_2\cdot\mathbf{m}(\mathbf{a}_8)\cdot\tau_2^{-1} \in P_{\mathbb{Y}}$ and hence

$$z_{\mathbb{Y}}(\tau_2\cdot\mathbf{m}(\mathbf{a}_8), \mathbf{n}(\mathbf{b}_9)\cdot\tau_2) = z_{\mathbb{Y}}(\tau_2, \mathbf{n}(\mathbf{b}_9)\cdot\tau_2) = \gamma_F(\frac{1}{2}\psi\circ q_9),$$

where q_9 is a non-degenerate symmetric bilinear form associated to

$$\frac{ad+bc}{bd}\cdot\begin{pmatrix} -J_2 & \\ & J \end{pmatrix}.$$

We have $\det q_9 \equiv -J_1 \bmod (F^\times)^2$ and

$$\begin{aligned} h_F(q_9) &= (-\frac{ad+bc}{bd}\cdot J_2, \frac{ad+bc}{bd}\cdot J)_F \\ &= (J_2, \frac{ad+bc}{bd})_F \\ &= (J_2, \frac{\nu}{ab-cdJ})_F. \end{aligned}$$

Hence we have

$$\begin{aligned} \gamma_F(\frac{1}{2}\psi\circ q_9) &= \gamma_F(\frac{1}{2}\psi)^2\cdot\gamma_F(-J_1, \frac{1}{2}\psi)\cdot(J_2, \frac{\nu}{ab-cdJ})_F \\ &= \gamma_F(J_1, \frac{1}{2}\psi)^{-1}\cdot(J_1, (ab-cdJ)\nu)_F. \end{aligned}$$

If $b+dt \neq 0$, then we have

$$\mathbf{m}(\mathbf{a}_8)\cdot\mathbf{n}(\mathbf{b}_9)\cdot\tau_2 \in \mathbf{m}(\mathbf{a}_9)\cdot\mathbf{n}(\mathbf{b}_{21})\cdot\tau''\cdot P_{\mathbb{Y}},$$

where

$$\mathbf{a}_9 = \begin{pmatrix} 1 & & & -1 \\ & 1 & -1 & \\ & & \frac{(b+dt)t}{bJ_1} & \\ & & & \frac{(b+dt)t}{b} \end{pmatrix},$$

$$\mathbf{b}_{21} = \frac{(ad+bc)b}{(b+dt)^2 d}\cdot\begin{pmatrix} 1 & & & 1 \\ & -J_1 & & \\ & & 0 & \\ 1 & & & 0 \end{pmatrix},$$

$$\tau'' = \begin{pmatrix} & & -\mathbf{1}_2 & \\ & \mathbf{1}_2 & & \\ \mathbf{1}_2 & & & \\ & & & \mathbf{1}_2 \end{pmatrix}.$$

Since $\tau_2 \cdot \mathbf{m}(\mathbf{a}_9) \cdot \mathbf{n}(\mathbf{b}_{21}) \cdot \tau_2^{-1} \in P_{\mathbb{Y}}$, we have

$$z_{\mathbb{Y}}(\tau_2 \cdot \mathbf{m}(\mathbf{a}_8), \mathbf{n}(\mathbf{b}_9) \cdot \tau_2) = z_{\mathbb{Y}}(\tau_2, \mathbf{m}(\mathbf{a}_9) \cdot \mathbf{n}(\mathbf{b}_{21}) \cdot \tau'') = z_{\mathbb{Y}}(\tau_2, \tau'') = 1.$$

Finally assume that $(b, d) \neq (0, 0)$ and $b^2 - d^2 J \neq 0$. We have $\mathbf{g} \cdot \mathbf{d}^{-1} \in \mathbf{n}(\mathbf{b}_{13}) \cdot \tau_4 \cdot P_{\mathbb{Y}}$, where $\mathbf{b}_{13}$ is as in the proof of Lemma C.2.7. Hence we have

$$z_{\mathbb{Y}}(\mathbf{h}_0, \mathbf{g} \cdot \mathbf{d}^{-1}) = z_{\mathbb{Y}}(\tau_2 \cdot \mathbf{m}(\mathbf{a}_6), \mathbf{n}(\mathbf{b}_{13}) \cdot \tau_4) = z_{\mathbb{Y}}(\tau_2, \mathbf{n}(\mathbf{b}_{22}) \cdot \tau_4),$$

where

$$\mathbf{b}_{22} = \mathbf{a}_6 \cdot \mathbf{b}_{13} \cdot {}^t\mathbf{a}_6 = \frac{1}{b^2 - d^2 J} \cdot \begin{pmatrix} l & & & -l't \\ & -lJ_1 & l't & \\ & l't & -2l'J_2 & \\ -l't & & & 2l'J \end{pmatrix}$$

with

$$l = ab - cdJ, \quad l' = (a - ct)(b + dt).$$

We write $\mathbf{b}_{22} = \mathbf{b}_{23} + \mathbf{b}_{24}$, where

$$\mathbf{b}_{23} = \frac{1}{b^2 - d^2 J} \cdot \begin{pmatrix} l & & & -l't \\ & -lJ_1 & l't & \\ & l't & & \\ -l't & & & \end{pmatrix},$$

$$\mathbf{b}_{24} = \frac{2(a - ct)}{b - dt} \cdot \begin{pmatrix} 0 & & & \\ & 0 & & \\ & & -J_2 & \\ & & & J \end{pmatrix}.$$

Since $\tau_2 \cdot \mathbf{b}_{23} \cdot \tau_2^{-1} \in P_{\mathbb{Y}}$, we have $z_{\mathbb{Y}}(\tau_2, \mathbf{n}(\mathbf{b}_{22}) \cdot \tau_4) = z_{\mathbb{Y}}(\tau_2, \mathbf{n}(\mathbf{b}_{24}) \cdot \tau_4)$. If $a - ct = 0$, then we have $z_{\mathbb{Y}}(\tau_2, \mathbf{n}(\mathbf{b}_{24}) \cdot \tau_4) = z_{\mathbb{Y}}(\tau_2, \tau_4) = 1$. If $a - ct \neq 0$, then we have $z_{\mathbb{Y}}(\tau_2, \mathbf{n}(\mathbf{b}_{24}) \cdot \tau_4) = \gamma_F(\frac{1}{2}\psi \circ q_{10})$, where q_{10} is a non-degenerate symmetric bilinear form associated to

$$\frac{2(a - ct)}{b - dt} \cdot \begin{pmatrix} -J_2 & \\ & J \end{pmatrix}.$$

We have $\det q_{10} \equiv -J_1 \bmod (F^\times)^2$ and

$$\begin{aligned} h_F(q_{10}) &= (-\frac{2(a - ct)}{b - dt} \cdot J_2, \frac{2(a - ct)}{b - dt} \cdot J)_F \\ &= (J_2, \frac{2(a - ct)}{b - dt})_F. \end{aligned}$$

Hence we have

$$\begin{aligned} \gamma_F(\frac{1}{2}\psi \circ q_{10}) &= \gamma_F(\frac{1}{2}\psi)^2 \cdot \gamma_F(-J_1, \frac{1}{2}\psi) \cdot (J_2, \frac{2(a - ct)}{b - dt})_F \\ &= \gamma_F(J_1, \frac{1}{2}\psi)^{-1} \cdot (J_1, \frac{2(a - ct)}{b - dt})_F. \end{aligned}$$

Thus we obtain

$$z_{\mathbb{Y}}(\mathbf{h}_0, \mathbf{g}) = \begin{cases} 1 & \text{if } b = d = 0, \\ \gamma_F(J_1, \frac{1}{2}\psi)^{-1} \cdot (J_1, (ab - cdJ)\nu)_F & \text{if } (b,d) \neq (0,0), \\ & \quad b^2 - d^2 J = 0,\ b + dt = 0, \\ 1 & \text{if } (b,d) \neq (0,0), \\ & \quad b^2 - d^2 J = 0,\ b + dt \neq 0, \\ 1 & \text{if } (b,d) \neq (0,0), \\ & \quad b^2 - d^2 J \neq 0,\ a - ct = 0, \\ \gamma_F(J_1, \frac{1}{2}\psi)^{-1} \cdot (J_1, \frac{2(a-ct)}{b-dt})_F & \text{if } (b,d) \neq (0,0), \\ & \quad b^2 - d^2 J \neq 0,\ a + ct \neq 0. \end{cases}$$

Now we compute $\mu(\mathbf{g}) = z_{\mathbb{Y}}(\mathbf{h}_0 \mathbf{g} \mathbf{h}_0^{-1}, \mathbf{h}_0) \cdot z_{\mathbb{Y}}(\mathbf{h}_0, \mathbf{g})^{-1}$. Recall that $J = t^2$ and $\nu = a^2 - b^2 u - c^2 J + d^2 uJ \neq 0$. First assume that $b = d = 0$. Then we have

$$\mu(\mathbf{g}) = (\nu, J_1)_F \cdot 1 = (\nu, J_1)_F.$$

Next assume that $(b,d) \neq (0,0)$ and $b^2 - d^2 J = 0$. Then we have $\nu = a^2 - c^2 J = (a + ct)(a - ct) \neq 0$. Since $b^2 - d^2 J = (b + dt)(b - dt)$, we have

$$b + dt = 0 \Longleftrightarrow b - dt \neq 0.$$

If $b + dt = 0$, then we have

$$\begin{aligned} \mu(\mathbf{g}) &= (\nu, J_1)_F \cdot \gamma_F(J_1, \frac{1}{2}\psi) \cdot (J_1, (ab - cdJ)\nu)_F \\ &= \gamma_F(J_1, \frac{1}{2}\psi) \cdot (J_1, ab - cdJ)_F. \end{aligned}$$

If $b + dt \neq 0$, then we have

$$(a - ct)(b + dt) = 2(a - ct)dt = 2(adt - cdJ) = 2(ab - cdJ).$$

Hence we have

$$\begin{aligned} \mu(\mathbf{g}) &= \gamma_F(J_1, \frac{1}{2}\psi) \cdot (2u \cdot \frac{a - ct}{b + dt}, J_1)_F \cdot 1 \\ &= \gamma_F(J_1, \frac{1}{2}\psi) \cdot (2(a - ct)(b + dt), J_1)_F \cdot (u, J_1)_F \\ &= \gamma_F(J_1, \frac{1}{2}\psi) \cdot (ab - cdJ, J_1)_F \cdot (u, J_1)_F. \end{aligned}$$

Finally assume that $(b,d) \neq (0,0)$ and $b^2 - d^2 J \neq 0$. Then we have $b + dt \neq 0$. If $a - ct = 0$, then we have $\nu = -b^2 u + d^2 uJ$ and hence

$$\begin{aligned} \mu(\mathbf{g}) &= (\nu, J_1)_F \cdot 1 \\ &= (-b^2 + d^2 J, J_1)_F \cdot (u, J_1)_F \\ &= (-(b^2 - d^2 J)J, J_1)_F \cdot (u, J_1)_F. \end{aligned}$$

If $a - ct \neq 0$, then we have

$$\begin{aligned}\mu(\mathbf{g}) &= \gamma_F(J_1, \frac{1}{2}\psi) \cdot (2u \cdot \frac{a-ct}{b+dt}, J_1)_F \cdot \gamma_F(J_1, \frac{1}{2}\psi) \cdot (J_1, \frac{2(a-ct)}{b-dt})_F \\ &= \gamma_F(J_1, \frac{1}{2}\psi)^2 \cdot (u(b+dt)(b-dt), J_1)_F \\ &= (-1, J_1)_F \cdot (b^2 - d^2 J, J_1)_F \cdot (u, J_1)_F \\ &= (-(b^2 - d^2 J)J, J_1)_F \cdot (u, J_1)_F.\end{aligned}$$

This completes the proof. □

C.3. The case $J_i \in (F^\times)^2$

We only consider the case $i = 1$; the other case is similar. Choose $t \in F^\times$ such that $J_1 = t^2$. We take an isomorphism

$$\mathfrak{i}_1 : B_1 \longrightarrow \mathrm{M}_2(F)$$

of F-algebras determined by

$$\mathfrak{i}_1(1) = \begin{pmatrix} 1 & \\ & 1 \end{pmatrix}, \quad \mathfrak{i}_1(\mathbf{i}) = \begin{pmatrix} & 2 \\ \frac{u}{2} & \end{pmatrix}, \quad \mathfrak{i}_1(\mathbf{j}_1) = \begin{pmatrix} t & \\ & -t \end{pmatrix}, \quad \mathfrak{i}_1(\mathbf{ij}_1) = \begin{pmatrix} & -2t \\ \frac{tu}{2} & \end{pmatrix}.$$

Note that

$$\mathfrak{i}_1(\boldsymbol{\alpha}_1^*) = \mathfrak{i}_1(\boldsymbol{\alpha}_1)^*$$

for $\boldsymbol{\alpha}_1 \in B_1$. Let

$$\mathbf{v} := \frac{1}{2}\mathbf{e}_1 + \frac{1}{2t}\mathbf{e}_2, \quad \mathbf{v}^* := \mathbf{e}_1^* + t\mathbf{e}_2^* = \frac{1}{u}\mathbf{e}_1\mathbf{i} - \frac{1}{tu}\mathbf{e}_2\mathbf{i}.$$

Then we have

$$V = \mathbf{v}B + \mathbf{v}^*B$$

and

$$\langle \mathbf{v}, \mathbf{v} \rangle = \langle \mathbf{v}^*, \mathbf{v}^* \rangle = 0, \quad \langle \mathbf{v}, \mathbf{v}^* \rangle = 1.$$

Moreover, we see that

$$\begin{bmatrix} \boldsymbol{\alpha}_1 \cdot \mathbf{v} & \boldsymbol{\alpha}_1 \cdot \mathbf{v}^* \end{bmatrix} = \begin{bmatrix} \mathbf{v} & \mathbf{v}^* \end{bmatrix} \cdot \mathfrak{i}_1(\boldsymbol{\alpha}_1)$$

for $\boldsymbol{\alpha}_1 \in B_1$, and

$$\begin{bmatrix} \boldsymbol{\alpha}_2 \cdot \mathbf{v} & \boldsymbol{\alpha}_2 \cdot \mathbf{v}^* \end{bmatrix} = \begin{bmatrix} \mathbf{v} & \mathbf{v}^* \end{bmatrix} \cdot (\alpha + \frac{\beta}{t}\mathbf{j})$$

for $\boldsymbol{\alpha}_2 = \alpha + \beta\mathbf{j}_2 \in B_2$ with $\alpha, \beta \in E$.

We regard $V' := V$ as a left B-space by putting

$$\boldsymbol{\alpha} \cdot x' := (x \cdot \boldsymbol{\alpha}^*)'$$

for $\boldsymbol{\alpha} \in B$ and $x' \in V'$. Here we let x' denote the element in V' corresponding to $x \in V$. We let $\mathrm{GL}(V')$ act on V' on the right. We define a skew-hermitian form

$$\langle \cdot, \cdot \rangle' : V' \times V' \longrightarrow B$$

by

$$\langle x', y' \rangle' := \langle x, y \rangle.$$

Note that

$$\langle \boldsymbol{\alpha} x', \boldsymbol{\beta} y' \rangle' = \boldsymbol{\alpha} \langle x', y' \rangle' \boldsymbol{\beta}^*$$

for $\boldsymbol{\alpha}, \boldsymbol{\beta} \in B$. For $x' \in V'$ and $g \in \mathrm{GL}(V)$, put

$$x' \cdot g := (g^{-1} \cdot x)'.$$

Then we have an isomorphism

$$\begin{aligned} \mathrm{GL}(V) &\longrightarrow \mathrm{GL}(V'), \\ g &\longmapsto [x' \mapsto x' \cdot g] \end{aligned}$$

so that we may identify $\mathrm{GU}(V)$ with $\mathrm{GU}(V')$ via this isomorphism. Let $V' = X' + Y'$ be a complete polarization given by

$$X' = B \cdot \mathbf{v}', \quad Y' = B \cdot (\mathbf{v}^*)'.$$

Note that

$$\begin{bmatrix} \mathbf{v}' \cdot \boldsymbol{\alpha} \\ (\mathbf{v}^*)' \cdot \boldsymbol{\alpha} \end{bmatrix} = {}^t\mathrm{i}_1(\boldsymbol{\alpha})^{-1} \cdot \begin{bmatrix} \mathbf{v}' \\ (\mathbf{v}^*)' \end{bmatrix}$$

for $\boldsymbol{\alpha} \in B_1$. We may identify V' with the space of row vectors B^2 so that

$$\langle x', y' \rangle' = x_1 y_2^* - x_2 y_1^*$$

for $x' = (x_1, x_2)$, $y' = (y_1, y_2) \in V'$. Then we may write

$$\mathrm{GU}(V') = \left\{ g \in \mathrm{GL}_2(B) \,\middle|\, g \begin{pmatrix} & 1 \\ -1 & \end{pmatrix} {}^t g^* = \nu(g) \cdot \begin{pmatrix} & 1 \\ -1 & \end{pmatrix} \right\}.$$

Similarly, we have a right B-space $W' := W$ with a hermitian form

$$\langle \cdot, \cdot \rangle' : W' \times W' \longrightarrow B.$$

We let $\mathrm{GL}(W')$ act on W' on the left. Now we consider an F-space

$$\mathbb{V}' := W' \otimes_B V'$$

with a symplectic form

$$\langle\!\langle \cdot, \cdot \rangle\!\rangle' := \frac{1}{2} \mathrm{tr}_{B/F}(\langle \cdot, \cdot \rangle' \otimes \langle \cdot, \cdot \rangle'^*).$$

We let $\mathrm{GL}(\mathbb{V}')$ act on $\mathbb{V}'$ on the right. For $\mathbf{x} = x \otimes y \in \mathbb{V}$ and $\mathbf{g} \in \mathrm{GL}(\mathbb{V})$, put

$$\mathbf{x}' := y' \otimes x' \in \mathbb{V}'$$

and

$$\mathbf{x}' \cdot \mathbf{g} := (\mathbf{x} \cdot \mathbf{g})'.$$

LEMMA C.3.1. *We have an isomorphism*

$$\begin{aligned} \mathrm{GSp}(\mathbb{V}) &\longrightarrow \mathrm{GSp}(\mathbb{V}'). \\ \mathbf{g} &\longmapsto [\mathbf{x}' \mapsto \mathbf{x}' \cdot \mathbf{g}] \end{aligned}$$

Moreover, this isomorphism induces a commutative diagram

$$\begin{array}{ccc} \mathrm{GU}(V) \times \mathrm{GU}(W) & \longrightarrow & \mathrm{GSp}(\mathbb{V}) \\ \downarrow & & \downarrow \\ \mathrm{GU}(W') \times \mathrm{GU}(V') & \longrightarrow & \mathrm{GSp}(\mathbb{V}') \end{array}.$$

PROOF. For $x_1, x_2 \in V$ and $y_1, y_2 \in W$, we have

$$\begin{aligned}\langle\!\langle y_1' \otimes x_1', y_2' \otimes x_2' \rangle\!\rangle' &= \frac{1}{2}\operatorname{tr}_{B/F}(\langle y_1', y_2'\rangle' \cdot \langle x_1', x_2'\rangle'^*) \\ &= \frac{1}{2}\operatorname{tr}_{B/F}(\langle y_1, y_2\rangle \cdot \langle x_1, x_2\rangle^*) \\ &= \frac{1}{2}\operatorname{tr}_{B/F}(\langle x_1, x_2\rangle \cdot \langle y_1, y_2\rangle^*) \\ &= \langle\!\langle x_1 \otimes y_1, x_2 \otimes y_2 \rangle\!\rangle.\end{aligned}$$

Also, for $\mathbf{g} = (g, h) \in \mathrm{GL}(V) \times \mathrm{GL}(W)$ and $\mathbf{x} = x \otimes y \in \mathbb{V}$, we have

$$\begin{aligned}\mathbf{x}' \cdot \mathbf{g} &= ((x \otimes y) \cdot (g, h))' = (g^{-1}x \otimes yh)' \\ &= (yh)' \otimes (g^{-1}x)' = h^{-1}y' \otimes x'g.\end{aligned}$$

This completes the proof. □

Thus we may identify $\mathrm{GSp}(\mathbb{V})$ with $\mathrm{GSp}(\mathbb{V}')$, and $\mathrm{GU}(V) \times \mathrm{GU}(W)$ with $\mathrm{GU}(W') \times \mathrm{GU}(V')$.

Let $\mathbb{V}' = \mathbb{X}' + \mathbb{Y}'$ be a complete polarization given by

$$\mathbb{X}' = W' \otimes_B X', \quad \mathbb{Y}' = W' \otimes_B Y'.$$

Put

$$s'(g) := \gamma^{j(g)}$$

for $g \in \mathrm{GU}(V')^0$, where

$$\gamma = \begin{cases} 1 & \text{if } B \text{ and } B_2 \text{ are split,} \\ -1 & \text{if } B \text{ and } B_2 \text{ are ramified,} \end{cases}$$

and

$$j(g) = \begin{cases} 0 & \text{if } g = \left(\begin{smallmatrix} * & * \\ 0 & * \end{smallmatrix}\right), \\ 1 & \text{othersiwe.} \end{cases}$$

LEMMA C.3.2. *We have*

$$z_{\mathbb{Y}'}(g, g') = s'(gg') \cdot s'(g)^{-1} \cdot s'(g')^{-1}$$

for $g, g' \in \mathrm{GU}(V)^0$.

PROOF. The proof is similar to that of Lemma C.2.3. If B is ramified, then we have

(C.2) $$z_{\mathbb{Y}'}(g, g') = s'(gg') \cdot s'(g)^{-1} \cdot s'(g')^{-1}$$

for $g, g' \in \mathrm{U}(V)^0$ by [**40**, Theorem 3.1, case 2_+]. If B is split, then we see that (C.2) also holds by using Morita theory as in §C.2 and [**40**, Theorem 3.1, case 1_-].

Let $g, g' \in \mathrm{GU}(V)^0$. For $\nu \in F^\times$, put

$$d(\nu) = \begin{pmatrix} 1 & \\ & \nu \end{pmatrix} \in \mathrm{GU}(V)^0.$$

We write

$$g = h \cdot d(\nu), \quad g' = h' \cdot d(\nu')$$

with $h, h' \in \mathrm{U}(V)^0$ and $\nu, \nu' \in F^\times$. Then we have

$$z_{\mathbb{Y}'}(g, g') = z_{\mathbb{Y}'}(h, h'') \cdot v_{\mathbb{Y}'}(h', \nu),$$

where

$$h'' = d(\nu) \cdot h' \cdot d(\nu)^{-1}.$$

By (C.2), we have

$$z_{\mathbb{Y}'}(h, h'') = s'(hh'') \cdot s'(h)^{-1} \cdot s'(h'')^{-1}.$$

We have $s'(h) = s'(g)$, and since $j(h'') = j(h')$, we have $s'(h'') = s'(h') = s'(g')$. Moreover, since $gg' = hh'' \cdot d(\nu\nu')$, we have $s'(hh'') = s'(gg')$. Thus we obtain

$$z_{\mathbb{Y}'}(h, h'') = s'(gg') \cdot s'(g)^{-1} \cdot s'(g')^{-1}.$$

By Lemma B.2.1, we have

$$v_{\mathbb{Y}'}(h', \nu) = (x_{\mathbb{Y}'}(h'), \nu)_F \cdot \gamma_F(\nu, \frac{1}{2}\psi)^{-j_{\mathbb{Y}'}(h')},$$

where $x_{\mathbb{Y}'}$ and $j_{\mathbb{Y}'}$ are as in §B.1 with respect to the complete polarization $\mathbb{V}' = \mathbb{X}' + \mathbb{Y}'$. Since the determinant over F of the automorphism $x \mapsto x \cdot \boldsymbol{\alpha}$ of B is $\nu(\boldsymbol{\alpha})^2$ for $\boldsymbol{\alpha} \in B^\times$, we have $x_{\mathbb{Y}'}(h') \equiv 1 \bmod (F^\times)^2$. Noting that either $c = 0$ or $c \in B^\times$ for $h' = \left(\begin{smallmatrix} a & b \\ c & d \end{smallmatrix}\right)$, one can see that $j_{\mathbb{Y}'}(h') = 4 \cdot j(h')$. Hence we have

$$v_{\mathbb{Y}'}(h', \nu) = 1.$$

This completes the proof. □

LEMMA C.3.3. *We have*

$$z_{\mathbb{Y}'}(h, h') = 1$$

for $h, h' \in \mathrm{GU}(W)$.

PROOF. The proof is similar to that of Lemma C.2.4.

For $g, g' \in \mathrm{GU}(W)$, we have

$$z_{\mathbb{Y}'}(g, g') = z_{\mathbb{Y}'}(h, h'') \cdot v_{\mathbb{Y}'}(h', \nu),$$

where

$$h = g \cdot d_{\mathbb{Y}'}(\nu)^{-1}, \qquad h' = g' \cdot d_{\mathbb{Y}'}(\nu')^{-1}, \qquad h'' = d_{\mathbb{Y}'}(\nu) \cdot h' \cdot d_{\mathbb{Y}'}(\nu)^{-1},$$
$$\nu = \nu(g), \qquad \nu' = \nu(g').$$

We have $h, h' \in P_{\mathbb{Y}'}$ and $z_{\mathbb{Y}'}(h, h'') = 1$. Since the determinant over F of the automorphism $x \mapsto \boldsymbol{\alpha} \cdot x$ of B is $\nu(\boldsymbol{\alpha})^2$ for $\boldsymbol{\alpha} \in B^\times$, we have $x_{\mathbb{Y}'}(h') \equiv 1 \bmod (F^\times)^2$, so that $v_{\mathbb{Y}'}(h', \nu) = 1$ by Lemma B.2.1. This completes the proof. □

LEMMA C.3.4. *We have*

$$z_{\mathbb{Y}'}(g, h) = z_{\mathbb{Y}'}(h, g) = 1$$

for $g \in \mathrm{GU}(V)^0$ *and* $h \in \mathrm{GU}(W)$.

PROOF. The proof is similar to that of Lemma C.2.5.

For $g \in \mathrm{GU}(V)^0$ and $h \in \mathrm{GU}(W)$, we have

$$z_{\mathbb{Y}'}(g, h) = z_{\mathbb{Y}'}(g', h'') \cdot v_{\mathbb{Y}'}(h', \nu), \quad z_{\mathbb{Y}'}(h, g) = z_{\mathbb{Y}'}(h', g'') \cdot v_{\mathbb{Y}'}(g', \nu'),$$

where

$$g' = g \cdot d(\nu)^{-1}, \qquad g'' = d(\nu') \cdot g' \cdot d(\nu')^{-1}, \qquad \nu = \nu(g),$$
$$h' = h \cdot d_{\mathbb{Y}'}(\nu')^{-1}, \qquad h'' = d_{\mathbb{Y}'}(\nu) \cdot h' \cdot d_{\mathbb{Y}'}(\nu)^{-1}, \qquad \nu' = \nu(h).$$

Since $h', h'' \in P_{\mathbb{Y}'}$, we have $z_{\mathbb{Y}'}(g', h'') = z_{\mathbb{Y}'}(h', g'') = 1$. As in the proof of Lemma C.3.3, we have $v_{\mathbb{Y}'}(h', \nu) = 1$. As in the proof of Lemma C.3.2, we have $v_{\mathbb{Y}'}(g', \nu') = 1$. This completes the proof. $\square$

We define a map $s' : \mathrm{GU}(V)^0 \times \mathrm{GU}(W) \to \mathbb{C}^1$ by

$$s'(\mathbf{g}) = \gamma^{j(g)}$$

for $\mathbf{g} = (g, h) \in \mathrm{GU}(V)^0 \times \mathrm{GU}(W)$. By Lemmas C.3.2, C.3.3, C.3.4, we see that

$$z_{\mathbb{Y}'}(\mathbf{g}, \mathbf{g}') = s'(\mathbf{g}\mathbf{g}') \cdot s'(\mathbf{g})^{-1} \cdot s'(\mathbf{g}')^{-1}$$

for $\mathbf{g}, \mathbf{g}' \in \mathrm{GU}(V)^0 \times \mathrm{GU}(W)$.

Recall that we may identify $\mathbb{V}$ with $\mathbb{V}'$, and we have two complete polarizations $\mathbb{V} = \mathbb{X} + \mathbb{Y} = \mathbb{X}' + \mathbb{Y}'$, where

$$\mathbb{X} = F\mathbf{e}_1 + F\mathbf{e}_2 + F\mathbf{e}_3 + F\mathbf{e}_4, \qquad \mathbb{Y} = F\mathbf{e}_1^* + F\mathbf{e}_2^* + F\mathbf{e}_3^* + F\mathbf{e}_4^*,$$
$$\mathbb{X}' = \mathbf{v} \cdot B, \qquad \mathbb{Y}' = \mathbf{v}^* \cdot B.$$

Put

$$\mathbf{h}_0 = \begin{pmatrix} \frac{1}{2} & \frac{1}{2t} & & & & & & \\ & & \frac{1}{2} & \frac{1}{2t} & & & & \\ & & & & -\frac{1}{2} & \frac{t}{2} & & \\ & & & & & & -\frac{1}{2} & \frac{t}{2} \\ & & & & 1 & t & & \\ & & & & & & 1 & t \\ 1 & -\frac{1}{t} & & & & & & \\ & & 1 & -\frac{1}{t} & & & & \end{pmatrix} \in \mathrm{Sp}(\mathbb{V}).$$

Then we have

$$\begin{bmatrix} \mathbf{v} \\ \frac{1}{t}\mathbf{v}\mathbf{j} \\ -\frac{1}{u}\mathbf{v}\mathbf{i} \\ -\frac{t}{uJ}\mathbf{v}\mathbf{i}\mathbf{j} \\ \mathbf{v}^* \\ -\frac{t}{J}\mathbf{v}^*\mathbf{j} \\ \mathbf{v}^*\mathbf{i} \\ -\frac{1}{t}\mathbf{v}^*\mathbf{i}\mathbf{j} \end{bmatrix} = \mathbf{h}_0 \begin{bmatrix} \mathbf{e}_1 \\ \mathbf{e}_2 \\ \mathbf{e}_3 \\ \mathbf{e}_4 \\ \mathbf{e}_1^* \\ \mathbf{e}_2^* \\ \mathbf{e}_3^* \\ \mathbf{e}_4^* \end{bmatrix},$$

and hence $\mathbb{X}' = \mathbb{X}\mathbf{h}_0$ and $\mathbb{Y}' = \mathbb{Y}\mathbf{h}_0$. Put

$$s(\mathbf{g}) := s'(\mathbf{g}) \cdot \mu(\mathbf{g}),$$

where

$$\mu(\mathbf{g}) := z_{\mathbb{Y}}(\mathbf{h}_0\mathbf{g}\mathbf{h}_0^{-1}, \mathbf{h}_0) \cdot z_{\mathbb{Y}}(\mathbf{h}_0, \mathbf{g})^{-1}$$

for $\mathbf{g} \in \mathrm{GU}(V)^0 \times \mathrm{GU}(W)$. Then we have

$$z_{\mathbb{Y}}(\mathbf{g}, \mathbf{g}') = s(\mathbf{g}\mathbf{g}') \cdot s(\mathbf{g})^{-1} \cdot s(\mathbf{g}')^{-1}$$

for $\mathbf{g}, \mathbf{g}' \in \mathrm{GU}(V)^0 \times \mathrm{GU}(W)$.

LEMMA C.3.5. *Let* $\mathbf{g}_1 := \boldsymbol{\alpha}_1^{-1} \in \mathrm{GU}(V)^0$ *with* $\boldsymbol{\alpha}_1 = a_1 + b_1\mathbf{i} + c_1\mathbf{j}_1 + d_1\mathbf{ij}_1 \in B_1^\times$. *Then we have*

$$\mu(\mathbf{g}_1)$$
$$= \begin{cases} 1 & \text{if } b_1 = d_1 = 0, \\ \gamma_F(J_2, \frac{1}{2}\psi) \cdot ((a_1b_1 + c_1d_1J_1)\nu_1J_1, J_2)_F & \text{if } (b_1, d_1) \neq (0,0),\ b_1^2 - d_1^2J_1 = 0, \\ (-(b_1^2 - d_1^2J_1)\nu_1J_1, J_2)_F & \text{if } (b_1, d_1) \neq (0,0),\ b_1^2 - d_1^2J_1 \neq 0, \end{cases}$$
$$\times \begin{cases} 1 & \text{if } b_1 - d_1t = 0, \\ (u, J)_F & \text{if } b_1 - d_1t \neq 0, \end{cases}$$

where $\nu_1 = \nu(\boldsymbol{\alpha}_1)$.

PROOF. Put $\mathbf{d} := d_{\mathbb{Y}}(\nu_1) \in \mathrm{GSp}(\mathbb{V})$. We have

$$z_{\mathbb{Y}}(\mathbf{h}_0\mathbf{g}_1\mathbf{h}_0^{-1}, \mathbf{h}_0) = z_{\mathbb{Y}}(\mathbf{h}_0\mathbf{g}_1\mathbf{h}_0^{-1} \cdot \mathbf{d}^{-1}, \mathbf{d} \cdot \mathbf{h}_0 \cdot \mathbf{d}^{-1}) \cdot v_{\mathbb{Y}}(\mathbf{h}_0, \nu_1).$$

We have $\mathbf{h}_0 = \tau_2 \cdot \mathbf{m}(\mathbf{a}_{10})$, where

$$\mathbf{a}_{10} = \begin{pmatrix} \frac{1}{2} & \frac{1}{2t} & & \\ & & \frac{1}{2} & \frac{1}{2t} \\ 1 & -\frac{1}{t} & & \\ & & 1 & -\frac{1}{t} \end{pmatrix},$$

so that $x_{\mathbb{Y}}(\mathbf{h}_0) \equiv -1 \bmod (F^\times)^2$ and $j_{\mathbb{Y}}(\mathbf{h}_0) = 2$. Hence we have

$$v_{\mathbb{Y}}(\mathbf{h}_0, \nu_1) = (-1, \nu_1)_F \cdot \gamma_F(\nu_1, \frac{1}{2}\psi)^{-2} = 1.$$

We have

$$\mathbf{h}_0\mathbf{g}_1\mathbf{h}_0^{-1} = \begin{pmatrix} (a_1 + c_1t) \cdot \mathbf{1}_4 & \frac{(b_1+d_1t)u}{2} \cdot \mathbf{y} \\ 2(b_1 - d_1t) \cdot \mathbf{y}^{-1} & (a_1 - c_1t) \cdot \mathbf{1}_4 \end{pmatrix},$$

where

$$\mathbf{y} = \begin{pmatrix} 1 & & & \\ & -J_2 & & \\ & & -\frac{1}{u} & \\ & & & \frac{1}{uJ_2} \end{pmatrix}.$$

If $b_1 - d_1t = 0$, then we have $\mathbf{h}_0\mathbf{g}_1\mathbf{h}_0^{-1} \cdot \mathbf{d}^{-1} \in P_{\mathbb{Y}}$ and hence $z_{\mathbb{Y}}(\mathbf{h}_0\mathbf{g}_1\mathbf{h}_0^{-1} \cdot \mathbf{d}^{-1}, \mathbf{d} \cdot \mathbf{h}_0 \cdot \mathbf{d}^{-1}) = 1$. If $b_1 - d_1t \neq 0$, then we have $\mathbf{h}_0\mathbf{g}_1\mathbf{h}_0^{-1} \cdot \mathbf{d}^{-1} \in P_{\mathbb{Y}} \cdot \tau_4 \cdot \mathbf{n}(\mathbf{b}_{25})$, where

$$\mathbf{b}_{25} = \frac{a_1 - c_1t}{2\nu_1(b_1 - d_1t)} \cdot \mathbf{y}.$$

Since $\mathbf{d} \cdot \mathbf{h}_0 \cdot \mathbf{d}^{-1} \in \tau_2 \cdot P_{\mathbb{Y}}$, we have

$$z_{\mathbb{Y}}(\mathbf{h}_0\mathbf{g}_1\mathbf{h}_0^{-1} \cdot \mathbf{d}^{-1}, \mathbf{d} \cdot \mathbf{h}_0 \cdot \mathbf{d}^{-1}) = z_{\mathbb{Y}}(\tau_4 \cdot \mathbf{n}(\mathbf{b}_{25}), \tau_2).$$

If $a_1 - c_1t = 0$, then we have $z_{\mathbb{Y}}(\tau_4 \cdot \mathbf{n}(\mathbf{b}_{25}), \tau_2) = z_{\mathbb{Y}}(\tau_4, \tau_2) = 1$. If $a_1 - c_1t \neq 0$, then we have $z_{\mathbb{Y}}(\tau_4 \cdot \mathbf{n}(\mathbf{b}_{25}), \tau_2) = \gamma_F(\frac{1}{2}\psi \circ q_{11})$, where q_{11} is a non-degenerate symmetric bilinear form associated to

$$\frac{a_1 - c_1t}{2\nu_1(b_1 - d_1t)} \cdot \begin{pmatrix} -\frac{1}{u} & \\ & \frac{1}{uJ_2} \end{pmatrix}.$$

We have $\det q_{11} \equiv -J_2 \bmod (F^\times)^2$ and

$$h_F(q_{11}) = (-\frac{a_1 - c_1 t}{2\nu_1 u(b_1 - d_1 t)}, \frac{a_1 - c_1 t}{2\nu_1 u J_2(b_1 - d_1 t)})_F$$
$$= (-\frac{a_1 - c_1 t}{2\nu_1 u(b_1 - d_1 t)}, J_2)_F.$$

Hence we have

$$\gamma_F(\frac{1}{2}\psi \circ q_{11}) = \gamma_F(\frac{1}{2}\psi)^2 \cdot \gamma_F(-J_2, \frac{1}{2}\psi) \cdot (-\frac{a_1 - c_1 t}{2\nu_1 u(b_1 - d_1 t)}, J_2)_F$$
$$= \gamma_F(J_2, \frac{1}{2}\psi) \cdot (\frac{a_1 - c_1 t}{2\nu_1 u(b_1 - d_1 t)}, J_2)_F.$$

Thus we obtain

$$z_{\mathbb{Y}}(\mathbf{h}_0 \mathbf{g}_1 \mathbf{h}_0^{-1}, \mathbf{h}_0)$$
$$= \begin{cases} 1 & \text{if } b_1 - d_1 t = 0, \\ 1 & \text{if } b_1 - d_1 t \neq 0,\ a_1 - c_1 t = 0, \\ \gamma_F(J_2, \frac{1}{2}\psi) \cdot (\frac{a_1 - c_1 t}{2\nu_1 u(b_1 - d_1 t)}, J_2)_F & \text{if } b_1 - d_1 t \neq 0,\ a_1 - c_1 t \neq 0. \end{cases}$$

Now we compute $z_{\mathbb{Y}}(\mathbf{h}_0, \mathbf{g}_1)$. We have

$$z_{\mathbb{Y}}(\mathbf{h}_0, \mathbf{g}_1) = z_{\mathbb{Y}}(\mathbf{h}_0, \mathbf{g}_1 \cdot \mathbf{d}^{-1}).$$

First assume that $b_1 = d_1 = 0$. Then we have $\mathbf{g}_1 \cdot \mathbf{d}^{-1} \in P_{\mathbb{Y}}$ and hence

$$z_{\mathbb{Y}}(\mathbf{h}_0, \mathbf{g}_1 \cdot \mathbf{d}^{-1}) = 1.$$

Next assume that $(b_1, d_1) \neq (0, 0)$ and $b_1^2 - d_1^2 J_1 = 0$. Then we have $b_1 \neq 0$ and $d_1 \neq 0$. As in the proof of Lemma C.2.6, we have

$$(a_1 d_1 - b_1 c_1) \cdot (a_1 b_1 + c_1 d_1 J_1) = \nu_1 b_1 d_1 \neq 0.$$

We have $\mathbf{g}_1 \cdot \mathbf{d}^{-1} \in \mathbf{m}(\mathbf{a}_1) \cdot \mathbf{n}(\mathbf{b}_3) \cdot \tau_2 \cdot P_{\mathbb{Y}}$, where $\mathbf{a}_1$ and $\mathbf{b}_3$ are as in the proof of Lemma C.2.6. Hence we have

$$z_{\mathbb{Y}}(\mathbf{h}_0, \mathbf{g}_1 \cdot \mathbf{d}^{-1}) = z_{\mathbb{Y}}(\tau_2 \cdot \mathbf{m}(\mathbf{a}_{10}), \mathbf{m}(\mathbf{a}_1) \cdot \mathbf{n}(\mathbf{b}_3) \cdot \tau_2)$$
$$= z_{\mathbb{Y}}(\tau_2 \cdot \mathbf{m}(\mathbf{a}_{11}), \mathbf{n}(\mathbf{b}_3) \cdot \tau_2),$$

where

$$\mathbf{a}_{11} = \mathbf{a}_{10} \cdot \mathbf{a}_1 = \begin{pmatrix} \frac{b_1 + d_1 t}{2} & & \frac{1}{2t} & \\ & \frac{b_1 + d_1 t}{2} & & \frac{1}{2t} \\ b_1 - d_1 t & & -\frac{1}{t} & \\ & b_1 - d_1 t & & -\frac{1}{t} \end{pmatrix}.$$

If $b_1 - d_1 t = 0$, then we have $\tau_2 \cdot \mathbf{m}(\mathbf{a}_{11}) \cdot \tau_2^{-1} \in P_{\mathbb{Y}}$ and hence

$$z_{\mathbb{Y}}(\tau_2 \cdot \mathbf{m}(\mathbf{a}_{11}), \mathbf{n}(\mathbf{b}_3) \cdot \tau_2) = z_{\mathbb{Y}}(\tau_2, \mathbf{n}(\mathbf{b}_3) \cdot \tau_2) = \gamma_F(\frac{1}{2}\psi \circ q_{12}),$$

where q_{12} is a non-degenerate symmetric bilinear form associated to

$$\frac{a_1 d_1 - b_1 c_1}{b_1 d_1} \cdot \begin{pmatrix} J_1 & \\ & -J \end{pmatrix}.$$

We have $\det q_{12} \equiv -J_2 \bmod (F^\times)^2$ and

$$\begin{aligned} h_F(q_{12}) &= (\frac{a_1d_1 - b_1c_1}{b_1d_1} \cdot J_1, -\frac{a_1d_1 - b_1c_1}{b_1d_1} \cdot J)_F \\ &= (\frac{a_1d_1 - b_1c_1}{b_1d_1}, J)_F \\ &= (\frac{\nu_1}{a_1b_1 + c_1d_1J_1}, J)_F. \end{aligned}$$

Hence we have

$$\begin{aligned} \gamma_F(\frac{1}{2}\psi \circ q_{12}) &= \gamma_F(\frac{1}{2}\psi)^2 \cdot \gamma_F(-J_2, \frac{1}{2}\psi) \cdot (\frac{\nu_1}{a_1b_1 + c_1d_1J_1}, J)_F \\ &= \gamma_F(J_2, \frac{1}{2}\psi)^{-1} \cdot ((a_1b_1 + c_1d_1J_1)\nu_1, J_2)_F. \end{aligned}$$

If $b_1 - d_1t \neq 0$, then we have

$$\mathbf{m}(\mathbf{a}_{11}) \cdot \mathbf{n}(\mathbf{b}_3) \cdot \tau_2 \in \mathbf{n}(\mathbf{b}_{26}) \cdot \tau'' \cdot P_{\mathbb{Y}},$$

where

$$\mathbf{b}_{26} = \frac{b_1}{2(b_1 - d_1t)} \cdot \begin{pmatrix} 1 & & & \\ & -J_2 & & \\ & & 0 & \\ & & & 0 \end{pmatrix}$$

and τ'' is as in the proof of Lemma C.2.9. Since $\tau_2 \cdot \mathbf{n}(\mathbf{b}_{26}) \cdot \tau_2^{-1} \in P_{\mathbb{Y}}$, we have

$$z_{\mathbb{Y}}(\tau_2 \cdot \mathbf{m}(\mathbf{a}_{11}), \mathbf{n}(\mathbf{b}_3) \cdot \tau_2) = z_{\mathbb{Y}}(\tau_2, \mathbf{n}(\mathbf{b}_{26}) \cdot \tau'') = z_{\mathbb{Y}}(\tau_2, \tau'') = 1.$$

Finally assume that $(b_1, d_1) \neq (0,0)$ and $b_1^2 - d_1^2J_1 \neq 0$. We have $\mathbf{g}_1 \cdot \mathbf{d}^{-1} \in \mathbf{n}(\mathbf{b}_5) \cdot \tau_4 \cdot P_{\mathbb{Y}}$, where $\mathbf{b}_5$ is as in the proof of Lemma C.2.6. Hence we have

$$z_{\mathbb{Y}}(\mathbf{h}_0, \mathbf{g} \cdot \mathbf{d}^{-1}) = z_{\mathbb{Y}}(\tau_2 \cdot \mathbf{m}(\mathbf{a}_{10}), \mathbf{n}(\mathbf{b}_5) \cdot \tau_4) = z_{\mathbb{Y}}(\tau_2, \mathbf{n}(\mathbf{b}_{27}) \cdot \tau_4),$$

where

$$\mathbf{b}_{27} = \mathbf{a}_{10} \cdot \mathbf{b}_5 \cdot {}^t\mathbf{a}_{10} = \frac{1}{b_1^2 - d_1^2J_1} \cdot \begin{pmatrix} \frac{l}{2} & & & \\ & -\frac{lJ_2}{2} & & \\ & & 2l' & \\ & & & -2l'J_2 \end{pmatrix}$$

with

$$l = (a_1 + c_1t)(b_1 + d_1t), \quad l' = (a_1 - c_1t)(b_1 - d_1t).$$

We write $\mathbf{b}_{27} = \mathbf{b}_{28} + \mathbf{b}_{29}$, where

$$\mathbf{b}_{28} = \frac{a_1 + c_1t}{2(b_1 - d_1t)} \cdot \begin{pmatrix} 1 & & & \\ & -J_2 & & \\ & & 0 & \\ & & & 0 \end{pmatrix},$$

$$\mathbf{b}_{29} = \frac{2(a_1 - c_1t)}{b_1 + d_1t} \cdot \begin{pmatrix} 0 & & & \\ & 0 & & \\ & & 1 & \\ & & & -J_2 \end{pmatrix}.$$

Since $\tau_2 \cdot \mathbf{b}_{28} \cdot \tau_2^{-1} \in P_{\mathbb{Y}}$, we have $z_{\mathbb{Y}}(\tau_2, \mathbf{n}(\mathbf{b}_{27}) \cdot \tau_4) = z_{\mathbb{Y}}(\tau_2, \mathbf{n}(\mathbf{b}_{29}) \cdot \tau_4)$. If $a_1 - c_1t = 0$, then we have $z_{\mathbb{Y}}(\tau_2, \mathbf{n}(\mathbf{b}_{29}) \cdot \tau_4) = z_{\mathbb{Y}}(\tau_2, \tau_4) = 1$. If $a_1 - c_1t \neq 0$, then we have $z_{\mathbb{Y}}(\tau_2, \mathbf{n}(\mathbf{b}_{29}) \cdot \tau_4) = \gamma_F(\frac{1}{2}\psi \circ q_{13})$, where q_{13} is a non-degenerate symmetric bilinear form associated to

$$\frac{2(a_1 - c_1 t)}{b_1 + d_1 t} \cdot \begin{pmatrix} 1 & \\ & -J_2 \end{pmatrix}.$$

We have $\det q_{13} \equiv -J_2 \bmod (F^\times)^2$ and

$$\begin{aligned} h_F(q_{13}) &= (\frac{2(a_1 - c_1 t)}{b_1 + d_1 t}, -\frac{2(a_1 - c_1 t)}{b_1 + d_1 t} \cdot J_2)_F \\ &= (\frac{2(a_1 - c_1 t)}{b_1 + d_1 t}, J_2)_F. \end{aligned}$$

Hence we have

$$\begin{aligned} \gamma_F(\frac{1}{2}\psi \circ q_{13}) &= \gamma_F(\frac{1}{2}\psi)^2 \cdot \gamma_F(-J_2, \frac{1}{2}\psi) \cdot (\frac{2(a_1 - c_1 t)}{b_1 + d_1 t}, J_2)_F \\ &= \gamma_F(J_2, \frac{1}{2}\psi)^{-1} \cdot (\frac{2(a_1 - c_1 t)}{b_1 + d_1 t}, J_2)_F. \end{aligned}$$

Thus we obtain

$$\begin{aligned} &z_{\mathbb{Y}}(\mathbf{h}_0, \mathbf{g}_1) \\ &= \begin{cases} 1 & \text{if } b_1 = d_1 = 0, \\ \gamma_F(J_2, \frac{1}{2}\psi)^{-1} \cdot ((a_1 b_1 + c_1 d_1 J_1)\nu_1, J_2)_F & \text{if } (b_1, d_1) \neq (0,0), \\ & \quad b_1^2 - d_1^2 J_1 = 0,\ b_1 - d_1 t = 0, \\ 1 & \text{if } (b_1, d_1) \neq (0,0), \\ & \quad b_1^2 - d_1^2 J_1 = 0,\ b_1 - d_1 t \neq 0, \\ 1 & \text{if } (b_1, d_1) \neq (0,0), \\ & \quad b_1^2 - d_1^2 J_1 \neq 0,\ a_1 - c_1 t = 0, \\ \gamma_F(J_2, \frac{1}{2}\psi)^{-1} \cdot (\frac{2(a_1 - c_1 t)}{b_1 + d_1 t}, J_2)_F & \text{if } (b_1, d_1) \neq (0,0), \\ & \quad b_1^2 - d_1^2 J_1 \neq 0,\ a_1 - c_1 t \neq 0. \end{cases} \end{aligned}$$

Now we compute $\mu(\mathbf{g}_1) = z_{\mathbb{Y}}(\mathbf{h}_0 \mathbf{g}_1 \mathbf{h}_0^{-1}, \mathbf{h}_0) \cdot z_{\mathbb{Y}}(\mathbf{h}_0, \mathbf{g}_1)^{-1}$. Recall that $J_1 = t^2$ and $\nu_1 = a_1^2 - b_1^2 u - c_1^2 J_1 + d_1^2 u J_1 \neq 0$. First assume that $b_1 = d_1 = 0$. Then we have

$$\mu(\mathbf{g}_1) = 1 \cdot 1 = 1.$$

Next assume that $(b_1, d_1) \neq (0,0)$ and $b_1^2 - d_1^2 J_1 = 0$. Then we have $\nu_1 = a_1^2 - c_1^2 J_1 = (a_1 + c_1 t)(a_1 - c_1 t) \neq 0$. Since $b_1^2 - d_1^2 J_1 = (b_1 + d_1 t)(b_1 - d_1 t)$, we have

$$b_1 - d_1 t = 0 \iff b_1 + d_1 t \neq 0.$$

If $b_1 - d_1 t = 0$, then we have

$$\begin{aligned} \mu(\mathbf{g}_1) &= 1 \cdot \gamma_F(J_2, \frac{1}{2}\psi) \cdot ((a_1 b_1 + c_1 d_1 J_1)\nu_1, J_2)_F \\ &= \gamma_F(J_2, \frac{1}{2}\psi) \cdot ((a_1 b_1 + c_1 d_1 J_1)\nu_1 J_1, J_2)_F. \end{aligned}$$

If $b_1 - d_1 t \neq 0$, then we have

$$\begin{aligned} (a_1 - c_1 t)(b_1 - d_1 t) &= -2(a_1 - c_1 t) d_1 t \\ &= 2(-a_1 d_1 t + c_1 d_1 J_1) \\ &= 2(a_1 b_1 + c_1 d_1 J_1). \end{aligned}$$

Hence we have

$$\begin{aligned}\mu(\mathbf{g}_1) &= \gamma_F(J_2, \frac{1}{2}\psi) \cdot (\frac{a_1 - c_1 t}{2\nu_1 u(b_1 - d_1 t)}, J_2)_F \cdot 1 \\ &= \gamma_F(J_2, \frac{1}{2}\psi) \cdot (2\nu_1(a_1 - c_1 t)(b_1 - d_1 t), J_2)_F \cdot (u, J_2)_F \\ &= \gamma_F(J_2, \frac{1}{2}\psi) \cdot ((a_1 b_1 + c_1 d_1 J_1)\nu_1, J_2)_F \cdot (u, J_2)_F \\ &= \gamma_F(J_2, \frac{1}{2}\psi) \cdot ((a_1 b_1 + c_1 d_1 J_1)\nu_1 J_1, J_2)_F \cdot (u, J)_F.\end{aligned}$$

Finally assume that $(b_1, d_1) \neq (0,0)$ and $b_1^2 - d_1^2 J_1 \neq 0$. Then we have $b_1 - d_1 t \neq 0$. If $a_1 - c_1 t = 0$, then we have

$$\mu(\mathbf{g}_1) = 1 \cdot 1 = 1.$$

On the other hand, since $\nu_1 = -b_1^2 u + d_1^2 u J_1$, we have

$$\begin{aligned}(-(b_1^2 - d_1^2 J_1)\nu_1 J_1, J_2)_F \cdot (u, J)_F &= (\frac{\nu_1}{u} \cdot \nu_1 J_1, J_2)_F \cdot (u, J_2)_F \\ &= (\nu_1^2 J_1, J_2)_F \\ &= 1.\end{aligned}$$

If $a_1 - c_1 t \neq 0$, then we have

$$\begin{aligned}\mu(\mathbf{g}_1) &= \gamma_F(J_2, \frac{1}{2}\psi) \cdot (\frac{a_1 - c_1 t}{2\nu_1 u(b_1 - d_1 t)}, J_2)_F \cdot \gamma_F(J_2, \frac{1}{2}\psi) \cdot (\frac{2(a_1 - c_1 t)}{b_1 + d_1 t}, J_2)_F \\ &= \gamma_F(J_2, \frac{1}{2}\psi)^2 \cdot (\nu_1 u(b_1 + d_1 t)(b_1 - d_1 t), J_2)_F \\ &= (-1, J_2) \cdot ((b_1^2 - d_1^2 J_1)\nu_1, J_2)_F \cdot (u, J_2)_F \\ &= (-(b_1^2 - d_1^2 J_1)\nu_1, J_2)_F \cdot (u, J_2)_F \\ &= (-(b_1^2 - d_1^2 J_1)\nu_1 J_1, J_2)_F \cdot (u, J)_F.\end{aligned}$$

This completes the proof. □

LEMMA C.3.6. *Let* $\mathbf{g}_2 := \boldsymbol{\alpha}_2^{-1} \in \mathrm{GU}(V)^0$ *with* $\boldsymbol{\alpha}_2 = a_2 + b_2\mathbf{i} + c_2\mathbf{j}_2 + d_2\mathbf{ij}_2 \in B_2^\times$. *Then we have*

$$\mu(\mathbf{g}_2) = \begin{cases} 1 & \text{if } b_2 = d_2 = 0, \\ \gamma_F(J_1, \frac{1}{2}\psi) \cdot ((a_2 b_2 + c_2 d_2 J_2)\nu_2 J_2, J_1)_F & \text{if } (b_2, d_2) \neq (0,0),\ b_2^2 - d_2^2 J_2 = 0, \\ (-(b_2^2 - d_2^2 J_2)\nu_2 J_2, J_1)_F & \text{if } (b_2, d_2) \neq (0,0),\ b_2^2 - d_2^2 J_2 \neq 0, \end{cases}$$

where $\nu_2 = \nu(\boldsymbol{\alpha}_2)$.

PROOF. Put $\mathbf{d} := d_{\mathbb{Y}}(\nu_2) \in \mathrm{GSp}(\mathbb{V})$. We have

$$z_{\mathbb{Y}}(\mathbf{h}_0 \mathbf{g}_2 \mathbf{h}_0^{-1}, \mathbf{h}_0) = z_{\mathbb{Y}}(\mathbf{h}_0 \mathbf{g}_2 \mathbf{h}_0^{-1} \cdot \mathbf{d}^{-1}, \mathbf{d} \cdot \mathbf{h}_0 \cdot \mathbf{d}^{-1}) \cdot v_{\mathbb{Y}}(\mathbf{h}_0, \nu_2).$$

Since $\mathbb{Y}' \mathbf{g}_2 = \mathbb{Y}'$, we have $\mathbf{h}_0 \mathbf{g}_2 \mathbf{h}_0^{-1} \cdot \mathbf{d}^{-1} \in P_{\mathbb{Y}}$ and hence

$$z_{\mathbb{Y}}(\mathbf{h}_0 \mathbf{g}_2 \mathbf{h}_0^{-1} \cdot \mathbf{d}^{-1}, \mathbf{d} \cdot \mathbf{h}_0 \cdot \mathbf{d}^{-1}) = 1.$$

As in the proof of Lemma C.3.5, we have $v_{\mathbb{Y}}(\mathbf{h}_0, \nu_2) = 1$. Thus we obtain

$$z_{\mathbb{Y}}(\mathbf{h}_0 \mathbf{g}_2 \mathbf{h}_0^{-1}, \mathbf{h}_0) = 1.$$

Now we compute $z_{\mathbb{Y}}(\mathbf{h}_0, \mathbf{g}_2)$. We have

$$z_{\mathbb{Y}}(\mathbf{h}_0, \mathbf{g}_2) = z_{\mathbb{Y}}(\mathbf{h}_0, \mathbf{g}_2 \cdot \mathbf{d}^{-1}).$$

First assume that $b_2 = d_2 = 0$. Then we have $\mathbf{g}_2 \cdot \mathbf{d}^{-1} \in P_{\mathbb{Y}}$ and hence

$$z_{\mathbb{Y}}(\mathbf{h}_0, \mathbf{g}_2 \cdot \mathbf{d}^{-1}) = 1.$$

Next assume that $(b_2, d_2) \neq (0,0)$ and $b_2^2 - d_2^2 J_2 = 0$. Then we have $b_2 \neq 0$ and $d_2 \neq 0$. As in the proof of Lemma C.2.6, we have $a_2 b_2 + c_2 d_2 J_2 \neq 0$. We have $\mathbf{g}_2 \cdot \mathbf{d}^{-1} \in \mathbf{m}(\mathbf{a}_{12}) \cdot \mathbf{n}(\mathbf{b}_{30}) \cdot \tau_2 \cdot P_{\mathbb{Y}}$, where

$$\mathbf{a}_{12} = \begin{pmatrix} d_2 & & & \\ & d_2 & & \\ b_2 & & 1 & \\ & b_2 & & 1 \end{pmatrix},$$

$$\mathbf{b}_{30} = \frac{a_2 d_2 - b_2 c_2}{b_2 d_2} \cdot \begin{pmatrix} 0 & & & \\ & 0 & & \\ & & J_2 & \\ & & & -J \end{pmatrix}.$$

Hence we have

$$\begin{aligned} z_{\mathbb{Y}}(\mathbf{h}_0, \mathbf{g}_2 \cdot \mathbf{d}^{-1}) &= z_{\mathbb{Y}}(\tau_2 \cdot \mathbf{m}(\mathbf{a}_{10}), \mathbf{m}(\mathbf{a}_{12}) \cdot \mathbf{n}(\mathbf{b}_{30}) \cdot \tau_2) \\ &= z_{\mathbb{Y}}(\tau_2, \mathbf{m}(\mathbf{a}_{10}) \cdot \mathbf{m}(\mathbf{a}_{12}) \cdot \mathbf{n}(\mathbf{b}_{30}) \cdot \tau_2), \end{aligned}$$

where $\mathbf{a}_{10}$ is as in the proof of Lemma C.3.5. We have

$$\mathbf{m}(\mathbf{a}_{10}) \cdot \mathbf{m}(\mathbf{a}_{12}) \cdot \mathbf{n}(\mathbf{b}_{30}) \cdot \tau_2 \in \mathbf{m}(\mathbf{a}_{13}) \cdot \mathbf{n}(\mathbf{b}_{31}) \cdot \tau' \cdot P_{\mathbb{Y}},$$

where

$$\mathbf{a}_{13} = \begin{pmatrix} d_2 & & & \\ b_2 & 1 & & \\ & & d_2 & \\ & & b_2 & 1 \end{pmatrix},$$

$$\mathbf{b}_{31} = \frac{(a_2 d_2 - b_2 c_2) J_2}{b_2 d_2} \cdot \begin{pmatrix} 0 & & & \\ & 0 & & 1 \\ & & 0 & \\ & 1 & & 0 \end{pmatrix},$$

and τ' is as in the proof of Lemma C.2.8. Hence we have

$$\begin{aligned} z_{\mathbb{Y}}(\tau_2, \mathbf{m}(\mathbf{a}_{10}) \cdot \mathbf{m}(\mathbf{a}_{12}) \cdot \mathbf{n}(\mathbf{b}_{30}) \cdot \tau_2) &= z_{\mathbb{Y}}(\tau_2, \mathbf{m}(\mathbf{a}_{13}) \cdot \mathbf{n}(\mathbf{b}_{31}) \cdot \tau') \\ &= z_{\mathbb{Y}}(\tau_2 \cdot \mathbf{m}(\mathbf{a}_{13}) \cdot \mathbf{n}(\mathbf{b}_{31}), \tau'). \end{aligned}$$

Since $\tau_2 \cdot \mathbf{m}(\mathbf{a}_{13}) \cdot \mathbf{n}(\mathbf{b}_{31}) \cdot \tau_2^{-1} \in P_{\mathbb{Y}}$, we have

$$z_{\mathbb{Y}}(\tau_2 \cdot \mathbf{m}(\mathbf{a}_{13}) \cdot \mathbf{n}(\mathbf{b}_{31}), \tau') = z_{\mathbb{Y}}(\tau_2, \tau') = 1.$$

On the other hand, since $J_1 \in (F^\times)^2$, we have $\gamma_F(J_1, \frac{1}{2}\psi) = 1$ and

$$((a_2 b_2 + c_2 d_2 J_2)\nu_2 J_2, J_1)_F = 1.$$

Finally assume that $(b_2, d_2) \neq (0,0)$ and $b_2^2 - d_2^2 J_2 \neq 0$. We have $\mathbf{g}_2 \cdot \mathbf{d}^{-1} \in \mathbf{n}(\mathbf{b}_{32}) \cdot \tau_4 \cdot P_{\mathbb{Y}}$, where

$$\mathbf{b}_{32} = \frac{1}{b_2^2 - d_2^2 J_2} \cdot \begin{pmatrix} l & & l' & \\ & -lJ_1 & & -l'J_1 \\ l' & & lJ_2 & \\ & -l'J & & -lJ \end{pmatrix}$$

with

$$l = a_2 b_2 + c_2 d_2 J_2, \quad l' = (a_2 d_2 + b_2 c_2) J_2.$$

Hence we have

$$\begin{aligned} z_{\mathbb{Y}}(\mathbf{h}_0, \mathbf{g}_2 \cdot \mathbf{d}^{-1}) &= z_{\mathbb{Y}}(\tau_2 \cdot \mathbf{m}(\mathbf{a}_{10}), \mathbf{n}(\mathbf{b}_{32}) \cdot \tau_4) \\ &= z_{\mathbb{Y}}(\tau_2 \cdot \mathbf{m}(\mathbf{a}_{10}), \mathbf{n}(\mathbf{b}_{32}) \cdot \mathbf{m}(\mathbf{a}_{10})^{-1} \cdot \tau_4) \\ &= z_{\mathbb{Y}}(\tau_2 \cdot \mathbf{m}(\mathbf{a}_{10}) \cdot \mathbf{n}(\mathbf{b}_{32}) \cdot \mathbf{m}(\mathbf{a}_{10})^{-1}, \tau_4). \end{aligned}$$

Since $\tau_2 \cdot \mathbf{m}(\mathbf{a}_{10}) \cdot \mathbf{n}(\mathbf{b}_{32}) \cdot \mathbf{m}(\mathbf{a}_{10})^{-1} \cdot \tau_2^{-1} \in P_{\mathbb{Y}}$, we have

$$z_{\mathbb{Y}}(\mathbf{h}_0, \mathbf{g}_2 \cdot \mathbf{d}^{-1}) = z_{\mathbb{Y}}(\tau_2, \tau_4) = 1.$$

On the other hand, since $J_1 \in (F^\times)^2$, we have

$$(-(b_2^2 - d_2^2 J_2)\nu_2 J_2, J_1)_F = 1.$$

This completes the proof. □

LEMMA C.3.7. *Let $\mathbf{g} := \boldsymbol{\alpha} \in \mathrm{GU}(W)$ with $\boldsymbol{\alpha} = a + b\mathbf{i} + c\mathbf{j} + d\mathbf{ij} \in B^\times$. Then, for any $i \in \{1,2\}$, we have*

$$\mu(\mathbf{g}) = \begin{cases} (\nu, J_i)_F & \text{if } b = d = 0, \\ \gamma_F(J_i, \frac{1}{2}\psi) \cdot (ab - cdJ, J_i)_F & \text{if } (b,d) \neq (0,0),\ b^2 - d^2 J = 0, \\ (-(b^2 - d^2 J)J, J_i)_F & \text{if } (b,d) \neq (0,0),\ b^2 - d^2 J \neq 0, \end{cases}$$

where $\nu = \nu(\boldsymbol{\alpha})$.

PROOF. Put $\mathbf{d} := d_{\mathbb{Y}}(\nu) \in \mathrm{GSp}(\mathbb{V})$. We have

$$z_{\mathbb{Y}}(\mathbf{h}_0 \mathbf{g} \mathbf{h}_0^{-1}, \mathbf{h}_0) = z_{\mathbb{Y}}(\mathbf{h}_0 \mathbf{g} \mathbf{h}_0^{-1} \cdot \mathbf{d}^{-1}, \mathbf{d} \cdot \mathbf{h}_0 \cdot \mathbf{d}^{-1}) \cdot v_{\mathbb{Y}}(\mathbf{h}_0, \nu).$$

Since $\mathbb{Y}'\mathbf{g} = \mathbb{Y}'$, we have $\mathbf{h}_0 \mathbf{g} \mathbf{h}_0^{-1} \cdot \mathbf{d}^{-1} \in P_{\mathbb{Y}}$ and hence

$$z_{\mathbb{Y}}(\mathbf{h}_0 \mathbf{g} \mathbf{h}_0^{-1} \cdot \mathbf{d}^{-1}, \mathbf{d} \cdot \mathbf{h}_0 \cdot \mathbf{d}^{-1}) = 1.$$

As in the proof of Lemma C.3.5, we have $v_{\mathbb{Y}}(\mathbf{h}_0, \nu) = 1$. Thus we obtain

$$z_{\mathbb{Y}}(\mathbf{h}_0 \mathbf{g} \mathbf{h}_0^{-1}, \mathbf{h}_0) = 1.$$

Now we compute $z_{\mathbb{Y}}(\mathbf{h}_0, \mathbf{g})$. We have

$$z_{\mathbb{Y}}(\mathbf{h}_0, \mathbf{g}) = z_{\mathbb{Y}}(\mathbf{h}_0, \mathbf{g} \cdot \mathbf{d}^{-1}).$$

First assume that $b = d = 0$. Then we have $\mathbf{g} \cdot \mathbf{d}^{-1} \in P_{\mathbb{Y}}$ and hence

$$z_{\mathbb{Y}}(\mathbf{h}_0, \mathbf{g} \cdot \mathbf{d}^{-1}) = 1.$$

On the other hand, since $J_1 \in (F^\times)^2$, we have $(\nu, J_1)_F = 1$ and

$$(\nu, J_2)_F = (\nu, J)_F = (a^2 - c^2 J, J)_F = 1.$$

Next assume that $(b,d) \neq (0,0)$ and $b^2 - d^2 J = 0$. Then we have $b \neq 0$ and $d \neq 0$. As in the proof of Lemma C.2.9, we have $ab - cdJ \neq 0$. We have $\mathbf{g} \cdot \mathbf{d}^{-1} \in$

$\mathbf{m}(\mathbf{a}_3) \cdot \mathbf{n}(\mathbf{b}_9) \cdot \tau_2 \cdot P_{\mathbb{Y}}$, where $\mathbf{a}_3$ and $\mathbf{b}_9$ are as in the proof of Lemma C.2.7. Hence we have

$$\begin{aligned} z_{\mathbb{Y}}(\mathbf{h}_0, \mathbf{g} \cdot \mathbf{d}^{-1}) &= z_{\mathbb{Y}}(\tau_2 \cdot \mathbf{m}(\mathbf{a}_{10}), \mathbf{m}(\mathbf{a}_3) \cdot \mathbf{n}(\mathbf{b}_9) \cdot \tau_2) \\ &= z_{\mathbb{Y}}(\tau_2, \mathbf{m}(\mathbf{a}_{10}) \cdot \mathbf{m}(\mathbf{a}_3) \cdot \mathbf{n}(\mathbf{b}_9) \cdot \tau_2), \end{aligned}$$

where $\mathbf{a}_{10}$ is as in the proof of Lemma C.3.5. We have

$$\mathbf{m}(\mathbf{a}_{10}) \cdot \mathbf{m}(\mathbf{a}_3) \cdot \mathbf{n}(\mathbf{b}_9) \cdot \tau_2 \in \mathbf{m}(\mathbf{a}_{14}) \cdot \mathbf{n}(\mathbf{b}_{33}) \cdot \tau' \cdot P_{\mathbb{Y}},$$

where

$$\mathbf{a}_{14} = \begin{pmatrix} bt & & & \\ -dJ & 1 & & \\ & & bt & \\ & & dJ & 1 \end{pmatrix},$$

$$\mathbf{b}_{33} = -\frac{(ad+bc)J_2}{bd} \cdot \begin{pmatrix} 0 & & & \\ & 0 & & 1 \\ & & 0 & \\ & 1 & & 0 \end{pmatrix},$$

and τ' is as in the proof of Lemma C.2.8. Hence we have

$$\begin{aligned} z_{\mathbb{Y}}(\tau_2, \mathbf{m}(\mathbf{a}_{10}) \cdot \mathbf{m}(\mathbf{a}_3) \cdot \mathbf{n}(\mathbf{b}_9) \cdot \tau_2) &= z_{\mathbb{Y}}(\tau_2, \mathbf{m}(\mathbf{a}_{14}) \cdot \mathbf{n}(\mathbf{b}_{33}) \cdot \tau') \\ &= z_{\mathbb{Y}}(\tau_2 \cdot \mathbf{m}(\mathbf{a}_{14}) \cdot \mathbf{n}(\mathbf{b}_{33}), \tau'). \end{aligned}$$

Since $\tau_2 \cdot \mathbf{m}(\mathbf{a}_{14}) \cdot \mathbf{n}(\mathbf{b}_{33}) \cdot \tau_2^{-1} \in P_{\mathbb{Y}}$, we have

$$z_{\mathbb{Y}}(\tau_2 \cdot \mathbf{m}(\mathbf{a}_{14}) \cdot \mathbf{n}(\mathbf{b}_{33}), \tau') = z_{\mathbb{Y}}(\tau_2, \tau') = 1.$$

On the other hand, since $J_1 \in (F^\times)^2$ and $J \in (F^\times)^2$, we have $\gamma_F(J_1, \frac{1}{2}\psi) = \gamma_F(J_2, \frac{1}{2}\psi) = 1$ and

$$(ab - cdJ, J_1)_F = (ab - cdJ, J_2)_F = 1.$$

Finally assume that $(b, d) \neq (0, 0)$ and $b^2 - d^2 J \neq 0$. We have $\mathbf{g} \cdot \mathbf{d}^{-1} \in \mathbf{n}(\mathbf{b}_{13}) \cdot \tau_4 \cdot P_{\mathbb{Y}}$, where $\mathbf{b}_{13}$ is as in the proof of Lemma C.2.7. Hence we have

$$\begin{aligned} z_{\mathbb{Y}}(\mathbf{h}_0, \mathbf{g} \cdot \mathbf{d}^{-1}) &= z_{\mathbb{Y}}(\tau_2 \cdot \mathbf{m}(\mathbf{a}_{10}), \mathbf{n}(\mathbf{b}_{13}) \cdot \tau_4) \\ &= z_{\mathbb{Y}}(\tau_2 \cdot \mathbf{m}(\mathbf{a}_{10}), \mathbf{n}(\mathbf{b}_{13}) \cdot \mathbf{m}(\mathbf{a}_{10})^{-1} \cdot \tau_4) \\ &= z_{\mathbb{Y}}(\tau_2 \cdot \mathbf{m}(\mathbf{a}_{10}) \cdot \mathbf{n}(\mathbf{b}_{13}) \cdot \mathbf{m}(\mathbf{a}_{10})^{-1}, \tau_4). \end{aligned}$$

Since $\tau_2 \cdot \mathbf{m}(\mathbf{a}_{10}) \cdot \mathbf{n}(\mathbf{b}_{13}) \cdot \mathbf{m}(\mathbf{a}_{10})^{-1} \cdot \tau_2^{-1} \in P_{\mathbb{Y}}$, we have

$$z_{\mathbb{Y}}(\mathbf{h}_0, \mathbf{g} \cdot \mathbf{d}^{-1}) = z_{\mathbb{Y}}(\tau_2, \tau_4) = 1.$$

On the other hand, since $J_1 \in (F^\times)^2$, we have $(-(b^2 - d^2 J)J, J_1)_F = 1$ and

$$(-(b^2 - d^2 J)J, J_2)_F = (-(b^2 - d^2 J)J, J)_F = (b^2 - d^2 J, J)_F = 1.$$

This completes the proof. $\square$

C.4. The product formula

Suppose that F is a number field. First we fix quaternion algebras B_1 and B_2 over F. Next we fix a quadratic extension E of F such that E embeds into B_1 and B_2. Let B be the quaternion algebra over F which is the product of B_1 and B_2 in the Brauer group. Then E also embeds into B.

Fix a finite set Σ of places of F containing

$$\Sigma_\infty \cup \Sigma_2 \cup \Sigma_E \cup \Sigma_B \cup \Sigma_{B_1} \cup \Sigma_{B_2}.$$

Here Σ_∞ is the set of archimedean places of F, Σ_2 is the set of places of F lying above 2, and $\Sigma_\bullet$ is the set of places v of F such that $\bullet_v$ is ramified over F_v for $\bullet = E, B, B_1, B_2$.

We write $B_i = E + E\mathbf{j}_i$. Put $J_i = \mathbf{j}_i^2$ and $J = J_1 J_2$. We may write $B = E + E\mathbf{j}$ such that $\mathbf{j}^2 = J$. Then, for each place v of F, we have

- $J \in \mathrm{N}_{E_v/F_v}(E_v^\times)$ if $v \notin \Sigma_B$,
- $J_1 \in \mathrm{N}_{E_v/F_v}(E_v^\times)$ if $v \notin \Sigma_{B_1}$,
- $J_2 \in \mathrm{N}_{E_v/F_v}(E_v^\times)$ if $v \notin \Sigma_{B_2}$.

By using the weak approximation theorem and replacing $\mathbf{j}_i$ by $\alpha_i \mathbf{j}_i$ with some $\alpha_i \in E^\times$ if necessary, we may assume that

$$J \in (F_v^\times)^2 \quad \text{or} \quad J_1 \in (F_v^\times)^2 \quad \text{or} \quad J_2 \in (F_v^\times)^2$$

for all $v \in \Sigma$.

LEMMA C.4.1. *We have*

$$u \in (F_v^\times)^2 \quad \textit{or} \quad J \in (F_v^\times)^2 \quad \textit{or} \quad J_1 \in (F_v^\times)^2 \quad \textit{or} \quad J_2 \in (F_v^\times)^2$$

for all $v \notin \Sigma$.

PROOF. Let $v \notin \Sigma$. We may assume that v is inert in E. Assume that $J_i \notin (F_v^\times)^2$ for $i = 1, 2$. Since $J_i \in \mathrm{N}_{E_v/F_v}(E_v^\times)$, we have $J_i \in \varepsilon \cdot (F_v^\times)^2$ for $i = 1, 2$, where $\varepsilon \in \mathfrak{o}_{F_v}^\times$ but $\varepsilon \notin (F_v^\times)^2$. Hence we have

$$J = J_1 J_2 \in (F_v^\times)^2.$$

This yields the lemma. □

Thus, for each place v of F, we can define a map

$$s_v : \mathrm{GU}(V_v)^0 \times \mathrm{GU}(W_v) \longrightarrow \mathbb{C}^1$$

by $s_v := s'_v \cdot \mu_v$, where s'_v and μ_v are as in §§C.2, C.3. Here, for $\bullet = u, J, J_1, J_2$ with $\bullet \in (F_v^\times)^2$, we have chosen $t \in F_v^\times$ such that $\bullet = t^2$. Recall that

$$z_{\mathbb{Y}_v}(\mathbf{g}, \mathbf{g}') = s_v(\mathbf{g}\mathbf{g}') \cdot s_v(\mathbf{g})^{-1} \cdot s_v(\mathbf{g}')^{-1}$$

for $\mathbf{g}, \mathbf{g}' \in \mathrm{GU}(V_v)^0 \times \mathrm{GU}(W_v)$.

PROPOSITION C.4.2. (i) *Let* $\mathbf{g}_i := \boldsymbol{\alpha}_i^{-1} \in \mathrm{GU}(V_v)^0$ *with* $\boldsymbol{\alpha}_i = a_i + b_i\mathbf{i} + c_i\mathbf{j}_i + d_i\mathbf{i}\mathbf{j}_i \in B_{i,v}^\times$. *Then we have*

$$s_v(\mathbf{g}_i) = \begin{cases} 1 & \textit{if } b_i = d_i = 0, \\ \gamma_{F_v}(J_j, \frac{1}{2}\psi_v) \cdot ((a_i b_i + c_i d_i J_i)\nu_i J_i, J_j)_{F_v} & \textit{if } (b_i, d_i) \neq (0,0),\ b_i^2 - d_i^2 J_i = 0, \\ (-(b_i^2 - d_i^2 J_i)\nu_i J_i, J_j)_{F_v} & \textit{if } (b_i, d_i) \neq (0,0),\ b_i^2 - d_i^2 J_i \neq 0, \end{cases}$$

where $\nu_i = \nu(\boldsymbol{\alpha}_i)$ *and* $\{i, j\} = \{1, 2\}$.

(ii) *Let* $\mathbf{g} := \boldsymbol{\alpha} \in \mathrm{GU}(W_v)$ *with* $\boldsymbol{\alpha} = a + b\mathbf{i} + c\mathbf{j} + d\mathbf{ij} \in B_v^\times$. *Then we have*

$$s_v(\mathbf{g}) = \begin{cases} (\nu, J_1)_{F_v} & \text{if } b = d = 0, \\ \gamma_{F_v}(J_1, \frac{1}{2}\psi_v) \cdot (ab - cdJ, J_1)_{F_v} & \text{if } (b,d) \neq (0,0),\ b^2 - d^2 J = 0, \\ (-(b^2 - d^2 J)J, J_1)_{F_v} & \text{if } (b,d) \neq (0,0),\ b^2 - d^2 J \neq 0, \end{cases}$$

where $\nu = \nu(\boldsymbol{\alpha})$.

PROOF. If $u \in (F_v^\times)^2$, then $B_{i,v}$ is split and the assertion follows from Lemmas C.2.6 and C.2.7.

Assume that $J \in (F_v^\times)^2$. Let $\mathfrak{i} : B_v \to \mathrm{M}_2(F_v)$ be the isomorphism as in §C.2. Since

$$\mathfrak{i}(\boldsymbol{\alpha}) = \begin{pmatrix} a + ct & b - dt \\ u(b + dt) & a - ct \end{pmatrix},$$

we have

$$j(\mathbf{g}) = \begin{cases} 0 & \text{if } b + dt = 0, \\ 1 & \text{if } b + dt \neq 0. \end{cases}$$

Since

$$(u, J_1)_{F_v} = \begin{cases} 1 & \text{if } B_{1,v} \text{ is split}, \\ -1 & \text{if } B_{1,v} \text{ is ramified}, \end{cases}$$

the assertion follows from Lemmas C.2.8 and C.2.9.

Assume that $J_i \in (F_v^\times)^2$. We only consider the case $i = 1$; the other case is similar. Let $\mathfrak{i}_1 : B_{1,v} \to \mathrm{M}_2(F_v)$ be the isomorphism as in §C.3. Since

$${}^t\mathfrak{i}_1(\boldsymbol{\alpha}_1) = \begin{pmatrix} a_1 + c_1 t & \frac{u}{2}(b_1 + d_1 t) \\ 2(b_1 - d_1 t) & a_1 - c_1 t \end{pmatrix},$$

we have

$$j(\mathbf{g}_1) = \begin{cases} 0 & \text{if } b_1 - d_1 t = 0, \\ 1 & \text{if } b_1 - d_1 t \neq 0. \end{cases}$$

Also, we have $j(\mathbf{g}_2) = 0$. Since

$$(u, J)_{F_v} = \begin{cases} 1 & \text{if } B_v \text{ is split}, \\ -1 & \text{if } B_v \text{ is ramified}, \end{cases}$$

the assertion follows from Lemmas C.3.5, C.3.6, and C.3.7. □

Recall that, for almost all v, we have a maximal compact subgroup K_v of $\mathrm{Sp}(\mathbb{V}_v)$ and a map $s_{\mathbb{Y}_v} : K_v \to \mathbb{C}^1$ such that

$$z_{\mathbb{Y}_v}(k, k') = s_{\mathbb{Y}_v}(kk') \cdot s_{\mathbb{Y}_v}(k)^{-1} \cdot s_{\mathbb{Y}_v}(k')^{-1}$$

for $k, k' \in K_v$. Put

$$\mathbf{K}_v := \mathrm{G}(\mathrm{U}(V_v) \times \mathrm{U}(W_v))^0 \cap K_v.$$

Then $\mathbf{K}_v$ is a maximal compact subgroup of $\mathrm{G}(\mathrm{U}(V_v) \times \mathrm{U}(W_v))^0$ for almost all v.

LEMMA C.4.3. *We have*

$$s_v|_{\mathbf{K}_v} = s_{\mathbb{Y}_v}|_{\mathbf{K}_v}$$

for almost all v.

PROOF. Recall that $s_v(\mathbf{g}) = s'_v(\mathbf{g}) \cdot \mu_v(\mathbf{g})$ for $\mathbf{g} \in \mathrm{GU}(V_v)^0 \times \mathrm{GU}(W_v)$, where

$$s'_v : \mathrm{GU}(V_v)^0 \times \mathrm{GU}(W_v) \longrightarrow \mathbb{C}^1$$

is the map as in §§C.2, C.3 and

$$\mu_v(\mathbf{g}) = z_{\mathbb{Y}_v}(\mathbf{h}_0\mathbf{g}\mathbf{h}_0^{-1}, \mathbf{h}_0) \cdot z_{\mathbb{Y}_v}(\mathbf{h}_0, \mathbf{g})^{-1}$$

for $\mathbf{g} \in \mathrm{GSp}(\mathbb{V}_v)$ with some $\mathbf{h}_0 \in \mathrm{Sp}(\mathbb{V}_v)$ such that $\mathbb{X}'_v = \mathbb{X}_v\mathbf{h}_0$ and $\mathbb{Y}'_v = \mathbb{Y}_v\mathbf{h}_0$. By the uniqueness of the splitting, we have

$$s_{\mathbb{Y}_v} = s_{\mathbb{Y}'_v} \cdot \mu_v|_{K_v}$$

for almost all v. On the other hand, by definition, one can see that

$$s'_v|_{\mathbf{K}_v} = s_{\mathbb{Y}'_v}|_{\mathbf{K}_v}$$

for almost all v. This yields the lemma. □

PROPOSITION C.4.4. *Let $\gamma \in \mathrm{GU}(V)^0(F) \times \mathrm{GU}(W)(F)$. Then we have $s_v(\gamma) = 1$ for almost all v and*

$$\prod_v s_v(\gamma) = 1.$$

PROOF. Let $\gamma_1, \gamma_2 \in \mathrm{GU}(V)^0(F) \times \mathrm{GU}(W)(F)$. Suppose that $s_v(\gamma_i) = 1$ for almost all v and $\prod_v s_v(\gamma_i) = 1$ for $i = 1, 2$. Since $s_v(\gamma_1\gamma_2) = s_v(\gamma_1) \cdot s_v(\gamma_2) \cdot z_{\mathbb{Y}_v}(\gamma_1, \gamma_2)$, the product formulas for the quadratic Hilbert symbol and the Weil index imply that $s_v(\gamma_1\gamma_2) = 1$ for almost all v and $\prod_v s_v(\gamma_1\gamma_2) = 1$. Hence the assertion follows from Proposition C.4.2. □

APPENDIX D

Splittings for the Doubling Method: Quaternionic Unitary Groups

In this section, we show that the splitting constructed in Appendix C is compatible with the one constructed in [**40**, §4] via the doubling method. This compatibility (Lemma D.4.2) is used in Chapter 4 to prove the Rallis inner product formula.

D.1. Setup

Let F be a number field and B a quaternion algebra over F. Recall that

- V is a 2-dimensional right skew-hermitian B-space with $\det V = 1$,
- W is a 1-dimensional left hermitian B-space,
- $\mathbb{V} := V \otimes_B W$ is an 8-dimensional symplectic F-space,
- $\mathbb{V} = \mathbb{X} \oplus \mathbb{Y}$ is a complete polarization over F.

We consider a 2-dimensional left B-space $W^\square := W \oplus W$ equipped with a hermitian form

$$\langle (x, x'), (y, y') \rangle := \langle x, y \rangle - \langle x', y' \rangle$$

for $x, x', y, y' \in W$. Put $W_+ := W \oplus \{0\}$ and $W_- := \{0\} \oplus W$. We regard $\mathrm{GU}(W_\pm)$ as a subgroup of $\mathrm{GL}(W)$ and identify it with $\mathrm{GU}(W)$ via the identity map. Note that the identity map $\mathrm{GU}(W_-) \to \mathrm{GU}(W)$ is induced by an anti-isometry $W_- \to W$. We have a natural map

$$\iota : \mathrm{G}(\mathrm{U}(W) \times \mathrm{U}(W)) \longrightarrow \mathrm{GU}(W^\square)$$

and seesaw dual pairs

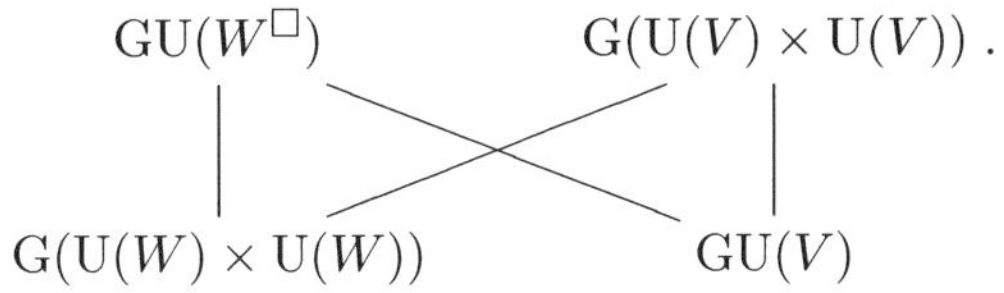

Put

$$W^\triangle := \{(x, x) \in W^\square \mid x \in W\},$$
$$W^\triangledown := \{(x, -x) \in W^\square \mid x \in W\}.$$

Then $W^\square = W^\triangledown \oplus W^\triangle$ is a complete polarization over B. Choosing a basis $\mathbf{w}, \mathbf{w}^*$ of $W^\square$ such that

$$W^\triangledown = B\mathbf{w}, \quad W^\triangle = B\mathbf{w}^*, \quad \langle \mathbf{w}, \mathbf{w}^* \rangle = 1,$$

we may write

$$\mathrm{GU}(W^\square) = \left\{ g \in \mathrm{GL}_2(B) \;\middle|\; g \begin{pmatrix} & 1 \\ 1 & \end{pmatrix} {}^t g^* = \nu(g) \cdot \begin{pmatrix} & 1 \\ 1 & \end{pmatrix} \right\}.$$

For $\nu \in F^\times$, put

$$d(\nu) = d_{W^\triangle}(\nu) := \begin{pmatrix} 1 & \\ & \nu \end{pmatrix} \in \mathrm{GU}(W^\square).$$

Similarly, we consider a 16-dimensional F-space $\mathbb{V}^\square := V \otimes_B W^\square = \mathbb{V} \oplus \mathbb{V}$ equipped with a symplectic form

$$\langle\!\langle (x, x'), (y, y') \rangle\!\rangle := \langle\!\langle x, y \rangle\!\rangle - \langle\!\langle x', y' \rangle\!\rangle \tag{D.1}$$

for $x, x', y, y' \in \mathbb{V}$. Put $\mathbb{V}_+ := \mathbb{V} \oplus \{0\}$ and $\mathbb{V}_- := \{0\} \oplus \mathbb{V}$. We regard $\mathrm{Sp}(\mathbb{V}_\pm)$ as a subgroup of $\mathrm{GL}(\mathbb{V})$ and identify it with $\mathrm{Sp}(\mathbb{V})$ via the identity map. Note that the identity map $\mathrm{Sp}(\mathbb{V}_-) \to \mathrm{Sp}(\mathbb{V})$ is induced by an anti-isometry $\mathbb{V}_- \to \mathbb{V}$. We have a natural map

$$\iota : \mathrm{Sp}(\mathbb{V}) \times \mathrm{Sp}(\mathbb{V}) \longrightarrow \mathrm{Sp}(\mathbb{V}^\square).$$

Put

$$\mathbb{V}^\triangle := V \otimes_B W^\triangle = \{(x, x) \in \mathbb{V}^\square \mid x \in \mathbb{V}\}, \qquad \mathbb{X}^\square := \mathbb{X} \oplus \mathbb{X},$$
$$\mathbb{V}^\nabla := V \otimes_B W^\nabla = \{(x, -x) \in \mathbb{V}^\square \mid x \in \mathbb{V}\}, \qquad \mathbb{Y}^\square := \mathbb{Y} \oplus \mathbb{Y}.$$

Then $\mathbb{V}^\square = \mathbb{V}^\nabla \oplus \mathbb{V}^\triangle = \mathbb{X}^\square \oplus \mathbb{Y}^\square$ are complete polarizations over F.

For the rest of this section, we fix a place v of F and suppress the subscript v from the notation. Thus $F = F_v$ will be a local field of characteristic zero. We may lift the natural map $\iota : \mathrm{Sp}(\mathbb{V}) \times \mathrm{Sp}(\mathbb{V}) \to \mathrm{Sp}(\mathbb{V}^\square)$ to a unique homomorphism

$$\tilde{\iota} : \mathrm{Mp}(\mathbb{V}) \times \mathrm{Mp}(\mathbb{V}) \longrightarrow \mathrm{Mp}(\mathbb{V}^\square)$$

such that $\tilde{\iota}(z_1, z_2) = z_1 z_2^{-1}$ for $z_1, z_2 \in \mathbb{C}^1$.

D.2. Splitting $z_{\mathbb{V}^\triangle}$

First assume that B is split. Fix an isomorphism $\mathfrak{i} : B \to \mathrm{M}_2(F)$. Put $e = \mathfrak{i}^{-1}\left(\begin{smallmatrix} 1 & 0 \\ 0 & 0 \end{smallmatrix}\right)$ and $e' = \mathfrak{i}^{-1}\left(\begin{smallmatrix} 0 & 1 \\ 0 & 0 \end{smallmatrix}\right)$. Then $W^{\square\dagger} := eW^\square$ is a 4-dimensional symplectic F-space and the restriction $\mathrm{GU}(W^\square) \to \mathrm{GSp}(W^{\square\dagger})$ is an isomorphism. Using a basis $e\mathbf{w}, e'\mathbf{w}, e'\mathbf{w}^*, -e\mathbf{w}^*$ of $W^{\square\dagger}$, we write

$$\mathrm{GSp}(W^{\square\dagger}) = \left\{ h \in \mathrm{GL}_4(F) \;\middle|\; h \begin{pmatrix} & \mathbf{1}_2 \\ -\mathbf{1}_2 & \end{pmatrix} {}^t h = \nu(h) \cdot \begin{pmatrix} & \mathbf{1}_2 \\ -\mathbf{1}_2 & \end{pmatrix} \right\}.$$

Then the restriction $\mathrm{GU}(W^\square) \to \mathrm{GSp}(W^{\square\dagger})$ is given by

$$\begin{pmatrix} a & b \\ c & d \end{pmatrix} \mapsto \begin{pmatrix} \mathbf{1}_2 & \\ & \tau^{-1} \end{pmatrix} \cdot \begin{pmatrix} \mathfrak{i}(a) & \mathfrak{i}(b) \\ \mathfrak{i}(c) & \mathfrak{i}(d) \end{pmatrix} \cdot \begin{pmatrix} \mathbf{1}_2 & \\ & \tau \end{pmatrix},$$

where $\tau = \tau_1 = \left(\begin{smallmatrix} & -1 \\ 1 & \end{smallmatrix}\right)$. Note that $x^* = \tau \cdot {}^t x \cdot \tau^{-1}$ for $x \in \mathrm{M}_2(F)$. Also, $W^\dagger := eW$ is a 2-dimensional symplectic F-space and $V^\dagger := Ve$ is a 4-dimensional quadratic F-space. We define a map

$$\hat{s} : \mathrm{G}(\mathrm{U}(V) \times \mathrm{U}(W^\square)) \longrightarrow \mathbb{C}^1$$

by

$$\hat{s}(\mathbf{g}) = \gamma^{\hat{\jmath}(h)}$$

for $\mathbf{g} = (g, h) \in \mathrm{G}(\mathrm{U}(V) \times \mathrm{U}(W^\square))$, where

$$\gamma = \begin{cases} 1 & \text{if } V^\dagger \text{ is isotropic,} \\ -1 & \text{if } V^\dagger \text{ is anisotropic,} \end{cases}$$

and

$$\jmath(h) = \begin{cases} 0 & \text{if } c = 0, \\ 1 & \text{if } c \neq 0,\ \det c = 0, \\ 2 & \text{if } \det c \neq 0, \end{cases} \qquad h = \begin{pmatrix} a & b \\ c & d \end{pmatrix} \in \mathrm{GSp}(W^{\square\dagger}).$$

Since $\dim_F V^\dagger = 4$ and $\det V^\dagger = 1$, we have

$$z_{\mathbb{V}^\triangle}(h, h') = \hat{s}(hh') \cdot \hat{s}(h)^{-1} \cdot \hat{s}(h')^{-1}$$

for $h, h' \in \mathrm{U}(W^\square)$ by [**40**, Theorem 3.1, case 1_+].

Next assume that B is ramified. We define a map

$$\hat{s} : \mathrm{G}(\mathrm{U}(V) \times \mathrm{U}(W^\square)) \longrightarrow \mathbb{C}^1$$

by

$$\hat{s}(\mathbf{g}) = 1$$

for $\mathbf{g} \in \mathrm{G}(\mathrm{U}(V) \times \mathrm{U}(W^\square))$. Since $\dim_B V = 2$ and $\det V = 1$, we have

$$z_{\mathbb{V}^\triangle}(h, h') = \hat{s}(hh') \cdot \hat{s}(h)^{-1} \cdot \hat{s}(h')^{-1}$$

for $h, h' \in \mathrm{U}(W^\square)$ by [**40**, Theorem 3.1, case 2_-].

LEMMA D.2.1. *We have*

$$z_{\mathbb{V}^\triangle}(\mathbf{g}, \mathbf{g}') = \hat{s}(\mathbf{g}\mathbf{g}') \cdot \hat{s}(\mathbf{g})^{-1} \cdot \hat{s}(\mathbf{g}')^{-1}$$

for $\mathbf{g}, \mathbf{g}' \in \mathrm{G}(\mathrm{U}(V) \times \mathrm{U}(W^\square))$.

PROOF. Let $\mathbf{g}_i = (g_i, h_i) \in \mathrm{G}(\mathrm{U}(V) \times \mathrm{U}(W^\square))$ and put $h_i' = h_i \cdot d(\nu(h_i))^{-1} \in \mathrm{U}(W^\square)$. Then we have $h_1 h_2 = h_1' h_2'' \cdot d(\nu(h_1 h_2))$, where $h_2'' = d(\nu(h_1)) \cdot h_2' \cdot d(\nu(h_1))^{-1}$. Since

$$\mathbb{V}^\triangle \cdot g = \mathbb{V}^\triangle, \quad \mathbb{V}^\triangle \cdot d(\nu) = \mathbb{V}^\triangle$$

for $g \in \mathrm{GU}(V)$ and $\nu \in F^\times$, we have

$$\begin{aligned} \mathbb{V}^\triangle \cdot \mathbf{g}_1^{-1} &= \mathbb{V}^\triangle \cdot h_1^{-1} = \mathbb{V}^\triangle \cdot h_1'^{-1}, \\ \mathbb{V}^\triangle \cdot \mathbf{g}_2^{-1}\mathbf{g}_1^{-1} &= \mathbb{V}^\triangle \cdot h_2^{-1} h_1^{-1} = \mathbb{V}^\triangle \cdot h_2''^{-1} h_1'^{-1}. \end{aligned}$$

Hence we have

$$\begin{aligned} q(\mathbb{V}^\triangle, \mathbb{V}^\triangle \cdot \mathbf{g}_2^{-1}, \mathbb{V}^\triangle \cdot \mathbf{g}_1) &= q(\mathbb{V}^\triangle \cdot \mathbf{g}_1^{-1}, \mathbb{V}^\triangle \cdot \mathbf{g}_2^{-1}\mathbf{g}_1^{-1}, \mathbb{V}^\triangle) \\ &= q(\mathbb{V}^\triangle \cdot h_1'^{-1}, \mathbb{V}^\triangle \cdot h_2''^{-1} h_1'^{-1}, \mathbb{V}^\triangle) \\ &= q(\mathbb{V}^\triangle, \mathbb{V}^\triangle \cdot h_2''^{-1}, \mathbb{V}^\triangle \cdot h_1'), \end{aligned}$$

so that

$$z_{\mathbb{V}^\triangle}(\mathbf{g}_1, \mathbf{g}_2) = z_{\mathbb{V}^\triangle}(h_1', h_2'') = \hat{s}(h_1' h_2'') \cdot \hat{s}(h_1')^{-1} \cdot \hat{s}(h_2'')^{-1}.$$

By definition, we have $\hat{s}(h_1') = \hat{s}(\mathbf{g}_1)$, $\hat{s}(h_2'') = \hat{s}(h_2') = \hat{s}(\mathbf{g}_2)$, and

$$\hat{s}(h_1' h_2'') = \hat{s}(h_1 h_2 \cdot d(\nu(h_1 h_2))^{-1}) = \hat{s}(\mathbf{g}_1 \mathbf{g}_2).$$

This completes the proof. □

D.3. Splitting $z_{\mathbb{Y}'^\square}$

Let $\mathbb{V} = \mathbb{X}' \oplus \mathbb{Y}'$ be the complete polarization given in §C.2, C.3. Put

$$\mathbb{X}'^\square := \mathbb{X}' \oplus \mathbb{X}', \quad \mathbb{Y}'^\square := \mathbb{Y}' \oplus \mathbb{Y}'.$$

Then $\mathbb{V}^\square = \mathbb{X}'^\square \oplus \mathbb{Y}'^\square$ is a complete polarization. Noting that the symplectic form on $\mathbb{V}^\square = \mathbb{V} \oplus \mathbb{V}$ is given by (D.1), we have

$$\begin{aligned} z_{\mathbb{Y}'^\square,\psi}(\iota(g_1, g_2), \iota(g_1', g_2')) &= z_{\mathbb{Y}',\psi}(g_1, g_1') \cdot z_{\mathbb{Y}',\psi^{-1}}(g_2, g_2') \\ &= z_{\mathbb{Y}',\psi}(g_1, g_1') \cdot z_{\mathbb{Y}',\psi}(g_2, g_2')^{-1} \end{aligned}$$

for $g_i, g_i' \in \mathrm{Sp}(\mathbb{V})$, where we write $z_{\mathbb{Y}'} = z_{\mathbb{Y}',\psi}$ to indicate the dependence of the 2-cocycle on ψ. The Weil representation $\omega_\psi^\square$ of $\mathrm{Mp}(\mathbb{V}^\square)$ can be realized on the Schwartz space

$$\mathcal{S}(\mathbb{X}'^\square) = \mathcal{S}(\mathbb{X}') \otimes \mathcal{S}(\mathbb{X}').$$

As representations of $\mathrm{Mp}(\mathbb{V})_{\mathbb{Y}'} \times \mathrm{Mp}(\mathbb{V})_{\mathbb{Y}'}$, we have

$$\omega_\psi^\square \circ \tilde{\iota} = \omega_\psi \otimes (\omega_\psi \circ \tilde{\mathrm{j}}_{\mathbb{Y}'}),$$

where $\tilde{\mathrm{j}}_{\mathbb{Y}'}$ is the automorphism of $\mathrm{Mp}(\mathbb{V})_{\mathbb{Y}'} = \mathrm{Sp}(\mathbb{V}) \times \mathbb{C}^1$ defined by

$$\tilde{\mathrm{j}}_{\mathbb{Y}'}(g, z) = (\mathrm{j}_{\mathbb{Y}'}(g), z^{-1}), \quad \mathrm{j}_{\mathbb{Y}'}(g) = d_{\mathbb{Y}'}(-1) \cdot g \cdot d_{\mathbb{Y}'}(-1).$$

Fix $\mathbf{h}_0' \in \mathrm{Sp}(\mathbb{V}^\square)$ such that $\mathbb{X}'^\square = \mathbb{V}^\nabla \cdot \mathbf{h}_0'$ and $\mathbb{Y}'^\square = \mathbb{V}^\triangle \cdot \mathbf{h}_0'$. Put

$$\mu'(g) = z_{\mathbb{V}^\triangle}(g, \mathbf{h}_0'^{-1}) \cdot z_{\mathbb{V}^\triangle}(\mathbf{h}_0'^{-1}, \mathbf{h}_0' g \mathbf{h}_0'^{-1})^{-1}$$

for $g \in \mathrm{Sp}(\mathbb{V}^\square)$. Then we have

$$z_{\mathbb{Y}'^\square}(g, g') = z_{\mathbb{V}^\triangle}(g, g') \cdot \mu'(gg') \cdot \mu'(g)^{-1} \cdot \mu'(g')^{-1}$$

for $g, g' \in \mathrm{Sp}(\mathbb{V}^\square)$. Put

$$\mathcal{G} := \{(g, h_1, h_2) \in \mathrm{GU}(V)^0 \times \mathrm{GU}(W) \times \mathrm{GU}(W) \mid \nu(g) = \nu(h_1) = \nu(h_2)\}.$$

We have natural maps

$$\begin{aligned} &\mathcal{G} \hookrightarrow \mathrm{G}(\mathrm{U}(V) \times \mathrm{U}(W^\square)), \\ &\mathcal{G} \hookrightarrow \mathrm{G}(\mathrm{U}(V)^0 \times \mathrm{U}(W)) \times \mathrm{G}(\mathrm{U}(V)^0 \times \mathrm{U}(W)). \end{aligned}$$

Lemma D.3.1. *We have*

$$\hat{s} \cdot \mu' = s' \otimes (s' \circ \mathrm{j}_{\mathbb{Y}'})$$

on $\mathcal{G}$, where $s' : \mathrm{GU}(V)^0 \times \mathrm{GU}(W) \to \mathbb{C}^1$ is the map defined in §C.2, C.3.

The proof of this lemma will be given in the next two sections.

D.3.1. The case $u \in (F^\times)^2$ or $J \in (F^\times)^2$. Recall that $\mathbb{X}' = V^\dagger \otimes_F X$ and $\mathbb{Y}' = V^\dagger \otimes_F Y$, where $X = Fe$ and $Y = Fe'$, and $W^\dagger = X + Y$ is a complete polarization over F. We have

$$\mathbb{X}'^\square = V^\dagger \otimes_F X^\square, \quad \mathbb{Y}'^\square = V^\dagger \otimes_F Y^\square,$$

where $X^\square = X \oplus X$ and $Y^\square = Y \oplus Y$. We have $d_{\mathbb{Y}'}(-1) = \mathrm{id} \otimes d_Y(-1)$ and $\mathrm{j}_{\mathbb{Y}'} = \mathrm{id} \otimes \mathrm{j}_Y$, where

$$d_Y(\nu) = \begin{pmatrix} 1 & \\ & \nu \end{pmatrix} \in \mathrm{GSp}(W^\dagger)$$

and $\mathrm{j}_Y(h) = d_Y(-1) \cdot h \cdot d_Y(-1)$ for $h \in \mathrm{GSp}(W^\dagger)$. In particular, we have

$$\mathrm{j}_{\mathbb{Y}'}(\mathrm{G}(\mathrm{U}(V)^0 \times \mathrm{U}(W))) = \mathrm{G}(\mathrm{U}(V)^0 \times \mathrm{U}(W)).$$

Let $\iota : \mathrm{G}(\mathrm{Sp}(W^\dagger) \times \mathrm{Sp}(W^\dagger)) \to \mathrm{GSp}(W^{\square\dagger})$ be the natural map. We may take

$$\mathbf{w} = \frac{1}{2}(1, -1), \quad \mathbf{w}^* = (1, 1).$$

Since

$$\begin{bmatrix} (e,0) \\ (0,e) \\ (e',0) \\ (0,-e') \end{bmatrix} = h_0 \cdot \begin{bmatrix} e\mathbf{w} \\ e'\mathbf{w} \\ e'\mathbf{w}^* \\ -e\mathbf{w}^* \end{bmatrix}, \quad h_0 = \begin{pmatrix} 1 & & & -\frac{1}{2} \\ -1 & & & -\frac{1}{2} \\ & 1 & \frac{1}{2} & \\ & 1 & -\frac{1}{2} & \end{pmatrix} \in \mathrm{Sp}(W^{\square\dagger}),$$

we have

$$\iota(h_1, h_2) = h_0^{-1} \cdot \iota^\natural(h_1, \mathrm{j}_Y(h_2)) \cdot h_0.$$

Here, using a basis e, e' of $W^\dagger$, we identify $\mathrm{GSp}(W^\dagger)$ with $\mathrm{GL}_2(F)$ and put

$$\iota^\natural(h_1, h_2) = \begin{pmatrix} a_1 & & b_1 & \\ & a_2 & & b_2 \\ c_1 & & d_1 & \\ & c_2 & & d_2 \end{pmatrix}, \quad h_i = \begin{pmatrix} a_i & b_i \\ c_i & d_i \end{pmatrix}.$$

Since $X^\square = eW^\nabla \cdot h_0$ and $Y^\square = eW^\triangle \cdot h_0$, we may take $\mathbf{h}'_0 = \mathrm{id} \otimes h_0$.

Proof of Lemma D.3.1. Let $\mathbf{g} = (g, h_1, h_2) \in \mathcal{G}$ and $\nu = \nu(\mathbf{g})$. Put $\mathbf{g}_i = (g, h_i) \in \mathrm{G}(\mathrm{U}(V)^0 \times \mathrm{U}(W))$. By definition, we have

$$s'(\mathbf{g}_1) \cdot s'(\mathrm{j}_{\mathbb{Y}'}(\mathbf{g}_2)) = \gamma^{j(h_1)} \cdot \gamma^{j(\mathrm{j}_Y(h_2))},$$

where j is as in §C.2.

Put $h = \iota(h_1, h_2) \in \mathrm{GU}(W^\square)$. We identify $\mathbf{g}$ with $(g, h) \in \mathrm{G}(\mathrm{U}(V)^0 \times \mathrm{U}(W^\square))$. Since $\hat{\jmath}(h) = \hat{\jmath}(d(\nu)^{-1} \cdot h)$, we have

$$\hat{s}(\mathbf{g}) = \hat{s}(d(\nu)^{-1} \cdot h).$$

Put $\mathbf{g}' = (g, d(\nu)) \in \mathrm{G}(\mathrm{U}(V)^0 \times \mathrm{U}(W^\square))$. Then we have $\mathbb{V}^\triangle \cdot \mathbf{g}' = \mathbb{V}^\triangle$ and

$$\begin{aligned} \mathbf{g} &= \mathbf{g}' \cdot d(\nu)^{-1} \cdot h, \\ \mathbf{h}'_0 \mathbf{g} \mathbf{h}'^{-1}_0 &= h_0 h h_0^{-1} \cdot g = h_0 h h_0^{-1} \cdot d(\nu)^{-1} \cdot \mathbf{g}'. \end{aligned}$$

Hence, by Lemma D.2.1, we have

$$\begin{aligned} z_{\mathbb{V}^\triangle}(\mathbf{g}, \mathbf{h}'^{-1}_0) &= z_{\mathbb{V}^\triangle}(d(\nu)^{-1} \cdot h, h_0^{-1}) \\ &= \hat{s}(d(\nu)^{-1} \cdot h h_0^{-1}) \cdot \hat{s}(d(\nu)^{-1} \cdot h)^{-1} \cdot \hat{s}(h_0^{-1})^{-1}, \\ z_{\mathbb{V}^\triangle}(\mathbf{h}'^{-1}_0, \mathbf{h}'_0 \mathbf{g} \mathbf{h}'^{-1}_0) &= z_{\mathbb{V}^\triangle}(h_0^{-1}, h_0 h h_0^{-1} \cdot d(\nu)^{-1}) \\ &= \hat{s}(h h_0^{-1} \cdot d(\nu)^{-1}) \cdot \hat{s}(h_0^{-1})^{-1} \cdot \hat{s}(h_0 h h_0^{-1} \cdot d(\nu)^{-1})^{-1}. \end{aligned}$$

Since $\hat{\jmath}(d(\nu)^{-1} \cdot h h_0^{-1}) = \hat{\jmath}(h h_0^{-1} \cdot d(\nu)^{-1})$, we have

$$\hat{s}(d(\nu)^{-1} \cdot h h_0^{-1}) = \hat{s}(h h_0^{-1} \cdot d(\nu)^{-1}).$$

Since $h_0 h h_0^{-1} = \iota^\natural(h_1, \mathrm{j}_Y(h_2))$, we have $\hat{\jmath}(h_0 h h_0^{-1} \cdot d(\nu)^{-1}) = j(h_1) + j(\mathrm{j}_Y(h_2))$ and

$$\hat{s}(h_0 h h_0^{-1} \cdot d(\nu)^{-1}) = \gamma^{j(h_1) + j(\mathrm{j}_Y(h_2))}.$$

Thus we obtain

$$\begin{aligned}\hat{s}(\mathbf{g}) \cdot \mu'(\mathbf{g}) &= \hat{s}(d(\nu)^{-1} \cdot h) \cdot z_{\mathbb{V}^{\triangle}}(\mathbf{g}, \mathbf{h}_0'^{-1}) \cdot z_{\mathbb{V}^{\triangle}}(\mathbf{h}_0'^{-1}, \mathbf{h}_0'\mathbf{g}\mathbf{h}_0'^{-1})^{-1} \\ &= \hat{s}(h_0 h h_0^{-1} \cdot d(\nu)^{-1}) \\ &= \gamma^{j(h_1)+j(\mathrm{j}_Y(h_2))}.\end{aligned}$$

This completes the proof. □

D.3.2. The case $J_i \in (F^\times)^2$. We only consider the case $i = 1$; the other case is similar. As in §C.3, we regard V and W as left and right B-spaces, respectively. Recall that $\mathbb{X}' = W \otimes_B X$ and $\mathbb{Y}' = W \otimes_B Y$, where $X = B\mathbf{v}$ and $Y = B\mathbf{v}^*$, and $V = X + Y$ is a complete polarization over B. As in §D.1, we define a 4-dimensional left skew-hermitian B-space $V^\square = V \oplus V$ and a complete polarization $V^\square = V^\triangledown \oplus V^\triangle$ over B. Using a basis

$$\mathbf{v}_1 := \frac{1}{2}(\mathbf{v}, -\mathbf{v}), \quad \mathbf{v}_2 := \frac{1}{2}(\mathbf{v}^*, -\mathbf{v}^*), \quad \mathbf{v}_1^* := (\mathbf{v}^*, \mathbf{v}^*), \quad \mathbf{v}_2^* := (-\mathbf{v}, -\mathbf{v})$$

of $V^\square$, we write

$$\mathrm{GU}(V^\square) = \left\{ g \in \mathrm{GL}_4(B) \;\middle|\; g \begin{pmatrix} & \mathbf{1}_2 \\ -\mathbf{1}_2 & \end{pmatrix} {}^t g^* = \nu(g) \cdot \begin{pmatrix} & \mathbf{1}_2 \\ -\mathbf{1}_2 & \end{pmatrix} \right\}.$$

We may identify $\mathbb{V}^\square$ with $W \otimes_B V^\square$. Under this identification, we have

$$\begin{aligned} \mathbb{V}^\triangledown &= W \otimes_B V^\triangledown, & \mathbb{X}'^\square &= W \otimes_B X^\square, \\ \mathbb{V}^\triangle &= W \otimes_B V^\triangle, & \mathbb{Y}'^\square &= W \otimes_B Y^\square, \end{aligned}$$

where $X^\square = X \oplus X$ and $Y^\square = Y \oplus Y$. We have $d_{\mathbb{Y}'}(-1) = \mathrm{id} \otimes d_Y(-1)$ and $\mathrm{j}_{\mathbb{Y}'} = \mathrm{id} \otimes \mathrm{j}_Y$, where

$$d_Y(\nu) = \begin{pmatrix} 1 & \\ & \nu \end{pmatrix} \in \mathrm{GU}(V)^0$$

and $\mathrm{j}_Y(g) = d_Y(-1) \cdot g \cdot d_Y(-1)$ for $g \in \mathrm{GU}(V)$. In particular, we have

$$\mathrm{j}_{\mathbb{Y}'}(\mathrm{G}(\mathrm{U}(V)^0 \times \mathrm{U}(W))) = \mathrm{G}(\mathrm{U}(V)^0 \times \mathrm{U}(W)).$$

Let $\iota : \mathrm{G}(\mathrm{U}(V) \times \mathrm{U}(V)) \to \mathrm{GU}(V^\square)$ be the natural map. Since

$$\begin{bmatrix} (\mathbf{v}, 0) \\ (0, \mathbf{v}) \\ (\mathbf{v}^*, 0) \\ (0, -\mathbf{v}^*) \end{bmatrix} = g_0 \cdot \begin{bmatrix} \mathbf{v}_1 \\ \mathbf{v}_2 \\ \mathbf{v}_1^* \\ \mathbf{v}_2^* \end{bmatrix}, \quad g_0 = \begin{pmatrix} 1 & & & -\frac{1}{2} \\ -1 & & & -\frac{1}{2} \\ & 1 & \frac{1}{2} & \\ & 1 & -\frac{1}{2} & \end{pmatrix} \in \mathrm{U}(V^\square),$$

we have

$$\iota(g_1, g_2) = g_0^{-1} \cdot \iota^\natural(g_1, \mathrm{j}_Y(g_2)) \cdot g_0.$$

Here, regarding V as a left B-space and using a basis $\mathbf{v}, \mathbf{v}^*$ of V, we identify $\mathrm{GU}(V)$ with a subgroup of $\mathrm{GL}_2(B)$ and put

$$\iota^\natural(g_1, g_2) = \begin{pmatrix} a_1 & & b_1 & \\ & a_2 & & b_2 \\ c_1 & & d_1 & \\ & c_2 & & d_2 \end{pmatrix}, \quad g_i = \begin{pmatrix} a_i & b_i \\ c_i & d_i \end{pmatrix}.$$

Since $X^\square = V^\triangledown \cdot g_0$ and $Y^\square = V^\triangle \cdot g_0$, we may take $\mathbf{h}_0' = \mathrm{id} \otimes g_0$.

When B is split, we define a map

$$\hat{s}' : \mathrm{U}(V^\square) \longrightarrow \mathbb{C}^1$$

by

$$\hat{s}'(g) = 1$$

for $g \in \mathrm{U}(V^{\square})$. Then we have

$$z_{\mathbb{V}^{\triangle}}(g, g') = \hat{s}'(gg') \cdot \hat{s}'(g)^{-1} \cdot \hat{s}'(g')^{-1}$$

for $g, g' \in \mathrm{U}(W^{\square})$ by [**40**, Theorem 3.1, case 1_{-}]. When B is ramified, we define a map

$$\hat{s}' : \mathrm{U}(V^{\square}) \longrightarrow \mathbb{C}^1$$

by

$$\hat{s}'(g) = (-1)^{\hat{\jmath}'(g)}$$

for $g \in \mathrm{U}(V^{\square})$, where

$$\hat{\jmath}'(g) = \begin{cases} 0 & \text{if } c = 0, \\ 1 & \text{if } c \neq 0, \, \nu(c) = 0, \\ 2 & \text{if } \nu(c) \neq 0, \end{cases} \qquad g = \begin{pmatrix} a & b \\ c & d \end{pmatrix} \in \mathrm{U}(V^{\square}),$$

and $\nu : \mathrm{M}_2(B) \to F$ is the reduced norm. Since $\dim_B W = 1$, we have

$$z_{\mathbb{V}^{\triangle}}(g, g') = \hat{s}'(gg') \cdot \hat{s}'(g)^{-1} \cdot \hat{s}'(g')^{-1}$$

for $g, g' \in \mathrm{U}(W^{\square})$ by [**40**, Theorem 3.1, case 2_{+}].

PROOF OF LEMMA D.3.1. Let $\mathbf{g} = (g, h_1, h_2) \in \mathcal{G}$ and $\nu = \nu(\mathbf{g})$. Put $\mathbf{g}_i = (g, h_i) \in \mathrm{G}(\mathrm{U}(V)^0 \times \mathrm{U}(W))$. By definition, we have $j(\mathrm{j}_Y(g)) = j(g)$, where j is as in §C.3. Hence we have

$$s'(\mathbf{g}_1) \cdot s'(\mathrm{j}_{\mathbb{Y}'}(\mathbf{g}_2)) = \gamma^{j(g)} \cdot \gamma^{j(\mathrm{j}_Y(g))} = 1.$$

Here

$$\gamma = \begin{cases} 1 & \text{if } B \text{ is split}, \\ -1 & \text{if } B \text{ is ramified}. \end{cases}$$

Also, by definition, we have

$$\hat{s}(\mathbf{g}) = 1.$$

Now we compute $z_{\mathbb{V}^{\triangle}}(\mathbf{g}, \mathbf{h}_0'^{-1})$. We identify $\mathbf{g}$ with $(g, \iota(h_1, h_2)) \in \mathrm{G}(\mathrm{U}(V)^0 \times \mathrm{U}(W^{\square}))$, where $\iota : \mathrm{G}(\mathrm{U}(W) \times \mathrm{U}(W)) \to \mathrm{GU}(W^{\square})$ is the natural map. Put $\mathbf{g}' = (g, \iota(h_2, h_2)) \in \mathrm{G}(\mathrm{U}(V)^0 \times \mathrm{U}(W^{\square}))$ and $h = h_1^{-1} h_2 \in \mathrm{U}(W)$. Then we have $\mathbf{g} = \mathbf{g}' \cdot \iota(h^{-1}, 1)$. Via the identification $\mathbb{V}^{\square} = V \otimes_B W^{\square} = W \otimes_B V^{\square}$, we identify $\mathbf{g}'$ with $(h_2, \iota(g, g)) \in \mathrm{G}(\mathrm{U}(W) \times \mathrm{U}(V^{\square}))$. Since $\mathbb{V}^{\triangle} \cdot \mathbf{g}' = \mathbb{V}^{\triangle}$, we have

$$z_{\mathbb{V}^{\triangle}}(\mathbf{g}, \mathbf{h}_0'^{-1}) = z_{\mathbb{V}^{\triangle}}(\iota(h^{-1}, 1), g_0^{-1}).$$

Put

$$\tau := \begin{pmatrix} 1 & & & \\ & & & -1 \\ & & 1 & \\ & 1 & & \end{pmatrix} \in \mathrm{U}(V^{\square}).$$

Then we have

$$g_0 \tau = \begin{pmatrix} 1 & -\frac{1}{2} & & \\ -1 & -\frac{1}{2} & & \\ & & \frac{1}{2} & -1 \\ & & -\frac{1}{2} & -1 \end{pmatrix}$$

and $\mathbb{V}^\triangle \cdot \tau^{-1} g_0^{-1} = \mathbb{V}^\triangle$, so that

$$z_{\mathbb{V}^\triangle}(\iota(h^{-1},1), g_0^{-1}) = z_{\mathbb{V}^\triangle}(\iota(h^{-1},1), \tau).$$

Under the identification $\mathbb{V}^\square = \mathbb{V} \oplus \mathbb{V} = W \otimes_B V^\square$, we have

$$\begin{aligned} & x \otimes (y, \pm y) \cdot \iota(h^{-1},1) \\ &= (x \otimes y, \pm x \otimes y) \cdot \iota(h^{-1},1) \\ &= (hx \otimes y, \pm x \otimes y) \\ &= \frac{1}{2}((h \pm 1)x \otimes y, (h \pm 1)x \otimes y) + \frac{1}{2}((h \mp 1)x \otimes y, -(h \mp 1)x \otimes y) \\ &= \frac{1}{2}(h \pm 1)x \otimes (y, y) + \frac{1}{2}(h \mp 1)x \otimes (y, -y) \end{aligned}$$

for $x \in W$ and $y \in V$. Thus we obtain

$$\begin{aligned} x \otimes \mathbf{v}_1 \cdot \iota(h^{-1},1) &= \frac{1}{2}(h+1)x \otimes \mathbf{v}_1 - \frac{1}{4}(h-1)x \otimes \mathbf{v}_2^*, \\ x \otimes \mathbf{v}_2 \cdot \iota(h^{-1},1) &= \frac{1}{2}(h+1)x \otimes \mathbf{v}_2 + \frac{1}{4}(h-1)x \otimes \mathbf{v}_1^*, \\ x \otimes \mathbf{v}_1^* \cdot \iota(h^{-1},1) &= \frac{1}{2}(h+1)x \otimes \mathbf{v}_1^* + (h-1)x \otimes \mathbf{v}_2, \\ x \otimes \mathbf{v}_2^* \cdot \iota(h^{-1},1) &= \frac{1}{2}(h+1)x \otimes \mathbf{v}_2^* - (h-1)x \otimes \mathbf{v}_1. \end{aligned}$$

First assume that B is split. Fix an isomorphism $\mathfrak{i} : B \to \mathrm{M}_2(F)$. Put $e = \mathfrak{i}^{-1}\left(\begin{smallmatrix} 1 & 0 \\ 0 & 0 \end{smallmatrix}\right)$, $e' = \mathfrak{i}^{-1}\left(\begin{smallmatrix} 0 & 1 \\ 0 & 0 \end{smallmatrix}\right)$, and $e'' = \mathfrak{i}^{-1}\left(\begin{smallmatrix} 0 & 0 \\ 1 & 0 \end{smallmatrix}\right)$. Then $W^\dagger := We$ is a 2-dimensional symplectic F-space and the restriction $\mathrm{GU}(W) \to \mathrm{GSp}(W^\dagger)$ is an isomorphism. Also, $V^{\square\dagger} := eV^\square$ is an 8-dimensional quadratic F-space. We identify $\mathbb{V}^\square$ with $W^\dagger \otimes_F V^{\square\dagger}$. Put $f := 2e''$ and

$$\begin{aligned} &\mathbf{x}_1 := e\mathbf{v}_1, && \mathbf{x}_2 := e'\mathbf{v}_1, && \mathbf{x}_3 := e\mathbf{v}_2, && \mathbf{x}_4 := e'\mathbf{v}_2, \\ &\mathbf{y}_1 := e'\mathbf{v}_1^*, && \mathbf{y}_2 := -e\mathbf{v}_1^*, && \mathbf{y}_3 := e'\mathbf{v}_2^*, && \mathbf{y}_4 := -e\mathbf{v}_2^*. \end{aligned}$$

Using a basis

$$e \otimes \mathbf{x}_1, f \otimes \mathbf{x}_1, \dots, e \otimes \mathbf{x}_4, f \otimes \mathbf{x}_4, f \otimes \mathbf{y}_1, -e \otimes \mathbf{y}_1, \dots, f \otimes \mathbf{y}_4, -e \otimes \mathbf{y}_4$$

of $\mathbb{V}^\square$, we identify $\mathrm{Sp}(\mathbb{V}^\square)$ with $\mathrm{Sp}_{16}(F)$. We define $\mathbf{h} \in \mathrm{GL}_2(F)$ by

$$\begin{bmatrix} he \\ hf \end{bmatrix} = \mathbf{h} \cdot \begin{bmatrix} e \\ f \end{bmatrix}$$

and put

$$\dot{\mathbf{h}} = \frac{1}{2}(\mathbf{h} + \mathbf{1}_2), \quad \ddot{\mathbf{h}} = \frac{1}{2}(\mathbf{h} - \mathbf{1}_2).$$

Then we have $\mathbf{h} \cdot \mathbf{a} \cdot {}^t\mathbf{h} = \mathbf{a}$, where

$$\mathbf{a} = \begin{pmatrix} & -1 \\ 1 & \end{pmatrix} \in \mathrm{GL}_2(F).$$

Moreover, we have $\iota(h^{-1}, 1) = \mathbf{d}^{-1} \cdot \mathbf{h}' \cdot \mathbf{d}$ and

$$\tau = \mathbf{d}^{-1} \cdot \begin{pmatrix} \mathbf{1}_2 &&&&&&& \\ & \mathbf{1}_2 &&&&&& \\ && \mathbf{0}_2 &&&&& \mathbf{1}_2 \\ &&& \mathbf{0}_2 &&& -\mathbf{1}_2 & \\ &&&& \mathbf{1}_2 &&& \\ &&&&& \mathbf{1}_2 && \\ &&& \mathbf{1}_2 &&& \mathbf{0}_2 & \\ && -\mathbf{1}_2 &&&&& \mathbf{0}_2 \end{pmatrix} \cdot \mathbf{d}$$

$$= \tau_4 \cdot \begin{pmatrix} \mathbf{1}_2 &&&&&&& \\ & \mathbf{1}_2 &&&&&& \\ &&& -\mathbf{a} &&&& \\ && \mathbf{a} &&&&& \\ &&&& \mathbf{1}_2 &&& \\ &&&&& \mathbf{1}_2 && \\ &&&&&&& -\mathbf{a} \\ &&&&&& \mathbf{a} & \end{pmatrix},$$

where

$$\mathbf{d} = \begin{pmatrix} \mathbf{1}_2 &&&&&&& \\ & \mathbf{1}_2 &&&&&& \\ && \mathbf{1}_2 &&&&& \\ &&& \mathbf{1}_2 &&&& \\ &&&& \mathbf{a} &&& \\ &&&&& \mathbf{a} && \\ &&&&&& \mathbf{a} & \\ &&&&&&& \mathbf{a} \end{pmatrix},$$

$$\mathbf{h}' = \begin{pmatrix} \dot{\mathbf{h}} &&&&&&& \frac{1}{2}\ddot{\mathbf{h}} \\ & \dot{\mathbf{h}} &&&&& -\frac{1}{2}\ddot{\mathbf{h}} & \\ && \dot{\mathbf{h}} &&& -\frac{1}{2}\ddot{\mathbf{h}} && \\ &&& \dot{\mathbf{h}} & \frac{1}{2}\ddot{\mathbf{h}} &&& \\ &&& 2\ddot{\mathbf{h}} & \dot{\mathbf{h}} &&& \\ && -2\ddot{\mathbf{h}} &&& \dot{\mathbf{h}} && \\ & -2\ddot{\mathbf{h}} &&&&& \dot{\mathbf{h}} & \\ 2\ddot{\mathbf{h}} &&&&&&& \dot{\mathbf{h}} \end{pmatrix},$$

$$\tau_4 = \begin{pmatrix} \mathbf{1}_4 &&& \\ &&& -\mathbf{1}_4 \\ && \mathbf{1}_4 & \\ & \mathbf{1}_4 && \end{pmatrix}.$$

If $h = 1$, then we have $z_{\mathbb{Y}^\triangle}(\iota(h^{-1}, 1), \tau) = z_{\mathbb{Y}^\triangle}(1, \tau) = 1$. Assume that $h \neq 1$ and $\det \ddot{\mathbf{h}} = 0$. Since $\det \mathbf{h} = 1$, we have $\operatorname{tr} \ddot{\mathbf{h}} = 0$. Hence we may take $\mathbf{a}_1 \in \mathrm{SL}_2(F)$ such that $\mathbf{a}_1 \cdot \ddot{\mathbf{h}} \cdot \mathbf{a}_1^{-1} = \mathbf{x}$, where

$$\mathbf{x} = \begin{pmatrix} 0 & x \\ 0 & 0 \end{pmatrix}$$

with some $x \neq 0$. Put

$$\mathbf{m}_1 = \begin{pmatrix} \mathbf{a}_1 &&&&&&& \\ & \mathbf{a}_1 &&&&&& \\ && \mathbf{a}_1 &&&&& \\ &&& \mathbf{a}_1 &&&& \\ &&&& {}^t\mathbf{a}_1^{-1} &&& \\ &&&&& {}^t\mathbf{a}_1^{-1} && \\ &&&&&& {}^t\mathbf{a}_1^{-1} & \\ &&&&&&& {}^t\mathbf{a}_1^{-1} \end{pmatrix}.$$

Noting that $\mathbf{a}_1 \cdot \dot{\mathbf{h}} \cdot \mathbf{a}_1^{-1} = \mathbf{x} + \mathbf{1}_2$ and $\mathbf{a} \cdot {}^t\mathbf{a}_1^{-1} \cdot \mathbf{a}^{-1} = \mathbf{a}_1$, we have

$$\mathbf{m}_1 \cdot \iota(h^{-1}, 1) \cdot \mathbf{m}_1^{-1} = \mathbf{d}^{-1} \cdot \mathbf{h}'' \cdot \mathbf{d},$$

where

$$\mathbf{h}'' = \begin{pmatrix} \mathbf{x}+\mathbf{1}_2 &&&&&&& \frac{1}{2}\mathbf{x} \\ & \mathbf{x}+\mathbf{1}_2 &&&&& -\frac{1}{2}\mathbf{x} & \\ && \mathbf{x}+\mathbf{1}_2 &&& -\frac{1}{2}\mathbf{x} && \\ &&& \mathbf{x}+\mathbf{1}_2 & \frac{1}{2}\mathbf{x} &&& \\ &&& 2\mathbf{x} & \mathbf{x}+\mathbf{1}_2 &&& \\ && -2\mathbf{x} &&& \mathbf{x}+\mathbf{1}_2 && \\ & -2\mathbf{x} &&&&& \mathbf{x}+\mathbf{1}_2 & \\ 2\mathbf{x} &&&&&&& \mathbf{x}+\mathbf{1}_2 \end{pmatrix}.$$

We have

$$\begin{aligned} z_{\mathbb{V}^\triangle}(\iota(h^{-1},1),\tau) &= z_{\mathbb{V}^\triangle}(\iota(h^{-1},1),\tau_4) \\ &= z_{\mathbb{V}^\triangle}(\mathbf{m}_1 \cdot \iota(h^{-1},1), \tau_4 \cdot \tau_4^{-1} \cdot \mathbf{m}_1^{-1} \cdot \tau_4) \\ &= z_{\mathbb{V}^\triangle}(\mathbf{m}_1 \cdot \iota(h^{-1},1) \cdot \mathbf{m}_1^{-1}, \tau_4) \\ &= z_{\mathbb{V}^\triangle}(\mathbf{d}^{-1} \cdot \mathbf{h}'' \cdot \mathbf{d}, \tau_4). \end{aligned}$$

Put

$$\mathbf{e} = \begin{pmatrix} 1 & \\ & 0 \end{pmatrix}, \quad \mathbf{e}^* = \begin{pmatrix} 0 & \\ & 1 \end{pmatrix}, \quad \mathbf{a}_2 = \begin{pmatrix} -x^{-1} & \\ & -x \end{pmatrix},$$

so that

$$\mathbf{a}_2 \cdot \mathbf{x} \cdot \mathbf{a} = -\mathbf{e}, \quad {}^t\mathbf{a}_2^{-1} \cdot \mathbf{a}^{-1} \cdot \mathbf{x} = \mathbf{e}^*.$$

Put

$$\mathbf{x}' = -\mathbf{e}\mathbf{a}^{-1} + \mathbf{a}_2, \quad \mathbf{x}'' = \mathbf{e}^*\mathbf{a} + {}^t\mathbf{a}_2^{-1},$$

and

$$\mathbf{m}_2 = \begin{pmatrix} &&& 2\mathbf{a}_2 &&&& \\ && -2\mathbf{a}_2 &&&&& \\ & -2\mathbf{a}_2 &&&&&& \\ 2\mathbf{a}_2 &&&&&&& \\ &&&&&&& \frac{1}{2}{}^t\mathbf{a}_2^{-1} \\ &&&&&& -\frac{1}{2}{}^t\mathbf{a}_2^{-1} & \\ &&&&& -\frac{1}{2}{}^t\mathbf{a}_2^{-1} && \\ &&&& \frac{1}{2}{}^t\mathbf{a}_2^{-1} &&& \end{pmatrix}.$$

Then we have $\mathbf{m}_2 \cdot \mathbf{d}^{-1} \cdot \mathbf{h}'' \cdot \mathbf{d} = \mathbf{h}'''$, where

$$\mathbf{h}''' = \begin{pmatrix} & & & 2\mathbf{x}' & -\mathbf{e} & & & \\ & & -2\mathbf{x}' & & & -\mathbf{e} & & \\ & -2\mathbf{x}' & & & & & -\mathbf{e} & \\ 2\mathbf{x}' & & & & & & & -\mathbf{e} \\ \mathbf{e}^* & & & & & & & \frac{1}{2}\mathbf{x}'' \\ & \mathbf{e}^* & & & & & -\frac{1}{2}\mathbf{x}'' & \\ & & \mathbf{e}^* & & & -\frac{1}{2}\mathbf{x}'' & & \\ & & & \mathbf{e}^* & \frac{1}{2}\mathbf{x}'' & & & \end{pmatrix}.$$

We have

$$z_{\mathbb{V}^\triangle}(\mathbf{d}^{-1} \cdot \mathbf{h}'' \cdot \mathbf{d}, \tau_4) = z_{\mathbb{V}^\triangle}(\mathbf{m}_2 \cdot \mathbf{d}^{-1} \cdot \mathbf{h}'' \cdot \mathbf{d}, \tau_4) = z_{\mathbb{V}^\triangle}(\mathbf{h}''', \tau_4).$$

Put

$$\mathbf{b} = \begin{pmatrix} 0 & \\ & x^{-1} \end{pmatrix}$$

and

$$\mathbf{n} = \begin{pmatrix} \mathbf{1}_2 & & & & & & & \frac{1}{2}\mathbf{b} \\ & \mathbf{1}_2 & & & & & -\frac{1}{2}\mathbf{b} & \\ & & \mathbf{1}_2 & & & -\frac{1}{2}\mathbf{b} & & \\ & & & \mathbf{1}_2 & \frac{1}{2}\mathbf{b} & & & \\ & & & & \mathbf{1}_2 & & & \\ & & & & & \mathbf{1}_2 & & \\ & & & & & & \mathbf{1}_2 & \\ & & & & & & & \mathbf{1}_2 \end{pmatrix}.$$

We have

$$\mathbf{h}''' \cdot \mathbf{n} = \begin{pmatrix} & & & 2\mathbf{x}' & \tilde{\mathbf{x}}' & & & \\ & & -2\mathbf{x}' & & & \tilde{\mathbf{x}}' & & \\ & -2\mathbf{x}' & & & & & \tilde{\mathbf{x}}' & \\ 2\mathbf{x}' & & & & & & & \tilde{\mathbf{x}}' \\ \mathbf{e}^* & & & & & & & \frac{1}{2}\tilde{\mathbf{x}}'' \\ & \mathbf{e}^* & & & & & -\frac{1}{2}\tilde{\mathbf{x}}'' & \\ & & \mathbf{e}^* & & & -\frac{1}{2}\tilde{\mathbf{x}}'' & & \\ & & & \mathbf{e}^* & \frac{1}{2}\tilde{\mathbf{x}}'' & & & \end{pmatrix},$$

where $\tilde{\mathbf{x}}' = \mathbf{x}'\mathbf{b} - \mathbf{e}$ and $\tilde{\mathbf{x}}'' = \mathbf{e}^*\mathbf{b} + \mathbf{x}''$. Since

$$\tilde{\mathbf{x}}'' = \begin{pmatrix} -x & 0 \\ 1 & 0 \end{pmatrix},$$

we have $\mathbb{V}^\triangle \cdot \mathbf{h}''' \cdot \mathbf{n} \cdot \boldsymbol{\tau}^{-1} = \mathbb{V}^\triangle$, where

$$\boldsymbol{\tau} = \begin{pmatrix} \mathbf{e} & & & & -\mathbf{e}^* & & & \\ & \mathbf{e} & & & & -\mathbf{e}^* & & \\ & & \mathbf{e} & & & & -\mathbf{e}^* & \\ & & & \mathbf{e} & & & & -\mathbf{e}^* \\ \mathbf{e}^* & & & & \mathbf{e} & & & \\ & \mathbf{e}^* & & & & \mathbf{e} & & \\ & & \mathbf{e}^* & & & & \mathbf{e} & \\ & & & \mathbf{e}^* & & & & \mathbf{e} \end{pmatrix}.$$

Hence we have

$$\begin{aligned} z_{\mathbb{V}^\triangle}(\mathbf{h}''', \tau_4) &= z_{\mathbb{V}^\triangle}(\mathbf{h}''', \tau_4 \cdot \tau_4^{-1} \cdot \mathbf{n} \cdot \tau_4) \\ &= z_{\mathbb{V}^\triangle}(\mathbf{h}''' \cdot \mathbf{n}, \tau_4) \\ &= z_{\mathbb{V}^\triangle}(\boldsymbol{\tau}, \tau_4) \\ &= 1. \end{aligned}$$

Assume that $h \neq 1$ and $\det \ddot{\mathbf{h}} \neq 0$. Put

$$\mathbf{k}_1 = {}^t\mathbf{a}^t\ddot{\mathbf{h}}^{-1}, \quad \mathbf{k}_2 = \mathbf{a}^{-1}\ddot{\mathbf{h}}, \quad \mathbf{k}_3 = \ddot{\mathbf{h}}^{-1}\dot{\mathbf{h}}\mathbf{a}.$$

We have

$$\iota(h^{-1}, 1) = \begin{pmatrix} & & & \frac{1}{2}\mathbf{k}_1 & \dot{\mathbf{h}} & & & \\ & & -\frac{1}{2}\mathbf{k}_1 & & & \dot{\mathbf{h}} & & \\ & -\frac{1}{2}\mathbf{k}_1 & & & & & \dot{\mathbf{h}} & \\ \frac{1}{2}\mathbf{k}_1 & & & & & & & \dot{\mathbf{h}} \\ & & & & & & & 2\mathbf{k}_2 \\ & & & & & & -2\mathbf{k}_2 & \\ & & & & & -2\mathbf{k}_2 & & \\ & & & & 2\mathbf{k}_2 & & & \end{pmatrix} \cdot \tau_8 \cdot \mathbf{n},$$

where

$$\tau_8 = \begin{pmatrix} & -\mathbf{1}_8 \\ \mathbf{1}_8 & \end{pmatrix},$$

$$\mathbf{n} = \begin{pmatrix} \mathbf{1}_2 & & & & & & & \frac{1}{2}\mathbf{k}_3 \\ & \mathbf{1}_2 & & & & & -\frac{1}{2}\mathbf{k}_3 & \\ & & \mathbf{1}_2 & & & -\frac{1}{2}\mathbf{k}_3 & & \\ & & & \mathbf{1}_2 & \frac{1}{2}\mathbf{k}_3 & & & \\ & & & & \mathbf{1}_2 & & & \\ & & & & & \mathbf{1}_2 & & \\ & & & & & & \mathbf{1}_2 & \\ & & & & & & & \mathbf{1}_2 \end{pmatrix}.$$

Hence we have

$$z_{\mathbb{V}^\triangle}(\iota(h^{-1}, 1), \tau) = z_{\mathbb{V}^\triangle}(\tau_8 \cdot \mathbf{n}, \tau_4) = z_{\mathbb{V}^\triangle}(\tau_8, \mathbf{n} \cdot \tau_4).$$

Since $\mathbb{V}^\triangle \cdot \tau_4^{-1}\mathbf{n}\tau_4 = \mathbb{V}^\triangle$, we have

$$z_{\mathbb{V}^\triangle}(\tau_8, \mathbf{n} \cdot \tau_4) = z_{\mathbb{V}^\triangle}(\tau_8, \tau_4) = 1.$$

Thus we obtain

$$z_{\mathbb{V}^\triangle}(\mathbf{g}, \mathbf{h}_0'^{-1}) = 1.$$

Next assume that B is ramified. Choose a basis e_1, e_2, e_3, e_4 of W over F. We may assume that $\frac{1}{2}\operatorname{tr}_{B/F}(\langle e_i, e_j \rangle) = a_i \cdot \delta_{ij}$ with some $a_i \in F^\times$. Put $e_i' := e_i \cdot a_i^{-1}$. Using a basis

$$e_1 \otimes \mathbf{v}_1, \ldots, e_4 \otimes \mathbf{v}_1, e_1 \otimes \mathbf{v}_2, \ldots, e_4 \otimes \mathbf{v}_2, e_1' \otimes \mathbf{v}_1^*, \ldots, e_4' \otimes \mathbf{v}_1^*, e_1' \otimes \mathbf{v}_2^*, \ldots, e_4' \otimes \mathbf{v}_2^*$$

of $\mathbb{V}^\square$, we identify $\mathrm{Sp}(\mathbb{V}^\square)$ with $\mathrm{Sp}_{16}(F)$. We define $\mathbf{h} \in \mathrm{GL}_4(F)$ by

$$\begin{bmatrix} he_1 \\ he_2 \\ he_3 \\ he_4 \end{bmatrix} = \mathbf{h} \cdot \begin{bmatrix} e_1 \\ e_2 \\ e_3 \\ e_4 \end{bmatrix}$$

and put

$$\dot{\mathbf{h}} = \frac{1}{2}(\mathbf{h} + \mathbf{1}_4), \quad \ddot{\mathbf{h}} = \frac{1}{2}(\mathbf{h} - \mathbf{1}_4).$$

Then we have $\mathbf{h} \cdot \mathbf{a} \cdot {}^t\mathbf{h} = \mathbf{a}$, where

$$\mathbf{a} := \begin{pmatrix} a_1 & & & \\ & a_2 & & \\ & & a_3 & \\ & & & a_4 \end{pmatrix} \in \mathrm{GL}_4(F).$$

Moreover, we have $\iota(h^{-1}, 1) = \mathbf{d}^{-1} \cdot \mathbf{h}' \cdot \mathbf{d}$ and

$$\tau = \mathbf{d}^{-1} \cdot \tau_4 \cdot \mathbf{d} = \tau_4 \cdot \begin{pmatrix} \mathbf{1}_4 & & & \\ & \mathbf{a}^{-1} & & \\ & & \mathbf{1}_4 & \\ & & & \mathbf{a} \end{pmatrix},$$

where

$$\mathbf{d} = \begin{pmatrix} \mathbf{1}_4 & & & \\ & \mathbf{1}_4 & & \\ & & \mathbf{a} & \\ & & & \mathbf{a} \end{pmatrix},$$

$$\mathbf{h}' = \begin{pmatrix} \dot{\mathbf{h}} & & & -\frac{1}{2}\ddot{\mathbf{h}} \\ & \dot{\mathbf{h}} & \frac{1}{2}\ddot{\mathbf{h}} & \\ & 2\ddot{\mathbf{h}} & \dot{\mathbf{h}} & \\ -2\ddot{\mathbf{h}} & & & \dot{\mathbf{h}} \end{pmatrix},$$

$$\tau_4 = \begin{pmatrix} \mathbf{1}_4 & & & \\ & & & -\mathbf{1}_4 \\ & & \mathbf{1}_4 & \\ & \mathbf{1}_4 & & \end{pmatrix}.$$

If $h = 1$, then we have $z_{\mathbb{V}\triangle}(\iota(h^{-1}, 1), \tau) = z_{\mathbb{V}\triangle}(1, \tau) = 1$. Assume that $h \neq 1$. Since B is ramified, $h - 1$ is given by the automorphism $x \mapsto \boldsymbol{\alpha} \cdot x$ of B with some $\boldsymbol{\alpha} \in B^\times$. In particular, we have $\ddot{\mathbf{h}} \in \mathrm{GL}_4(F)$. Put

$$\mathbf{k}_1 = \mathbf{a}{}^t\ddot{\mathbf{h}}^{-1}, \quad \mathbf{k}_2 = \mathbf{a}^{-1}\ddot{\mathbf{h}}, \quad \mathbf{k}_3 = \ddot{\mathbf{h}}^{-1}\dot{\mathbf{h}}\mathbf{a}.$$

We have

$$\iota(h^{-1}, 1) = \begin{pmatrix} & \frac{1}{2}\mathbf{k}_1 & \dot{\mathbf{h}} & \\ -\frac{1}{2}\mathbf{k}_1 & & & \dot{\mathbf{h}} \\ & & & 2\mathbf{k}_2 \\ & & -2\mathbf{k}_2 & \end{pmatrix} \cdot \tau_8 \cdot \mathbf{n},$$

where

$$\tau_8 = \begin{pmatrix} & -\mathbf{1}_8 \\ \mathbf{1}_8 & \end{pmatrix},$$

$$\mathbf{n} = \begin{pmatrix} \mathbf{1}_4 & & & -\frac{1}{2}\mathbf{k}_3 \\ & \mathbf{1}_4 & \frac{1}{2}\mathbf{k}_3 & \\ & & \mathbf{1}_4 & \\ & & & \mathbf{1}_4 \end{pmatrix}.$$

Hence we have

$$z_{\mathbb{V}\triangle}(\iota(h^{-1}, 1), \tau) = z_{\mathbb{V}\triangle}(\tau_8 \cdot \mathbf{n}, \tau_4) = z_{\mathbb{V}\triangle}(\tau_8, \mathbf{n} \cdot \tau_4).$$

Since $\mathbb{V}^{\triangle} \cdot \tau_4^{-1}\mathbf{n}\tau_4 = \mathbb{V}^{\triangle}$, we have

$$z_{\mathbb{V}^{\triangle}}(\tau_8, \mathbf{n} \cdot \tau_4) = z_{\mathbb{V}^{\triangle}}(\tau_8, \tau_4) = 1.$$

Thus we obtain

$$z_{\mathbb{V}^{\triangle}}(\mathbf{g}, \mathbf{h}_0'^{-1}) = 1.$$

Now we compute $z_{\mathbb{V}^{\triangle}}(\mathbf{h}_0'^{-1}, \mathbf{h}_0'\mathbf{g}\mathbf{h}_0'^{-1})$. Put $\mathbf{g}'' = (d_Y(\nu), \iota(h_1, h_2)) \in \mathrm{G}(\mathrm{U}(V)^0 \times \mathrm{U}(W^{\square}))$ and $g' = g \cdot d_Y(\nu)^{-1} \in \mathrm{U}(V)^0$. Then we have $\mathbf{g} = g' \cdot \mathbf{g}''$. Via the identification $\mathbb{V}^{\square} = V \otimes_B W^{\square} = W \otimes_B V^{\square}$, we identify g' with $\iota(g', g') \in \mathrm{U}(V^{\square})$. Since $\mathbb{V}^{\triangle} \cdot \mathbf{h}_0' = \mathbb{Y}'^{\square} = Y \otimes_B W^{\square}$, we have $\mathbb{V}^{\triangle} \cdot \mathbf{h}_0'\mathbf{g}''\mathbf{h}_0'^{-1} = \mathbb{V}^{\triangle}$ and hence

$$\begin{aligned} z_{\mathbb{V}^{\triangle}}(\mathbf{h}_0'^{-1}, \mathbf{h}_0'\mathbf{g}\mathbf{h}_0'^{-1}) &= z_{\mathbb{V}^{\triangle}}(g_0^{-1}, g_0 \cdot \iota(g', g') \cdot g_0^{-1}) \\ &= \hat{s}'(\iota(g', g') \cdot g_0^{-1}) \cdot \hat{s}'(g_0^{-1})^{-1} \cdot \hat{s}'(g_0 \cdot \iota(g', g') \cdot g_0^{-1})^{-1}. \end{aligned}$$

Hence, if B is split, then we have

$$z_{\mathbb{V}^{\triangle}}(\mathbf{h}_0'^{-1}, \mathbf{h}_0'\mathbf{g}\mathbf{h}_0'^{-1}) = 1.$$

Assume that B is ramified. We write $g' = \left(\begin{smallmatrix} a & b \\ c & d \end{smallmatrix}\right)$. Since

$$g_0^{-1} = \begin{pmatrix} \frac{1}{2} & -\frac{1}{2} & & \\ & & \frac{1}{2} & \frac{1}{2} \\ & & 1 & -1 \\ -1 & -1 & & \end{pmatrix},$$

$$g_0 \cdot \iota(g', g') \cdot g_0^{-1} = \iota^{\natural}(g', \mathfrak{j}_Y'(g')) = \begin{pmatrix} a & & b & \\ & a & & -b \\ c & & d & \\ & -c & & d \end{pmatrix},$$

$$\iota(g', g') \cdot g_0^{-1} = \begin{pmatrix} \frac{a}{2} & -\frac{a}{2} & \frac{b}{2} & \frac{b}{2} \\ \frac{c}{2} & -\frac{c}{2} & \frac{d}{2} & \frac{d}{2} \\ c & c & d & -d \\ -a & -a & -b & b \end{pmatrix},$$

we have

$$\begin{aligned} \hat{s}'(g_0^{-1}) &= -1, \\ \hat{s}'(g_0 \cdot \iota(g', g') \cdot g_0^{-1}) &= 1, \\ \hat{s}'(\iota(g', g') \cdot g_0^{-1}) &= -1. \end{aligned}$$

Hence we have

$$z_{\mathbb{V}^{\triangle}}(\mathbf{h}_0'^{-1}, \mathbf{h}_0'\mathbf{g}\mathbf{h}_0'^{-1}) = 1.$$

Thus we obtain

$$\mu'(\mathbf{g}) = z_{\mathbb{V}^{\triangle}}(\mathbf{g}, \mathbf{h}_0'^{-1}) \cdot z_{\mathbb{V}^{\triangle}}(\mathbf{h}_0'^{-1}, \mathbf{h}_0'\mathbf{g}\mathbf{h}_0'^{-1})^{-1} = 1.$$

This completes the proof. $\square$

D.4. Splitting $z_{\mathbb{Y}^\square}$

Let $\mathbb{V} = \mathbb{X} \oplus \mathbb{Y}$ be the complete polarization given in §C.1. Put

$$\mathbb{X}^\square := \mathbb{X} \oplus \mathbb{X}, \quad \mathbb{Y}^\square := \mathbb{Y} \oplus \mathbb{Y}.$$

Then $\mathbb{V}^\square = \mathbb{X}^\square \oplus \mathbb{Y}^\square$ is a complete polarization. As in §D.3, we have

$$z_{\mathbb{Y}^\square}(\iota(g_1, g_2), \iota(g_1', g_2')) = z_{\mathbb{Y}}(g_1, g_1') \cdot z_{\mathbb{Y}}(g_2, g_2')^{-1}$$

for $g_i, g_i' \in \mathrm{Sp}(\mathbb{V})$. The Weil representation $\omega_\psi^\square$ of $\mathrm{Mp}(\mathbb{V}^\square)$ can be realized on the Schwartz space

$$\mathcal{S}(\mathbb{X}^\square) = \mathcal{S}(\mathbb{X}) \otimes \mathcal{S}(\mathbb{X}).$$

As representations of $\mathrm{Mp}(\mathbb{V})_\mathbb{Y} \times \mathrm{Mp}(\mathbb{V})_\mathbb{Y}$, we have

$$\omega_\psi^\square \circ \tilde{\iota} = \omega_\psi \otimes (\omega_\psi \circ \tilde{\mathrm{j}}_\mathbb{Y}),$$

where $\tilde{\mathrm{j}}_\mathbb{Y}$ is the automorphism of $\mathrm{Mp}(\mathbb{V})_\mathbb{Y} = \mathrm{Sp}(\mathbb{V}) \times \mathbb{C}^1$ defined by

$$\tilde{\mathrm{j}}_\mathbb{Y}(g, z) = (\mathrm{j}_\mathbb{Y}(g), z^{-1}), \quad \mathrm{j}_\mathbb{Y}(g) = d_\mathbb{Y}(-1) \cdot g \cdot d_\mathbb{Y}(-1).$$

Put $\mathbf{J} := ((\mathbf{j}_1, \mathbf{j}_2), \mathbf{j})$. Here we view $(\mathbf{j}_1, \mathbf{j}_2) \in \mathrm{GU}(V)^0$ and $\mathbf{j} \in \mathrm{GU}(W)$.

LEMMA D.4.1. *We have*

$$\mathrm{j}_\mathbb{Y}(\mathbf{g}) = \mathbf{J} \cdot \mathbf{g} \cdot \mathbf{J}^{-1}$$

for $\mathbf{g} \in \mathrm{GU}(V)^0 \times \mathrm{GU}(W)$. *In particular, we have*

$$\mathrm{j}_\mathbb{Y}(\mathrm{G}(\mathrm{U}(V)^0 \times \mathrm{U}(W))) = \mathrm{G}(\mathrm{U}(V)^0 \times \mathrm{U}(W)).$$

PROOF. Let $\mathbf{g} = ((\boldsymbol{\alpha}_1^{-1}, \boldsymbol{\alpha}_2^{-1}), \boldsymbol{\alpha}) \in \mathrm{GU}(V)^0 \times \mathrm{GU}(W)$ with $\boldsymbol{\alpha}_i = a_i + b_i\mathbf{i} + c_i\mathbf{j}_i + d_i\mathbf{i}\mathbf{j}_i \in B_i^\times$ and $\boldsymbol{\alpha} = a + b\mathbf{i} + c\mathbf{j} + d\mathbf{i}\mathbf{j} \in B^\times$. By §C.1, we see that $\mathrm{j}_\mathbb{Y}(\mathbf{g}) = ((\boldsymbol{\beta}_1^{-1}, \boldsymbol{\beta}_2^{-1}), \boldsymbol{\beta})$, where

$$\boldsymbol{\beta}_i = a_i - b_i\mathbf{i} + c_i\mathbf{j}_i - d_i\mathbf{i}\mathbf{j}_i, \quad \boldsymbol{\beta} = a - b\mathbf{i} + c\mathbf{j} - d\mathbf{i}\mathbf{j}.$$

On the other hand, since $\mathbf{j}_i\mathbf{i} = -\mathbf{i}\mathbf{j}_i$ and $\mathbf{j}\mathbf{i} = -\mathbf{i}\mathbf{j}$, we have $\mathbf{j}_i \cdot \boldsymbol{\alpha}_i \cdot \mathbf{j}_i^{-1} = \boldsymbol{\beta}_i$ and $\mathbf{j} \cdot \boldsymbol{\alpha} \cdot \mathbf{j}^{-1} = \boldsymbol{\beta}$. This yields the lemma. □

As in §C.2, C.3, fix $\mathbf{h}_0 \in \mathrm{Sp}(\mathbb{V})$ such that $\mathbb{X}' = \mathbb{X}\mathbf{h}_0$ and $\mathbb{Y}' = \mathbb{Y}\mathbf{h}_0$, and define a map $s : \mathrm{GU}(V)^0 \times \mathrm{GU}(W) \to \mathbb{C}^1$ by $s := s' \cdot \mu$, where

$$\mu(g) = z_\mathbb{Y}(\mathbf{h}_0 g \mathbf{h}_0^{-1}, \mathbf{h}_0) \cdot z_\mathbb{Y}(\mathbf{h}_0, g)^{-1}.$$

Put $\hat{\mathbf{h}}_0 := \mathbf{h}_0' \cdot \iota(\mathbf{h}_0, \mathbf{h}_0)^{-1} \in \mathrm{Sp}(\mathbb{V}^\square)$. Then we have

$$\mathbb{V}^\nabla \cdot \hat{\mathbf{h}}_0 = \mathbb{X}'^\square \cdot \iota(\mathbf{h}_0, \mathbf{h}_0)^{-1} = \mathbb{X}^\square,$$
$$\mathbb{V}^\triangle \cdot \hat{\mathbf{h}}_0 = \mathbb{Y}'^\square \cdot \iota(\mathbf{h}_0, \mathbf{h}_0)^{-1} = \mathbb{Y}^\square.$$

Put

$$\hat{\mu}(g) = z_{\mathbb{V}^\triangle}(g, \hat{\mathbf{h}}_0^{-1}) \cdot z_{\mathbb{V}^\triangle}(\hat{\mathbf{h}}_0^{-1}, \hat{\mathbf{h}}_0 g \hat{\mathbf{h}}_0^{-1})^{-1}$$

for $g \in \mathrm{Sp}(\mathbb{V}^\square)$. Then we have

$$z_{\mathbb{Y}^\square}(g, g') = z_{\mathbb{V}^\triangle}(g, g') \cdot \hat{\mu}(gg') \cdot \hat{\mu}(g)^{-1} \cdot \hat{\mu}(g')^{-1}$$

for $g, g' \in \mathrm{Sp}(\mathbb{V}^\square)$.

LEMMA D.4.2. *We have*

$$\hat{s} \cdot \hat{\mu} = s \otimes (s \circ \mathrm{j}_\mathbb{Y})$$

on $\mathcal{G}$.

PROOF. For $\mathbf{g} = (g, h_1, h_2) \in \mathcal{G}$, we identify $\mathbf{g}$ with $\iota(\mathbf{g}_1, \mathbf{g}_2) \in \mathrm{Sp}(\mathbb{V}^\square)$, where $\mathbf{g}_i = (g, h_i) \in \mathrm{G}(\mathrm{U}(V)^0 \times \mathrm{U}(W)) \subset \mathrm{Sp}(\mathbb{V})$. Then, by a direct calculation, one can see that

$$\begin{aligned}
&\hat{\mu}(\mathbf{g}) \cdot \mu'(\mathbf{g})^{-1} \\
&= z_{\mathbb{Y}^\square}(\iota(\mathbf{h}_0, \mathbf{h}_0) \cdot \mathbf{g} \cdot \iota(\mathbf{h}_0, \mathbf{h}_0)^{-1}, \iota(\mathbf{h}_0, \mathbf{h}_0)) \cdot z_{\mathbb{Y}^\square}(\iota(\mathbf{h}_0, \mathbf{h}_0), \mathbf{g})^{-1} \\
&= z_{\mathbb{Y}}(\mathbf{h}_0 \mathbf{g}_1 \mathbf{h}_0^{-1}, \mathbf{h}_0) \cdot z_{\mathbb{Y}}(\mathbf{h}_0 \mathbf{g}_2 \mathbf{h}_0^{-1}, \mathbf{h}_0)^{-1} \cdot z_{\mathbb{Y}}(\mathbf{h}_0, \mathbf{g}_1)^{-1} \cdot z_{\mathbb{Y}}(\mathbf{h}_0, \mathbf{g}_2) \\
&= \mu(\mathbf{g}_1) \cdot \mu(\mathbf{g}_2)^{-1}.
\end{aligned}$$

Hence, by Lemma D.3.1, we have

$$\hat{s} \cdot \hat{\mu} = \hat{s} \cdot \mu' \cdot (\mu \otimes \mu^{-1}) = (s' \cdot \mu) \otimes ((s' \circ \mathrm{j}_{\mathbb{Y}'}) \cdot \mu^{-1})$$

on $\mathcal{G}$. Since $s' \circ \mathrm{j}_{\mathbb{Y}'} = s' = s'^{-1}$, we have

$$\hat{s} \cdot \hat{\mu} = s \otimes s^{-1}$$

on $\mathcal{G}$. By Proposition C.4.2 and Lemma D.4.1, we have $s(\mathrm{j}_{\mathbb{Y}}(\mathbf{g})) = s(\mathbf{g})^{-1}$ for $\mathbf{g} = \boldsymbol{\alpha}_i^{-1} \in \mathrm{GU}(V)^0$ with $\boldsymbol{\alpha}_i \in B_i^\times$ and $\mathbf{g} = \boldsymbol{\alpha} \in \mathrm{GU}(W)$ with $\boldsymbol{\alpha} \in B^\times$. Also, we have $z_{\mathbb{Y}}(\mathrm{j}_{\mathbb{Y}}(g_1), \mathrm{j}_{\mathbb{Y}}(g_2)) = z_{\mathbb{Y}}(g_1, g_2)^{-1}$ for $g_1, g_2 \in \mathrm{GSp}(\mathbb{V})$. Since $s(\mathbf{g}_1 \mathbf{g}_2) = s(\mathbf{g}_1) \cdot s(\mathbf{g}_2) \cdot z_{\mathbb{Y}}(\mathbf{g}_1, \mathbf{g}_2)$ for $\mathbf{g}_1, \mathbf{g}_2 \in \mathrm{GU}(V)^0 \times \mathrm{GU}(W)$, we have

$$s \circ \mathrm{j}_{\mathbb{Y}} = s^{-1}$$

on $\mathrm{GU}(V)^0 \times \mathrm{GU}(W)$. This completes the proof. $\square$

Bibliography

[1] Ahmed Abbes and Emmanuel Ullmo, *À propos de la conjecture de Manin pour les courbes elliptiques modulaires* (French, with French summary), Compositio Math. **103** (1996), no. 3, 269–286. MR1414591

[2] Laure Barthel, *Local Howe correspondence for groups of similitudes*, J. Reine Angew. Math. **414** (1991), 207–220, DOI 10.1515/crll.1991.414.207. MR1092630

[3] Massimo Bertolini, Henri Darmon, and Kartik Prasanna, *Generalized Heegner cycles and p-adic Rankin L-series*, Duke Math. J. **162** (2013), no. 6, 1033–1148, DOI 10.1215/00127094-2142056. With an appendix by Brian Conrad. MR3053566

[4] Don Blasius and Jonathan D. Rogawski, *Motives for Hilbert modular forms*, Invent. Math. **114** (1993), no. 1, 55–87, DOI 10.1007/BF01232663. MR1235020

[5] Henri Carayol, *Sur les représentations l-adiques associées aux formes modulaires de Hilbert* (French), Ann. Sci. École Norm. Sup. (4) **19** (1986), no. 3, 409–468. MR870690

[6] William Casselman, *On some results of Atkin and Lehner*, Math. Ann. **201** (1973), 301–314, DOI 10.1007/BF01428197. MR0337789

[7] W. Casselman, *The unramified principal series of* $\mathfrak{p}$*-adic groups. I. The spherical function*, Compositio Math. **40** (1980), no. 3, 387–406. MR571057

[8] W. Casselman and J. Shalika, *The unramified principal series of p-adic groups. II. The Whittaker function*, Compositio Math. **41** (1980), no. 2, 207–231. MR581582

[9] Pierre Deligne, *Formes modulaires et représentations l-adiques* (French), Séminaire Bourbaki. Vol. 1968/69: Exposés 347–363, Lecture Notes in Math., vol. 175, Springer, Berlin, 1971, pp. Exp. No. 355, 139–172. MR3077124

[10] Pierre Deligne, *Variétés de Shimura: interprétation modulaire, et techniques de construction de modèles canoniques* (French), Automorphic forms, representations and L-functions (Proc. Sympos. Pure Math., Oregon State Univ., Corvallis, Ore., 1977), Proc. Sympos. Pure Math., XXXIII, Amer. Math. Soc., Providence, R.I., 1979, pp. 247–289. MR546620

[11] Pierre Deligne and Kenneth A. Ribet, *Values of abelian L-functions at negative integers over totally real fields*, Invent. Math. **59** (1980), no. 3, 227–286, DOI 10.1007/BF01453237. MR579702

[12] Pierre Deligne and Jean-Pierre Serre, *Formes modulaires de poids* 1 (French), Ann. Sci. École Norm. Sup. (4) **7** (1974), 507–530 (1975). MR0379379

[13] Bas Edixhoven, *On the Manin constants of modular elliptic curves*, Arithmetic algebraic geometry (Texel, 1989), Progr. Math., vol. 89, Birkhäuser Boston, Boston, MA, 1991, pp. 25–39. MR1085254

[14] M. Eichler, *Über die Idealklassenzahl total definiter Quaternionenalgebren* (German), Math. Z. **43** (1938), no. 1, 102–109, DOI 10.1007/BF01181088. MR1545717

[15] G. Faltings, *Endlichkeitssätze für abelsche Varietäten über Zahlkörpern* (German), Invent. Math. **73** (1983), no. 3, 349–366, DOI 10.1007/BF01388432. MR718935

[16] Wee Teck Gan and Atsushi Ichino, *On endoscopy and the refined Gross-Prasad conjecture for* $(\mathrm{SO}_5, \mathrm{SO}_4)$, J. Inst. Math. Jussieu **10** (2011), no. 2, 235–324, DOI 10.1017/S1474748010000198. MR2787690

[17] Paul B. Garrett and Michael Harris, *Special values of triple product L-functions*, Amer. J. Math. **115** (1993), no. 1, 161–240, DOI 10.2307/2374726. MR1209238

[18] Roger Godement and Hervé Jacquet, *Zeta functions of simple algebras*, Lecture Notes in Mathematics, Vol. 260, Springer-Verlag, Berlin-New York, 1972. MR0342495

[19] Benedict H. Gross and Don B. Zagier, *Heegner points and derivatives of L-series*, Invent. Math. **84** (1986), no. 2, 225–320, DOI 10.1007/BF01388809. MR833192

[20] Michael Harris, *Arithmetic vector bundles and automorphic forms on Shimura varieties. I*, Invent. Math. **82** (1985), no. 1, 151–189, DOI 10.1007/BF01394784. MR808114

[21] Michael Harris, *Arithmetic vector bundles and automorphic forms on Shimura varieties. II*, Compositio Math. **60** (1986), no. 3, 323–378. MR869106

[22] Michael Harris, *L-functions of 2×2 unitary groups and factorization of periods of Hilbert modular forms*, J. Amer. Math. Soc. **6** (1993), no. 3, 637–719, DOI 10.2307/2152780. MR1186960

[23] Michael Harris, *Hodge-de Rham structures and periods of automorphic forms*, Motives (Seattle, WA, 1991), Proc. Sympos. Pure Math., vol. 55, Amer. Math. Soc., Providence, RI, 1994, pp. 573–624. MR1265564

[24] Michael Harris, *Cohomological automorphic forms on unitary groups. I. Rationality of the theta correspondence*, Automorphic forms, automorphic representations, and arithmetic (Fort Worth, TX, 1996), Proc. Sympos. Pure Math., vol. 66, Amer. Math. Soc., Providence, RI, 1999, pp. 103–200. MR1703759

[25] Michael Harris, *Cohomological automorphic forms on unitary groups. II. Period relations and values of L-functions*, Harmonic analysis, group representations, automorphic forms and invariant theory, Lect. Notes Ser. Inst. Math. Sci. Natl. Univ. Singap., vol. 12, World Sci. Publ., Hackensack, NJ, 2007, pp. 89–149, DOI 10.1142/9789812770790_0005. MR2401812

[26] Michael Harris and Stephen S. Kudla, *The central critical value of a triple product L-function*, Ann. of Math. (2) **133** (1991), no. 3, 605–672, DOI 10.2307/2944321. MR1109355

[27] Michael Harris and Stephen S. Kudla, *Arithmetic automorphic forms for the nonholomorphic discrete series of* GSp(2), Duke Math. J. **66** (1992), no. 1, 59–121, DOI 10.1215/S0012-7094-92-06603-8. MR1159432

[28] Michael Harris, Stephen S. Kudla, and William J. Sweet, *Theta dichotomy for unitary groups*, J. Amer. Math. Soc. **9** (1996), no. 4, 941–1004, DOI 10.1090/S0894-0347-96-00198-1. MR1327161

[29] Haruzo Hida, *Congruence of cusp forms and special values of their zeta functions*, Invent. Math. **63** (1981), no. 2, 225–261, DOI 10.1007/BF01393877. MR610538

[30] Haruzo Hida, *On congruence divisors of cusp forms as factors of the special values of their zeta functions*, Invent. Math. **64** (1981), no. 2, 221–262, DOI 10.1007/BF01389169. MR629471

[31] Atsushi Ichino and Kartik Prasanna, *Periods of quaternionic Shimura varieties. II.* In preparation.

[32] Atsushi Ichino and Kartik Prasanna, *Periods of quaternionic Shimura varieties. III.* In preparation.

[33] Atsushi Ichino and Kartik Prasanna, *Cycles for Rankin-Selberg motives and p-adic L-functions.* In preparation.

[34] Atsushi Ichino and Kartik Prasanna, *Hodge classes and the Jacquet-Langlands correspondence.* Preprint. `arXiv:1806.10563`

[35] H. Jacquet and R. P. Langlands, *Automorphic forms on* GL(2), Lecture Notes in Mathematics, Vol. 114, Springer-Verlag, Berlin-New York, 1970. MR0401654

[36] H. Jacquet and J. A. Shalika, *On Euler products and the classification of automorphic representations. I*, Amer. J. Math. **103** (1981), no. 3, 499–558, DOI 10.2307/2374103. MR618323

[37] Frazer Jarvis, *On Galois representations associated to Hilbert modular forms*, J. Reine Angew. Math. **491** (1997), 199–216, DOI 10.1515/crll.1997.491.199. MR1476093

[38] Mark Kisin, *Integral models for Shimura varieties of abelian type*, J. Amer. Math. Soc. **23** (2010), no. 4, 967–1012, DOI 10.1090/S0894-0347-10-00667-3. MR2669706

[39] Max-Albert Knus, *Quadratic and Hermitian forms over rings*, Grundlehren der Mathematischen Wissenschaften [Fundamental Principles of Mathematical Sciences], vol. 294, Springer-Verlag, Berlin, 1991. With a foreword by I. Bertuccioni. MR1096299

[40] Stephen S. Kudla, *Splitting metaplectic covers of dual reductive pairs*, Israel J. Math. **87** (1994), no. 1-3, 361–401, DOI 10.1007/BF02773003. MR1286835

[41] Stephen S. Kudla, *Integrals of Borcherds forms*, Compositio Math. **137** (2003), no. 3, 293–349, DOI 10.1023/A:1024127100993. MR1988501

[42] Stephen Kudla, *Notes on the local theta correspondence.* Available online at: `http://www.math.toronto.edu/~skudla/castle.pdf`.

[43] Stephen S. Kudla and Stephen Rallis, *Poles of Eisenstein series and L-functions*, Festschrift in honor of I. I. Piatetski-Shapiro on the occasion of his sixtieth birthday, Part II (Ramat Aviv, 1989), Israel Math. Conf. Proc., vol. 3, Weizmann, Jerusalem, 1990, pp. 81–110. MR1159110

[44] Stephen S. Kudla and Stephen Rallis, *Ramified degenerate principal series representations for* $\mathrm{Sp}(n)$, Israel J. Math. **78** (1992), no. 2-3, 209–256, DOI 10.1007/BF02808058. MR1194967

[45] Stephen S. Kudla and Stephen Rallis, *A regularized Siegel-Weil formula: the first term identity*, Ann. of Math. (2) **140** (1994), no. 1, 1–80, DOI 10.2307/2118540. MR1289491

[46] Stephen S. Kudla, Michael Rapoport, and Tonghai Yang, *Modular forms and special cycles on Shimura curves*, Annals of Mathematics Studies, vol. 161, Princeton University Press, Princeton, NJ, 2006. MR2220359

[47] Erez Lapid and Omer Offen, *Compact unitary periods*, Compos. Math. **143** (2007), no. 2, 323–338, DOI 10.1112/S0010437X06002612. MR2309989

[48] Jian-Shu Li, *Nonvanishing theorems for the cohomology of certain arithmetic quotients*, J. Reine Angew. Math. **428** (1992), 177–217, DOI 10.1515/crll.1992.428.177. MR1166512

[49] Tom Lovering, *Integral canonical models for automorphic vector bundles of abelian type*, Algebra Number Theory **11** (2017), no. 8, 1837–1890, DOI 10.2140/ant.2017.11.1837. MR3720933

[50] I. G. Macdonald, *Spherical functions on a group of p-adic type*, Ramanujan Institute, Centre for Advanced Study in Mathematics,University of Madras, Madras, 1971. Publications of the Ramanujan Institute, No. 2. MR0435301

[51] I. G. Macdonald, *The volume of a compact Lie group*, Invent. Math. **56** (1980), no. 2, 93–95, DOI 10.1007/BF01392542. MR558859

[52] B. Mazur, *Rational isogenies of prime degree (with an appendix by D. Goldfeld)*, Invent. Math. **44** (1978), no. 2, 129–162, DOI 10.1007/BF01390348. MR482230

[53] J. S. Milne, *Canonical models of (mixed) Shimura varieties and automorphic vector bundles*, Automorphic forms, Shimura varieties, and L-functions, Vol. I (Ann Arbor, MI, 1988), Perspect. Math., vol. 10, Academic Press, Boston, MA, 1990, pp. 283–414. MR1044823

[54] Calvin C. Moore, *Group extensions of p-adic and adelic linear groups*, Inst. Hautes Études Sci. Publ. Math. **35** (1968), 157–222. MR0244258

[55] Jan Nekovář, *Kolyvagin's method for Chow groups of Kuga-Sato varieties*, Invent. Math. **107** (1992), no. 1, 99–125, DOI 10.1007/BF01231883. MR1135466

[56] Takayuki Oda, *Periods of Hilbert modular surfaces*, Progress in Mathematics, vol. 19, Birkhäuser, Boston, Mass., 1982. MR670069

[57] Takayuki Oda, *Hodge structures of Shimura varieties attached to the unit groups of quaternion algebras*, Galois groups and their representations (Nagoya, 1981), Adv. Stud. Pure Math., vol. 2, North-Holland, Amsterdam, 1983, pp. 15–36. MR732459

[58] I. I. Piatetski-Shapiro and Stephen Rallis, *L-functions for the classical groups.* Explicit constructions of automorphic L-functions. Lecture Notes in Mathematics **1254**, Springer-Verlag, 1987, pp. 1–52.

[59] Robert Pollack and Tom Weston, *On anticyclotomic μ-invariants of modular forms*, Compos. Math. **147** (2011), no. 5, 1353–1381, DOI 10.1112/S0010437X11005318. MR2834724

[60] Kartik Prasanna, *Integrality of a ratio of Petersson norms and level-lowering congruences*, Ann. of Math. (2) **163** (2006), no. 3, 901–967, DOI 10.4007/annals.2006.163.901. MR2215136

[61] Kartik Prasanna, *Arithmetic aspects of the theta correspondence and periods of modular forms*, Eisenstein series and applications, Progr. Math., vol. 258, Birkhäuser Boston, Boston, MA, 2008, pp. 251–269, DOI 10.1007/978-0-8176-4639-4_9. MR2402687

[62] S. Rallis, *On the Howe duality conjecture*, Compositio Math. **51** (1984), no. 3, 333–399. MR743016

[63] R. Ranga Rao, *On some explicit formulas in the theory of Weil representation*, Pacific J. Math. **157** (1993), no. 2, 335–371. MR1197062

[64] K. A. Ribet, *On modular representations of* $\mathrm{Gal}(\overline{\mathbf{Q}}/\mathbf{Q})$ *arising from modular forms*, Invent. Math. **100** (1990), no. 2, 431–476, DOI 10.1007/BF01231195. MR1047143

[65] Kenneth A. Ribet and Shuzo Takahashi, *Parametrizations of elliptic curves by Shimura curves and by classical modular curves*, Proc. Nat. Acad. Sci. U.S.A. **94** (1997), no. 21, 11110–11114, DOI 10.1073/pnas.94.21.11110. Elliptic curves and modular forms (Washington, DC, 1996). MR1491967

[66] Brooks Roberts, *The theta correspondence for similitudes*, Israel J. Math. **94** (1996), 285–317, DOI 10.1007/BF02762709. MR1394579

[67] J. D. Rogawski and J. B. Tunnell, *On Artin L-functions associated to Hilbert modular forms of weight one*, Invent. Math. **74** (1983), no. 1, 1–42, DOI 10.1007/BF01388529. MR722724

[68] Hideo Shimizu, *Theta series and automorphic forms on* GL_2, J. Math. Soc. Japan **24** (1972), 638–683, DOI 10.2969/jmsj/02440638. MR0333081

[69] Goro Shimura, *Introduction to the arithmetic theory of automorphic functions*, Publications of the Mathematical Society of Japan, vol. 11, Princeton University Press, Princeton, NJ, 1994. Reprint of the 1971 original; Kanô Memorial Lectures, 1. MR1291394

[70] Goro Shimura, *Algebraic relations between critical values of zeta functions and inner products*, Amer. J. Math. **105** (1983), no. 1, 253–285, DOI 10.2307/2374388. MR692113

[71] Goro Shimura, *On the critical values of certain Dirichlet series and the periods of automorphic forms*, Invent. Math. **94** (1988), no. 2, 245–305, DOI 10.1007/BF01394326. MR958833

[72] Carl Ludwig Siegel, *The volume of the fundamental domain for some infinite groups*, Trans. Amer. Math. Soc. **39** (1936), no. 2, 209–218, DOI 10.2307/1989745. MR1501843

[73] Richard Taylor, *On Galois representations associated to Hilbert modular forms*, Invent. Math. **98** (1989), no. 2, 265–280, DOI 10.1007/BF01388853. MR1016264

[74] Marie-France Vignéras, *Arithmétique des algèbres de quaternions* (French), Lecture Notes in Mathematics, vol. 800, Springer, Berlin, 1980. MR580949

[75] André Weil, *Sur certains groupes d'opérateurs unitaires* (French), Acta Math. **111** (1964), 143–211, DOI 10.1007/BF02391012. MR0165033

[76] A. Wiles, *On ordinary λ-adic representations associated to modular forms*, Invent. Math. **94** (1988), no. 3, 529–573, DOI 10.1007/BF01394275. MR969243

[77] Andrew Wiles, *Modular elliptic curves and Fermat's last theorem*, Ann. of Math. (2) **141** (1995), no. 3, 443–551, DOI 10.2307/2118559. MR1333035

[78] Shunsuke Yamana, *Degenerate principal series representations for quaternionic unitary groups*, Israel J. Math. **185** (2011), 77–124, DOI 10.1007/s11856-011-0102-9. MR2837129

[79] Shunsuke Yamana, *On the Siegel-Weil formula for quaternionic unitary groups*, Amer. J. Math. **135** (2013), no. 5, 1383–1432, DOI 10.1353/ajm.2013.0045. MR3117309

[80] Hiroyuki Yoshida, *On a conjecture of Shimura concerning periods of Hilbert modular forms*, Amer. J. Math. **117** (1995), no. 4, 1019–1038, DOI 10.2307/2374957. MR1342839

[81] Chia-Fu Yu, *Variations of mass formulas for definite division algebras*, J. Algebra **422** (2015), 166–186, DOI 10.1016/j.jalgebra.2014.08.047. MR3272073

[82] Wei Zhang, *Automorphic period and the central value of Rankin-Selberg L-function*, J. Amer. Math. Soc. **27** (2014), no. 2, 541–612, DOI 10.1090/S0894-0347-2014-00784-0. MR3164988